Feature Selection and Ensemble Methods for Bioinformatics:

Algorithmic Classification and Implementations

Oleg Okun
SMARTTECCO, Sweden

T0338721

Medical Information Science
REFERENCE

Senior Editorial Director:	Kristin Klinger
Director of Book Publications:	Julia Mosemann
Editorial Director:	Lindsay Johnston
Acquisitions Editor:	Erika Carter
Development Editor:	Julia Mosemann
Production Editor:	Sean Woznicki
Typesetters:	Mike Brehm, Keith Glazewski, Natalie Pronio, Jennifer Romanchak, Milan Vracarich Jr.
Print Coordinator:	Jamie Snavely
Cover Design:	Nick Newcomer

Published in the United States of America by
Medical Information Science Reference (an imprint of IGI Global)
701 E. Chocolate Avenue
Hershey PA 17033
Tel: 717-533-8845
Fax: 717-533-8661
E-mail: cust@igi-global.com
Web site: http://www.igi-global.com/reference

Library of Congress Cataloging-in-Publication Data

Okun, Oleg.
  Feature selection and ensemble methods for bioinformatics: algorithmic classification and implementations / by Oleg Okun.
       p. cm.
  Includes bibliographical references and index.
  Summary: "This book offers a unique perspective on machine learning aspects of microarray gene expression based cancer classification, combining computer science, and biology"--Provided by publisher.
  ISBN 978-1-60960-557-5 (hardcover) -- ISBN 978-1-60960-558-2 (ebook) 1. Bioinformatics. 2. Bioinformatics--Methodology. I. Title.
  QH324.2.O58 2011
  572.80285--dc22
                                    2010052304

British Cataloguing in Publication Data
A Cataloguing in Publication record for this book is available from the British Library.

All work contributed to this book is new, previously-unpublished material. The views expressed in this book are those of the authors, but not necessarily of the publisher.

Table of Contents

Preface viii

Chapter 1
Biological Background.. 1
A Little Bit of Biology... 1
References... 4
Endnotes... 5

Chapter 2
Gene Expression Data Sets... 6
Biological Data and Their Characteristics.. 6
References 9
Endnotes... 9

Chapter 3
Introduction to Data Classification 10
Problem of Data Classification.. 10
References.. 12
Endnotes.. 12

Chapter 4
Naïve Bayes.. 13
Bayes and Naïve Bayes 13
References 28
Endnotes.. 30

Chapter 5

Nearest Neighbor... **32**

Nearest Neighbor Classification 32

References 48

Endnotes.. 51

Chapter 6

Classification Tree... **53**

Tree-Like Classifier 53

References 66

Endnotes.. 67

Chapter 7

Support Vector Machines **68**

Support Vector Machines 68

References 112

Endnotes...114

Chapter 8

Introduction to Feature and Gene Selection.................................**117**

Problem of Feature Selection..117

References 121

Endnotes... 121

Chapter 9

Feature Selection Based on Elements of Game Theory.............. **123**

Feature Selection Based on the Shapley Value 123

References 139

Endnote 139

Chapter 10

Kernel-Based Feature Selection with the Hilbert-Schmidt Independence Criterion... **141**

Kernel Methods and Feature Selection.. 141

References 155

Endnotes... 157

Chapter 11

Extreme Value Distribution Based Gene Selection **159**

Blend of Elements of Extreme Value Theory and Logistic Regression 159

References 174

Endnotes... 175

Chapter 12
Evolutionary Algorithm for Identifying Predictive Genes **177**
Evolutionary Search for Optimal or Near-Optimal Set of Genes 177
References 199
Endnotes 201

Chapter 13
Redundancy-Based Feature Selection **203**
Redundancy of Features 203
References 220
Endnotes 222

Chapter 14
Unsupervised Feature Selection **223**
Unsupervised Feature Filtering 223
References 234

Chapter 15
Differential Evolution for Finding Predictive Gene Subsets **236**
Differential Evolution: Global, Evolution Strategy Based Optimization Method 236
References 249
Endnotes 251

Chapter 16
Ensembles of Classifiers **252**
Ensemble Learning 252
References 257
Endnotes 258

Chapter 17
Classifier Ensembles Built on Subsets of Features **260**
Shaking Stable Classifiers 260
References 291
Endnotes 294

Chapter 18
Bagging and Random Forests **296**
Bootstrap and its Use in Classifier Ensembles 296
References 311
Endnotes 313

Chapter 19

Boosting and AdaBoost... **314**

Weighted Learning, Boosting and AdaBoost 314

References 326

Endnotes.. 328

Chapter 20

Ensemble Gene Selection.. **329**

Getting Important Genes out of a Pool.. 329

References 333

Endnotes.. 333

Chapter 21

Introduction to Classification Error Estimation **334**

Problem of Classification Error Estimation 334

References 338

Endnotes.. 340

Chapter 22

ROC Curve, Area under it, Other Classification Performance

Characteristics and Statistical Tests... **341**

Classification Performance Evaluation... 341

References 379

Endnotes.. 381

Chapter 23

Bolstered Resubstitution Error... **383**

Alternative to Traditional Error Estimators 383

References 404

Endnotes.. 405

Chapter 24

Performance Evaluation: Final Check... **406**

Bayesian Confidence (Credible) Interval... 406

References 412

Endnotes.. 413

Chapter 25

Application Examples.. **414**

Joining All Pieces Together 414

References 434

Endnote 435

Chapter 26
End Remarks... **436**
A Few Words in the End.. 436
References 437

About the Author **439**

Index.. **440**

Preface

A FEW WORDS ABOUT THE BOOK

This book represents my attempt to create a unique book on machine learning aspects of one of the important tasks in bioinformatics – microarray gene expression based cancer classification. That is, the input called a training set consists of expression level values measured for many (order of thousands or even tens of thousands) genes at once. Such measurements were however done only for a few patients, thus making the number of features (genes) much larger than the number of instances. Given such information, the task is to correctly assign class labels to the instances outside the training set. In this book, only binary or two-class classification problems are treated. Two classes of the data are 'tumor' ('cancer', 'diseased') and 'normal' ('healthy').

Although this book considers gene expression based cancer classification as an application, this does not imply that the methods described in this book cannot be used as they are for other bioinformatics tasks, where a data set is small but high dimensional; hence, the word 'bioinformatics' in the book title.

The uniqueness of this book stems from the combination of three topics:

- machine learning,
- bioinformatics,
- MATLAB®.

There are a plenty of books on machine learning (check, e.g., the web site of IGI Global http://www.idea-group.com/ and search there for the phrase "machine learning"). There are a few books covering both machine learning and bioinformatics. However, to my best knowledge this is one of very few books that cover all three topics above in one volume.

As follows from its topic, the subject of the book lies on a crossroad between computer science and biology. Hence, the main intention was to write a book that could be used either as a textbook or a reference book by researchers and students from both fields. Also my purpose was to write a book that could be suitable both for novices and seasoned practitioners, for people from both academia and industry. So, it was a very challenging and ambitious undertaking, and you, the readers of this book, will decide whether it was successfully accomplished or not. In this book I often use 'we' when addressing to you, dear readers, since I always assume your invisible presence and do not want to turn the whole process exclusively into my monolog. Everything we do on this "journey", we do together.

To meet all diverse goals and the needs of broad audience, I decided that each chapter shall comprise an independent or almost independent containment of knowledge that can be read and understood independently of other chapters. Each chapter aims at explaining one machine learning method only. In addition, the organization of all chapters tries to follow the same structure. Namely, each chapter begins with the main idea and theory, which a given method is based on, followed by algorithm description in pseudo code, demonstrating how to implement the method step-by-step. After that, MATLAB® code is given together with detailed comments on it. MATLAB® was chosen because it is widely used by the research community; it includes many useful toolboxes such as Statistics Toolbox™ and Bioinformatics Toolbox™.

Thus, as a whole, each chapter describes the process of algorithm design from the very beginning to the very end as it was my humble hope to teach the readers the best practices that they could apply in their everyday research work. In other words, I tried to provide a kind of a standard of how algorithms shall be described (though I am far from imposing the rules I deem to be good on readers) in giving lectures for students and researchers and in research reports/theses prepared by both students and researchers.

Of course, many people may object me at this point. For example, some (but not all, I believe) adepts of extreme programming (for a quick introduction to this promising programming technique to write good software, read (chromatic, 2003); 'chromatic' is not a mistake but a nickname) may say that documentation is unnecessary for already heavily commented code. From my industrial and academic experience, I disagree with this statement as professionally composed documentation can save many days and weeks of work both for users/customers and for researchers that did not take part in the development of that piece of software. For example, imagine a situation when all people who developed software suddenly left a company (optimists may say that they got better job offers elsewhere, while pessimists (and many realists nowadays) may say that all of them were laid off because of a worsened financial situation) and a new staff needs to quickly advance further while utilizing code developed by their predecessors. How do you think this could happen if the previous staff members did not leave any traces of their thoughts or their code comments are scarce and not very informative? This is, of course, not an impossible mission but anyway difficult.

I am not an inventor of the principles I advocate here. For instance, Prof. David Donoho from Stanford University (http://www-stat.stanford.edu/~donoho/index. html) some time ago suggested and actively promotes open code (in many cases it is MATLAB®) that can be used to reproduce research results described in scientific articles. I see it as a good practice to adhere. Besides, several MATLAB®-based books influenced me while I was writing my own book. Among them are (Nabney, 2002), (Martinez & Martinez, Computational statistics handbook with MATLAB®, 2002), (Stork & Yom-Tov, 2004), (van der Heijden, Duin, de Ridder, & Tax, 2004), and (Martinez & Martinez, Exploratory data analysis with MATLAB®, 2005). These books as well as MATLAB® code taught me best practices to utilize in my own book. As a quick reference guide to MATLAB®, I can recommend a little book by Davis and Sigmon (Davis & Sigmon, 2005).

Other books that are relevant to the statistical aspects of bioinformatics and that target biologists include but not limited to (Zar, 1999), (van Emden, 2008), (Lee, 2010). The book of van Emden has a funny title "Statistics for the terrified biologists", lol. Does this mean that all or at least a majority of statisticians are cold-blooded (neither in the biological, nor in the criminal sense of these words) folks with un-shaken resolution? Based on the laws of statistics, this is, of course, not true. But it would be interesting to see a book titled "Biology for the terrified statisticians". Hey, biologists out there, who of you is ready to write such a sweet revenge book?

Other books that may be complementary to the subject of my book are (Cohen, 2007), (Cristianini & Hann, 2007), (Alterovitz & Ramoni, 2007) (the last two books also concentrate on MATLAB® as the programming environment).

Although I mentioned earlier that each chapter can be read independently of other chapters, it does not mean that there are absolutely no connections between chapters. The book is structurally divided into five parts. Each part (except for the last one) begins with an introductory chapter providing a compressed summary of the topics and problems discussed in the chapters that follow.

The first group of chapters concerns the classification algorithms (or simply classifiers) most commonly employed in bioinformatics research working with gene expression data. The second group of chapters deals with feature or gene selection algorithms. Due to the huge number of algorithms, I tried to choose the algorithms built on as diverse principles as possible, though covering all existing types would certainly be unrealistic (e.g., searching with Google Scholar for the exact phrase "feature selection" returned 96,200 links while searching for "gene selection" resulted in 11,700 links (searches were performed in May of 2009)). The third group of chapters concentrates on classifier ensembles, i.e. several clas-sifiers whose predictions are combined together in order to form the final vote.

Figure 1. General scheme involving a single classifier

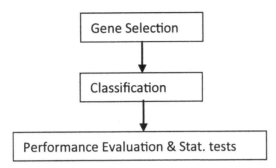

Compared to a single classifier, a classifier ensemble can deliver better and more stable performance. The fourth group of chapters describes advanced performance evaluation methods for single classifiers and classifier ensembles and statistical tests related to this evaluation. These methods, unlike many currently employed, are especially tuned to small-sample size problems such as classification of microarray gene expression data. Finally, the last part comprises a single chapter demonstrating how to utilize code spread across other four groups of chapters. It includes four examples putting feature selection, data classification, ensemble of classifiers, and performance evaluation techniques into one piece.

Such a book structure well matches to the general scheme used to solve problems like gene expression based cancer classification, where the number of features far exceeds the number of available instances (samples taken from patients). In the case when a single classifier is used, this scheme is shown in Figure 1, while in the case of a classifier ensemble, it is given in Figure 2.

Since there are too many genes compared to the number of instances, it is obvious that not all of them are related to cancer. In other words, many of them can be

Figure 2. General scheme involving an ensemble of classifiers

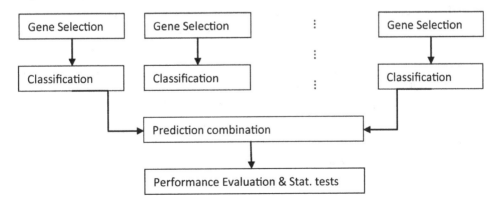

safely removed from the classification model so that they do not participate in the classification process. Their removal is also necessitated by the fact that such redundant genes, if left in a data set, will degrade the generalization ability of a classifier, i.e. the classifier will perfectly classify the training data but will suddenly degrade in performance on new, previously unseen (out-of-sample, test) instances.

As the task is not to perfectly classify the training data that are utilized for learning how to assign class labels but to correctly classify test data, generalization of the trained classifier is of great importance. With poor generalization, the classification mode is unusable as biologists and doctors cannot trust to the reliability of results. In other words, poor generalization is associated with 'no trustful outcome'. The removal of irrelevant and redundant genes out of a data set or alternatively the selection of highly relevant for disease prediction genes is called feature or gene selection. Usually, a small subset of the original genes remains after this procedure.

Once relevant genes have been selected, the next stage is the classification with selected genes, which typically consists of two steps: classifier training (optional if a classifier does not need it) on the training data and testing of the trained classifier on the test data. As the available data are scarce, the good solution is to artificially generate test instances based on the training instances. This is a rather new approach (see, e.g., (Braga-Neto & Dougherty, Bolstered error estimation, 2004), (Braga-Neto, Fads and fallacies in the name of small-sample microarray classification - A highlight of misunderstanding and erroneous usage in the applications of genomic signal processing, 2007), (Li, Fang, Lai, & Hu, 2009)), but I consider it very promising and appealing, compared to the other known techniques that reserve a part of the training instances for testing, thus reducing both training and test set sizes. When every extra instance is of importance, reducing either training or test data can easily bias classification results, i.e. to make them over-optimistic.

After a classifier or an ensemble of classifiers assigned class labels and/or class probability estimates to test instances, the final stage is the computation of the performance evaluation characteristics and running statistical tests in order to discover the statistically significant difference in classification performance (or the absence of such a difference) between several competing classification models.

Thus, the book covers the entire classifier design when using microarray gene expression data as input. What may look missing is probably the links to gene expression data sets to be used in experiments with the algorithms in this book. The paper by Statnikov et al. (Statnikov, Wang, & Aliferis, 2008) cites 22 data sets. Another paper (Yoon, Lee, Park, Bien, Chung, & Rha, 2008) refers to three prostate cancer data sets. Links to some data sets can be found at http://www.genecbr.org/links.htm. As active researchers, I suppose that you, dear readers, also know well where the microarray data are located on the web.

I hope readers will find this book useful for their work and education. Any questions, suggestions or comments as well as bug reports can be sent to me (olegokun@ yahoo.com).

This book would not appear without several people whom I would like to express my gratitude. I would like to thank Dr. Mehdi Khosrow-Pour, President of IGI Global, for his kind invitation to write this book. I deeply appreciate patience and professional support of Julia Mosemann, Director of Book Publications who was always eager to help in difficult situations. Critical comments of two anonymous reviewers provided the valuable and objective opinion about my work and thus helped me to dramatically improve the final book draft before sending it to IGI. I am also indebted to my parents, Raisa and Gregory Okun, for their never-ending support, wise advice and love through my entire life.

This book is dedicated to my parents and my son Antoshka.

Oleg Okun
Malmo, Sweden, 9 May 2010

REFERENCES

Alterovitz, G., & Ramoni, M. F. (Eds.). (2007). *Systems bioinformatics: an engineering case-based approach*. Norwood, MA: Artech House.

Braga-Neto, U. M. (2007). Fads and fallacies in the name of small-sample microarray classification - A highlight of misunderstanding and erroneous usage in the applications of genomic signal processing. *IEEE Signal Processing Magazine, 24*(1), 91–99. doi:10.1109/MSP.2007.273062

Braga-Neto, U. M., & Dougherty, E. R. (2004). Bolstered error estimation. *Pattern Recognition, 36*(7), 1267–1281. doi:10.1016/j.patcog.2003.08.017

Chromatic. (2003). *Extreme programming pocket guide.* Sebastopol, CA: O'Reilly Media.

Cohen, W. W. (2007). *A computer scientist's guide to cell biology.* New York: Springer Science+Business Media.

Cristianini, N., & Hann, M. W. (2007). *Introduction to computational genomics: a case studies approach*. Cambridge, UK: Cambridge University Press.

Davis, T. A., & Sigmon, K. (2005). *MATLAB® Primer* (7th ed.). Boca Raton, FL: Chapman & Hall/CRC Press.

Lee, J. K. (Ed.). (2010). *Statistical bioinformatics: for biomedical and life science researchers*. Hoboken, NJ: Wiley-Blackwell.

Li, D.-C., Fang, Y.-H., Lai, Y.-Y., & Hu, S. C. (2009). Utilization of virtual samples to facilitate cancer identification for DNA microarray data in the early stages of an investigation. *Information Sciences*, *179*(16), 2740–2753. doi:10.1016/j.ins.2009.04.003

Martinez, W. L., & Martinez, A. R. (2002). *Computational statistics handbook with MATLAB*. Boca Raton, FL: Chapman & Hall/CRC Press.

Martinez, W. L., & Martinez, A. R. (2005). *Exploratory data analysis with MATLAB*. Boca Raton, FL: Chapman & Hall/CRC Press.

Nabney, I. T. (2002). *NETLAB: algorithms for pattern recognition*. London: Springer-Verlag.

Statnikov, A., Wang, L., & Aliferis, C. F. (2008). A comprehensive comparison of random forests and support vector machines for microarray-based cancer classification. *BMC Bioinformatics*, *9*(319).

Stork, D. G., & Yom-Tov, E. (2004). *Computer manual in MATLAB to accompany Pattern Classification* (2nd ed.). Hoboken, NJ: John Wiley & Sons.

van der Heijden, F., Duin, R., de Ridder, D., & Tax, D. M. (2004). *Classification, parameter estimation and state estimation: an engineering approach using MATLAB*. Hoboken, NJ: John Wiley & Sons. doi:10.1002/0470090154

van Emden, H. (2008). *Statistics for the terrified biologists*. Hoboken, NJ: John Wiley & Sons.

Yoon, Y., Lee, J., Park, S., Bien, S., Chung, H. C., & Rha, S. Y. (2008). Direct integration of microarrays for selecting informative genes and phenotype classification. *Information Sciences*, *178*(1), 88–105. doi:10.1016/j.ins.2007.08.013

Zar, J. H. (1999). *Biostatistical analysis* (4th ed.). Upper Saddle River, NJ: Prentice Hall/Pearson Education International.

Chapter 1
Biological Background

A LITTLE BIT OF BIOLOGY

As this book treats microarray gene expression based cancer classification as an application, let us briefly consider biological aspects of this task.

Let us first recall the "central dogma" of biology as it is described in (Cohen, 2007): deoxyribonucleic acid or DNA is a nucleic acid that stores genetic information needed for the development and functioning of all living beings; DNA is used to construct proteins in the following way: a section of DNA called a gene is transcribed to a molecule called a messenger[1] RNA (ribonucleic acid) or mRNA and then translated into a protein by a ribosome; proteins carry out most functions of cells such as regulation of translation and transcription and DNA replication. DNA and mRNA molecules are sequences of four different nucleotides. Proteins are sequences of twenty different amino acids.

After the protein is constructed, the gene is said to be expressed. Transcription and translation can be considered as a kind of transformations, one of which is applied to DNA while another one is applied to mRNA. The entire process can be expressed as follows: DNA → mRNA → protein. As you can see, genes being the

DOI: 10.4018/978-1-60960-557-5.ch001

essential parts of DNA play the important role in this process. Gene expression is often viewed as the process of the protein synthesis (though proteins are not the only possible products of gene expression). By monitoring gene expression, one can get an indirect estimate of protein abundance, which is important for determining biological function.

The expressed genes within mammalian cells can be divided into housekeeping and tissue-specific ones (Weinberg, 2007). Housekeeping genes are responsible for maintaining viability of all cell types in the body; they carry out biological functions common to all cell types. On the other hand, the tissue-specific genes produce proteins that are specifically associated with a given tissue.

Microarray technology helps to get the expression levels of many genes at once. It is thanks to this technology[2] that we are flooded nowadays with a plenty of such measurements, however, done for a few samples (the reason for this will be discussed in the next chapter).

A DNA microarray is an array of thousands of locations, each containing DNA for a different gene (Cohen, 2007), (Li, Tseng, & Wong, 2003). This array or (glass, plastic) slide contains a large library of thousands of single stranded cDNA (complementary DNA, i.e. DNA complementary to mRNA) clones (probes), corresponding to different genes, i.e. each spot on the slide corresponds to a specific gene. A typical use of microarrays is to extract two mRNA samples from two cell cultures or tissues (e.g., normal and cancerous), separately reverse transcribe them into cDNA, and using fluorescence labeling, dye the cDNA in these samples red (for the sample extracted from cancerous tissue) and green (for the sample extracted from healthy tissue), respectively (Cohen, 2007). Both samples are then spread across the microarray and left to hybridize to their corresponding complementary cDNA (labeled cDNA try to bind to their complementary cDNA on the microarray in order to form a double stranded molecule in the process called hybridization). Hybridization thus acts like a detector of the presence of a certain gene. The slide is then scanned to obtained numerical intensities of each dye. The result of scanning is an image. Finally, image processing is used to find out the color at each location of the array. The four variants are possible: if genes are expressed in both samples, the color will be yellow; if genes are expressed in neither sample, the color will be black; if genes are only expressed in one sample, the color will be either red or green. The intensity of a color indicates the level of expression, i.e. the number of mRNA transcribed. Given two differently dyed cDNA, the goal is to compare the intensity values I_R and I_G of the red and green channels at each spot of the microarray. The most popular statistic is the intensity log-ratio: $\log_2\left(I_R / I_G\right)$ (Kohane, Kho, & Butte, 2003), (Speed, 2003), (Drăghici, 2003).

DNA microarrays are described in detail in several good books (see, e.g., (Speed, 2003), (Drăghici, 2003), (Simon, Korn, McShane, Radmacher, Wright, & Zhao, 2003), (Kohane, Kho, & Butte, 2003), (Lee, 2004), (Lund, Nielsen, Lundegaard, Keşmir, & Brunak, 2005), (Zhang, 2006), (Mallick, Gold, & Baladandayuthepani, 2009)). In addition, the following books include a description of the most commonly applied bioinformatics algorithms: (Baldi & Brunak, 2001), (Kohane, Kho, & Butte, 2003), (Lund, Nielsen, Lundegaard, Keşmir, & Brunak, 2005), (Speed, 2003), (Lee, 2004), (Mallick, Gold, & Baladandayuthepani, 2009), (Simon, Korn, McShane, Radmacher, Wright, & Zhao, 2003), (Zhang, 2006).

Cancer or a malignant (invasive, metastatic) tumor can be seen as a disease of DNA due to gene alterations and mutations, which results in uncontrolled growth of cells (cell proliferation). That is, tumor does not appear from nowhere (it is not a foreign body): it arises from mutated normal cells (Weinberg, 2007), (Schulz, 2007). Once a cancerous cell has been created, it undergoes clonal expansion via cell division. In other words, "parents" of the first generation cancerous cells are mutated and altered normal cells while "parents" of the next generation cancerous cells are the first generation cancerous cells, etc.

Detailed biology of cancer is described in (Weinberg, 2007), (Schulz, 2007). The second book (Schulz, 2007) includes chapters on biology of different cancer types such as colon, bladder, prostate, breast.

As genes undergo changes during progression of cancer, so do gene expressions. That is, for a given gene or a set of genes, the expression level or levels in the normal (healthy) state can be different from those in the cancerous state. Based on this fact, the idea of cancer classification sprang. In other words, it is assumed that comparing microarray gene expression levels of healthy and cancerous cells, it is possible to distinguish between these two states and to diagnose cancer. However, in some cancers and for some tumor progression states, this difference is more profound than for the others. Besides, as cancerous cells originate from mutated normal cells, these two types of cells may share many genes for which expression levels are almost identical. In addition, during cancer development tumor may rapidly mutate, which affects tissue-specific gene expression values. Some cancers change their phenotype to resemble cells from a different tissue in a process called metaplasia (Schulz, 2007). One also needs to take into account the fact that different cancer types generally have different sets of tissue-specific expressed genes. Therefore, the task of assigning a patient to one of the two classes (healthy or diseased) is not as straightforward as it may seem. The sheer number of genes makes this task even more challenging. This is where machine learning is typically called for help.

REFERENCES

Baldi, P., & Brunak, S. (2001). *Bioinformatics: the machine learning approach* (2nd ed.). Cambridge, MA: MIT Press.

Brown, P. O., & Botstein, D. (1999). Exploring the new world of the genome with DNA microarrays. *Nature Genetics, 21*(Suppl.1), 33–37. doi:10.1038/4462

Cohen, W. W. (2007). *A computer scientist's guide to cell biology.* New York, NY: Springer Science+Business Media.

Drăghici, S. (2003). *Data analysis tools for DNA microarrays.* Boca Raton, FL: Chapman & Hall/CRC Press.

Kohane, I. S., Kho, A. T., & Butte, A. J. (2003). *Microarrays for an integrative genomics.* Cambridge, MA: MIT Press.

Lee, M.-L. T. (2004). *Analysis of microarray gene expression data.* Boston, MA: Kluwer Academic Publishers.

Li, C., Tseng, G. C., & Wong, W. H. (2003). Model-based analysis of oligonucleotide arrays and issues in cDNA microarray analysis. In T. Speed (Ed.), *Statistical analysis of gene expression microarray data* (pp. 1-34). Boca Raton, FL: Chapman & Hall/CRC Press. Lund, O., Nielsen, M., Lundegaard, Keşmir, C., & Brunak, S. (2005). *Immunological bioinformatics.* Cambridge, MA: MIT Press.

Mallick, B. K., Gold, D., & Baladandayuthepani, V. (2009). *Bayesian analysis of gene expression data.* Hoboken, NJ: John Wiley & Sons. doi:10.1002/9780470742785

Schulz, W. A. (2007). *Molecular biology of human cancers: an advanced student's textbook.* Dordrecht, the Netherlands: Springer.

Simon, R. M., Korn, E. L., McShane, L. M., Radmacher, M. D., Wright, G. W., & Zhao, Y. (2003). *Design and analysis of DNA microarray investigations.* New York, NY: Springer-Verlag.

Speed, T. (Ed.). (2003). *Statistical analysis of gene expression microarray data.* Boca Raton, FL: Chapman & Hall/CRC Press.

Velculescu, V. E., Zhang, L., Vogelstein, B., & Kinzler, K. W. (1995). Serial analysis of gene expression. *Science, 270*(5235), 484–487. doi:10.1126/science.270.5235.484

Weinberg, R. A. (2007). *The biology of cancer.* New York, NY: Garland Science.

Zhang, A. (2006). *Advanced analysis of gene expression microarray data.* Singapore: World Scientific.

ENDNOTES

[1]   The mRNA transports a copy of the information in the DNA from the nucleus of the cell into the cytoplasm of the cell, where the machinery for protein making resides. In the cytoplasm, the mRNA is translated into a protein, using a different chemical mechanism (Drăghici, 2003).

[2]   There are also other technologies for monitoring gene expressions, e.g., such as Affymetrix oligonucleotide expression array (Brown & Botstein, 1999) or serial analysis of gene expression (SAGE) (Velculescu, Zhang, Vogelstein, & Kinzler, 1995). The former is microarray technology, too, while the latter is not.

Chapter 2
Gene Expression Data Sets

BIOLOGICAL DATA AND THEIR CHARACTERISTICS

Before embarking on a long tour across different machine learning methods, it is useful to look at some popular and their characteristics.

First of all, a typical gene expression data set contains a matrix X of real numbers. Let D and N be the number of its rows and columns, respectively. Then X is represented as

$$X = \begin{bmatrix} x_{11} & \cdots & x_{1N} \\ \vdots & \ddots & \vdots \\ x_{D1} & \cdots & x_{DN} \end{bmatrix},$$

where x_{ij} represents the value in the ith row and jth column.

DOI: 10.4018/978-1-60960-557-5.ch002

As there are thousands of gene expressions and only a few dozens of samples, D (the number of genes) is of order 1,000-10,000 while N (the number of biological samples) is somewhere between 10 and 100. Such a condition makes the application of many traditional statistical methods impossible as those techniques were developed under the assumption that $N \gg D$. You may ask: what's a problem?

The problem is in an underdetermined system where there are only a few equations versus many more unknown variables (Kohane, Kho, & Butte, 2003). Hence, the solution of such a system is not unique. In other words, multiple solutions exist. By translating this into the biological language of the applied problem treated in this book, this means that multiple subsets of genes may be equally relevant to cancer classification (Ein-Dor, Kela, Getz, Givol, & Domany, 2005), (Díaz-Uriarte & Alvarez de Andrés, 2006). However, in order to reduce a chance for noisy and/or irrelevant genes to be included into one of such subsets, one needs to eliminate irrelevant genes before the actual classification.

You may also wonder why it is impossible to increase N. The answer is that this is difficult as the measurement of gene expression requires a functionally relevant tissue taken under the right conditions, which is sadly rare due to impossibility to meet all requirements at once in practice (read more about these problems in (Kohane, Kho, & Butte, 2003)). So, we are left with the necessity to live and to deal with high dimensional data.

Below five popular gene expression data sets are briefly described in order to give a realistic picture of what gene expression data are.

Examples of Gene Expression Data Sets

For each data set below, all the data are stored in a text file, though the file extension may not be necessarily .txt. Often gene names are stored in a separate text file; hence, it is useful and recommended to study the content of all text files associated with a given data set. Downloading such files is straightforward but extracting numerical information is not so as different files stores gene expression data mixed with textual headers and other text information. In order to access gene expression data, one needs to write a separate script or program for each data set after studying the data structure in a file storing them. Any programming language or environment has commands/functions for input/output file operations. Sometimes, the entire file can be read into RAM memory during a single reading operation. However, as textual and numerical data are mixed together, further efforts are usually necessary in order to separate text from non-text. If you care of the standard way of storing and exchange gene expression data, then the MicroArray and Gene Expression (MAGE) group (http://www.mged.org/Workgroups/MAGE/mage.html) provides all the standardization information, which facilitates the exchange of data between different data systems.

Colon Data Set

This oligonucleotide data set[1], introduced in (Alon, et al., 1999), contains expression of 2,000 genes for 62 cases (22 normal and 40 colon tumor). Preprocessing includes the logarithmic transformation to base 10, followed by normalization to zero mean and unit variance.

Brain Data Set 1

This oligonucleotide data set[2], introduced in (Pomeroy, et al., 2002), contains cases of two classes of brain tumor. The data set, also known as Dataset B, contains 34 medulloblastoma cases, 9 of which are desmoplastic and 25 are classic. Preprocessing consists of thresholding of gene expression with a floor of 20 and ceiling of 16,000; filtering with exclusion of genes with $max / min \leq 3$ or $max - min < 100$, where max and min refer to the maximum and minimum expression of a certain gene across the 34 cases, respectively; base 10 logarithmic transformation; normalization across genes to zero mean and unit variance. As a result, 5,893 out of 7,129 original genes are only retained.

Brain Data Set 2

This oligonucleotide data set[3], known as Dataset C in (Pomeroy, et al., 2002), contains 60 medulloblastoma cases, corresponding to 39 survivors and 21 nonsurvivors according to the patient status. Preprocessing consists of thresholding of gene expression with a floor of 100 and ceiling of 16,000; filtering with exclusion of genes with $max / min \leq 5$ or $max - min < 500$, where max and min refer to the maximum and minimum expression of a certain gene across the 60 cases, respectively; base 10 logarithmic transformation; normalization across genes to zero mean and unit variance. As a result, 4,459 out of 7,129 original genes are only retained.

Prostate Data Set 1

This oligonucleotide data set[4], introduced in (Singh, et al., 2002), includes expression of 12,600 genes in 52 prostate and 50 normal cases.

Prostate Data Set 2

As the previous data set, this one[5] has been introduced in (Singh, et al., 2002). However, it was obtained independently of Prostate Data Set 1. It has 25 prostate and 9 normal cases with 12,600 expressed genes.

REFERENCES

Alon, U., Barkai, N., Notterman, D., Gish, K., Ybarra, S., & Mack, D. (1999). Broad patterns of gene expression revealed by clustering analysis of tumor and normal colon tissues probed by oligonucleotide arrays. *Proceedings of the National Academy of Sciences of the United States of America*, *96*(12), 6745–6750. doi:10.1073/pnas.96.12.6745

Díaz-Uriarte, R., & Alvarez de Andrés, S. (2006). Gene selection and classification of microarray data using random forest. *BMC Bioinformatics*, *7*(3).

Ein-Dor, L., Kela, I., Getz, G., Givol, D., & Domany, E. (2005). Outcome signature genes in breast cancer: is there a unique set? *Bioinformatics (Oxford, England)*, *21*(2), 171–178. doi:10.1093/bioinformatics/bth469

Kohane, I. S., Kho, A. T., & Butte, A. J. (2003). *Microarrays for an integrative genomics*. Cambridge, MA: MIT Press.

Pomeroy, S., Tamayo, P., Gaasenbeek, M., Sturla, L., Angelo, M., & McLaughlin, M. (2002). Prediction of central nervous system embryonal tumour outcome based on gene expression. *Nature*, *415*(6870), 436–442. doi:10.1038/415436a

Singh, D., Febbo, P., Ross, K., Jackson, D., Manola, J., & Ladd, C. (2002). Gene expression correlates of clinical prostate cancer behavior. *Cancer Cell*, *1*(2), 203–209. doi:10.1016/S1535-6108(02)00030-2

ENDNOTES

[1]   It can be downloaded from http://genomics-pubs.princeton.edu/oncology/.
[2]   It can be downloaded from http://www.broadinstitute.org/mpr/CNS/.
[3]   See the link in note 2.
[4]   It can be downloaded from http://www.broadinstitute.org/cgi-bin/cancer/publications/pub_paper.cgi?mode=view&paper_id=75.
[5]   See the link in note 4.

Chapter 3
Introduction to Data Classification

PROBLEM OF DATA CLASSIFICATION

Data classification is a scientific discipline researching the ways of assigning class membership values (labels) to unknown (in the sense that they have not previously be seen to an observer) observations or samples, based on a set of observations or samples provided with class membership values (labels). Each observation is represented by a feature vector associated with it. Unknown observations form a test set while their labeled counterparts together with class labels compose a training set. Labeling unknown observations is done by means of a classifier, which is a data classification algorithm implementing a mapping of feature vectors to class labels. An algorithm is a sequence of steps necessary for the solution of a data classification task at hand. Let us restrict ourselves to two-class problems.

Thus, the process of data classification involves either one or two following steps:

- Training a classifier (optional),
- Testing the trained classifier.

DOI: 10.4018/978-1-60960-557-5.ch003

Many classifiers have one or several parameters to be pre-defined before classification will start. Without knowing the optimal values of these parameters, data classification would be akin to random walk in search for the right solution. The words 'optimal values' mean such parameter values that allow a classifier to learn the correct mapping from features, describing each observation, to class labels. This learning done from the training data is possible, because the learner can always check the answer: for this, it needs to compare its output and the correct result as specified by class labels assigned to the observations from the training set. If there is a mismatch (classification error) between the two, the learner knows there should be some work to do.

Typically, classification error on the training data is not precisely 0% and sometimes it simply cannot be so due to the finite (limited) size of the training set. On the contrary, the zero error rate may indicate that you over-trained the classifier so that it learned every minute detail, which is often nothing but noise (garbage)[1]. Such a classifier will be unable to properly generalize. In other words, when presented with previously unseen data, its classification performance will be very bad. The smaller the training set is, the higher your chances to over-train a classifier, because different classes are likely to be under-represented. The more sophisticated classifier is, the higher chances are for its over-training, since sophisticated classifiers are capable of partitioning data classes in more complex ways (decision boundaries) than simpler classifiers. So, the training could be both evil and blessing. Sometimes, a classifier does not need training at all[2], which, however, does not automatically imply that this classifier will do its job well in all cases.

Independently of the fact whether classifier training is required or not, the testing phase has still to be carried out to complete the data classification task. This is done with the trained classifier applied to the test data. If no training was needed, then the word 'trained' is omitted before 'classifier' in the last sentence. As a result of testing, test error and other performance characteristics such Area Under the Receiver Operating Characteristic (ROC) Curve are computed, which can be further compared (by means of statistical tests) with errors/characteristics of other classifiers attained on the same test set.

Given that microarray gene expression data are high dimensional, it is advisable and even required reducing the number of features prior to data classification in order to alleviate the effect of classifier over-training. That is, dimensionality reduction should always precede classification when dealing with gene expression data sets.

There are many different classifiers that can be applied to gene expression data. In a series of chapters that follow we pay attention to the most common of them: Naïve Bayes, Nearest Neighbor, Decision Tree, and Support Vector Machine. These classifiers were also recently named among the top 10 algorithms in Data Mining. In addition, they form a bulk of base classifiers used in building classifier ensembles.

Finally, it is always good to know how to build a good classifier. We believe that the paper of Braga-Neto (Braga-Neto, 2007) can help you to avoid some common misunderstandings in the classifier design related to microarray data.

REFERENCES

Braga-Neto, U. M. (2007). Fads and fallacies in the name of small-sample microarray classification - A highlight of misunderstanding and erroneous usage in the applications of genomic signal processing. *IEEE Signal Processing Magazine, 24*(1), 91–99. doi:10.1109/MSP.2007.273062

Dudoit, S., & Fridlyand, J. (2003). Classification in microarray experiments. In Speed, T. (Ed.), *Statistical analysis of gene expression microarray data* (pp. 93–158). Boca Raton, FL: Chapman & Hall/CRC Press. doi:10.1201/9780203011232.ch3

Lu, Y., & Han, J. (2003). Cancer classification using gene expression data. *Information Systems, 28*(4), 243–268. doi:10.1016/S0306-4379(02)00072-8

Pepe, M. S. (2004). *The statistical evaluation of medical tests for classification and prediction* (1st paperback ed.). Oxford, UK: Oxford University Press.

Wu, X., & Kumar, V. (Eds.). (2009). *The top ten algorithms in data mining*. Boca Raton, FL: Chapman & Hall/CRC Press.

ENDNOTES

[1]   This effect is also known as overfitting.
[2]   It can happen when the optimal or close to optimal parameter values are known in advance, so that there is no need to learn them. Hence, the training step can be skipped even for classifiers that have parameters to set up.

Chapter 4
Naïve Bayes

BAYES AND NAÏVE BAYES

The name for this classifier looks a bit strange. You may ask: who is Bayes and why this person is considered to be naïve. You may also question the power of such a "naïve" algorithm. But do not rush to make conclusions as we will see later in this chapter that Naïve Bayes is not helpless in practice.

This classifier got its name after Thomas Bayes (1702-1761), the Englishman who was a clergyman as well as a mathematician. His contribution "Essay towards solving a problem in the doctrine of chances" was published in 1764 three years after his death in the Philosophical Transactions of the Royal Society. His work concerned so called inverse probability in connection with gambling and insurance and contained a version of what is today known as Bayes' Theorem. This theorem lays foundations of Naïve Bayes.

DOI: 10.4018/978-1-60960-557-5.ch004

From Probability to Bayes' Theorem

We begin with a number of definitions that will be used in this chapter and across the book. As a primal source of our definitions, we will rely on (Everitt, 2006).

The notion of probability lies in the heart of Naïve Bayes. Hence, let us introduce it first. Thus, probability is a measure associated with an event A and denoted by $P(A)$ which takes a value such that $0 \leq P(A) \leq 1$ (Everitt, 2006). Probability is the quantitative expression of the chance that an event will occur. The higher $P(A)$, the more likely it is that the event A will occur. If an event cannot happen, $P(A) = 0$; if an event is certainly to happen, $P(A) = 1$. Suppose that there are n equally likely outcomes and the event A is associated with $r(0 \leq r \leq n)$ of them, then $P(A) = r / n$. This formula for calculating $P(A)$ is based on the assumption that an experiments can be repeated a large number of times, n, and in r cases the event A occurs. The quantity r / n is called the relative frequency of A. As $n \to \infty$, $\lim_{n \to \infty}(r / n) \to P(A)$, i.e., the relative frequency converges to probability in the limit.

$P(A)$ is often called unconditional probability, because it does not depend on the outcome of other events (Everitt, 2006). This implies that there is also conditional probability, which is the probability that an event A occurs given the outcome of some other event B. This conditional probability is denoted by $P(A|B)$ and should be read as the probability of A given B. It is not true that $P(A|B) = P(B|A)$. If events A and B are independent of each other, then $P(A|B) = P(A)$.

If two events A and B are mutually exclusive (i.e., they cannot occur at the same time), the probability of either event occurring is the sum of the individual probabilities, i.e.

$$P(A\,or\,B) = P(A) + P(B).$$

It is straightforwardly to extend this formula to k mutually exclusive events $A_1, A_2, ..., A_k$:

$$P(A_1\,or\,A_2\,...or\,A_k) = P(A_1) + P(A_2) + ... + P(A_k) = \sum_{i=1}^{k} P(A_i).$$

This formula is the sum or addition rule for probabilities.

For events A and B that are independent, the probability that both occur is the product of the individual probabilities:

$$P\left(A\,and\,B\right) = P\left(A,B\right) = P\left(A\right)P\left(B\right).$$

If there are k independent events, then the last formula can be extended to

$$P\left(A_1\,and\,A_2\,...\,and\,A_k\right) = P\left(A_1,A_2,...,A_k\right) = P\left(A_1\right)P\left(A_2\right)...P\left(A_k\right) = \prod_{i=1}^{k}P\left(A_i\right).$$

As you might already guess, this formula is called the product or multiplication rule for probabilities.

In the last two formulas it was assumed that events are independent of each other. If this is not the case, then

$$P\left(A,B\right) = P\left(A|B\right)P\left(B\right).$$

Since $P\left(A,B\right) = P\left(B,A\right)$,

$$P\left(A,B\right) = P\left(A|B\right)P\left(B\right) = P\left(B,A\right) = P\left(B|A\right)P\left(A\right).$$

From this equation, we can get the famous Bayes' Theorem as follows

$$P\left(B|A\right) = \frac{P\left(A|B\right)P\left(B\right)}{P\left(A\right)}.$$

This theorem gives us a procedure for revising and updating the probability of some event in the light of new evidence[1]. Given k mutually exclusive and exhaustive events $C_1, C_2, ..., C_k$, we rewrite the last formula as

$$P\left(C_j|A\right) = \frac{P\left(A|C_j\right)P\left(C_j\right)}{\sum_{i=1}^{k}P\left(A|C_i\right)P\left(C_i\right)},$$

where the sum rule in the form $P(A) = \sum_B P(A, B)$ (marginalization) was used in order to express the denominator in terms of the quantities appearing in the numerator.

This gives the probabilities of the C_j when A is known to have occurred. The quantity $P(C_j)$ is termed the prior probability and $P(C_j|A)$ the posterior probability, i.e. the probability of C_j after A has become known. As you can see, there is a subtle but important difference between prior and posterior probabilities. In the former case, we judge on the probability of C_j without knowing anything about A while in the latter case, we estimate the probability of C_j after observing A.

Now let us come back to the classification problem. Let us assume that $C_1, \ldots, C_k$ are k different classes of data and A is a feature vector of gene expression values. In this interpretation the posterior probability $P(C_j|A)$ is the probability of assigning an example x to class C_j, based on the feature vector A.

Let us suppose that $k = 2$. Then

$$P(C_1|A) = \frac{P(A|C_1)P(C_1)}{P(A|C_1)P(C_1) + P(A|C_2)P(C_2)},$$

$$P(C_2|A) = \frac{P(A|C_2)P(C_2)}{P(A|C_1)P(C_1) + P(A|C_2)P(C_2)}.$$

The natural classification rule for classifying x given these posterior probabilities is $C^* = arg \max_{j=1,2} P(C_j|A)$. In other words, we assign x to the class that has the highest posterior probability. One can notice that when comparing the posteriors, one can safely omit the denominator from calculations as it does not affect the result. Therefore, the posterior probabilities are

$$P(C_1|A) \propto P(A|C_1)P(C_1),$$

$$P(C_2|A) \propto P(A|C_2)P(C_2),$$

where the symbol $\propto$ stands for 'proportional to'.

The quantity $P\left(A|C_j\right)$ is termed the likelihood, which is the probability of A given C_j. It can be seen that both the prior probabilities and likelihoods can be found from the data. Thus, construction of the classification based on Bayes' Theorem does not seem to be difficult. But where is 'Naïve'?

From Bayes' Theorem to Naïve Bayes

The vector of features A consists of individual features $A_i, i = 1, \ldots, D$. In general, features are not conditionally independent given class. However, in Naïve Bayes, it is assumed that they are conditionally independent, which leads to

$$P\left(C_j|A\right) \propto P\left(C_j\right)\prod_{i=1}^{D}P\left(A_i|C_j\right).$$

Assuming features independent allows reducing the D dimensional multivariate problem (estimating $P\left(A_1, \ldots, A_D|C_j\right)$) to the D univariate problems (estimating $P\left(A_1|C_j\right), \ldots, P\left(A_D|C_j\right)$). This shift leads to simple computations and requires a smaller training set to obtain accurate estimates than does the estimation of multivariate distribution. Besides, for microarray data, the accurate estimation of multivariate probabilities is impossible due to the small training set size.

Feature independence also facilitates the assessment of importance of each feature as shown in (Hand, 2009). However, one needs to remember that for such bioinformatics tasks as cancer classification and prediction, features are likely to be important in the group rather than individually. Hence, using Naïve Bayes for assessing feature importance should be treated as a preliminary step to more sophisticated analysis. The feature independence assumption protects against overfitting (over-trained classifier) and therefore, Naïve Bayes is one of the natural choices for microarray data classification (Keller, Schummer, Hood, & Ruzzo, 2000), (Lu & Han, 2003), (Blanco, Larrañaga, Inza, & Sierra, 2004), (Paul & Iba, 2004), (Wang, Makedon, Ford, & Pearlman, 2005), (De Ferrari & Aitken, 2006), (Finak, et al., 2008) as it can outperform more sophisticated classifiers. Feature selection prior to classification would likely improve the classification accuracy of Naïve Bayes (Hand, 2009).

Because of this deliberate ignorance of feature dependence, the classifier based on the formula above was named Naïve Bayes. However, as was shown in (Domingos & Pazzani, 1997), this did not prevent it to perform remarkably well even when feature dependence was known to exist. Besides, searching for feature dependencies is not necessarily the best approach to improving the performance of Naïve Bayes (Domingos & Pazzani, 1997). More information about these remarkable aspects of Naïve Bayes can be found in (Domingos & Pazzani, 1997) and (Hand, 2009).

Algorithm Description

One can notice that Naïve Bayes does not have parameters to set up, which makes it very attractive.

It is clear how to calculate $P\left(C_j\right)$ from the training data: N_j / N, where N is the total number of examples in a data set and N_j is the number of examples with label C_j. See also (Hand, 2009) for discussion if the classes are unbalanced, i.e. one class is much larger than the other. But how to find $P\left(A_i|C_j\right)$ given the fact that features (gene expression levels) are continuous? The answer is feature discretization when each feature is assumed to take a limited set of discrete (categorical) values, e.g., -1, 0, and 1. This change results in the following probabilities to be computed for the ith feature and the jth class:

$$P\left(A_i = -1|C_j\right),\ P\left(A_i = 0|C_j\right),\ P\left(A_i = +1|C_j\right).$$

For instance, if $i = 1, 2, 3$ and $j = 1, 2$, in theory one needs to estimate the 18 conditional probabilities:

$$P\left(A_1 = -1|C_1\right),\ P\left(A_1 = 0|C_1\right),\ P\left(A_1 = +1|C_1\right),$$

$$P\left(A_2 = -1|C_1\right),\ P\left(A_2 = 0|C_1\right),\ P\left(A_2 = +1|C_1\right),$$

$$P\left(A_3 = -1|C_1\right),\ P\left(A_3 = 0|C_1\right),\ P\left(A_3 = +1|C_1\right),$$

$$P\left(A_1 = -1|C_2\right),\ P\left(A_1 = 0|C_2\right),\ P\left(A_1 = +1|C_2\right),$$

$$P\left(A_2 = -1|C_2\right),\ P\left(A_2 = 0|C_2\right),\ P\left(A_2 = +1|C_2\right),$$

$$P\left(A_3 = -1|C_2\right),\ P\left(A_3 = 0|C_2\right),\ P\left(A_3 = +1|C_2\right).$$

In general, for D features, d categories per feature and k classes of data, the number of conditional probabilities to estimate based on training data is equal to $D \times d \times k$. As D is rather large for microarray data, it is recommended that d was small, i.e. 2 or 3. The computed conditional probabilities form a look-up table to be used during the test phase.

When classifying test data, we, however, do not need all conditional probabilities from this table. Suppose that after discretization, a 3-dimensional test example contains the following values: $A_1 = 0, A_2 = -1, A_3 = 0$. As a result, we need only 6 probabilities to retrieve from the look-up table:

$$P\left(A_1 = 0 | C_1\right),\ P\left(A_2 = -1 | C_1\right),\ P\left(A_3 = 0 | C_1\right),$$

$$P\left(A_1 = 0 | C_2\right),\ P\left(A_2 = -1 | C_2\right),\ P\left(A_3 = 0 | C_2\right).$$

Once this is done, we readily obtain the posterior probabilities

$$P\left(C_1 | A\right) \propto P\left(C_1\right) P\left(A_1 = 0 | C_1\right) P\left(A_2 = -1 | C_1\right) P\left(A_3 = 0 | C_1\right),$$

$$P\left(C_2 | A\right) \propto P\left(C_2\right) P\left(A_1 = 0 | C_2\right) P\left(A_2 = -1 | C_2\right) P\left(A_3 = 0 | C_2\right),$$

and choose the class corresponding to the highest posterior probability.

Thus, during the training phase, the look-up table of conditional probabilities is created, which is then used for classifying test examples. One should always remember to convert continuous feature values into categorical (discrete) ones. The same feature transformation rule has to be applied to both training and test data. Good examples with detailed explanations of how Naïve Bayes works are also presented in (Bramer, 2007), (Hand, 2009).

To summarize, pseudo code of Naïve Bayes is given in Figure 1.

MATLAB Code

As can be seen from pseudo code in Figure 1, feature discretization is applied to both training and test data before all probabilities are estimated. Although there are many algorithms to convert continuous values into discrete ones (Fayyad & Irani, 1993), (Dougherty, Kohavi, & Sahami, 1995), (Ho & Scott, 1997), (Liu, Hussain, Tan, & Dash, 2002), (Zhou, Wang, & Dougherty, 2003), we opted for the algorithm in (Ding & Peng, 2005) as it was designed with gene expression data in view.

Figure 1. Pseudo code of Naive Bayes

Naïve Bayes

Make both training and test data discrete;

Estimate the prior probabilities $P(C_j), j = 1, \ldots, k$ from the training data, where k is the number of classes;
Estimate the conditional probabilities $P(A_i = a_\ell | C_j), i = 1, \ldots, D, \ j = 1, \ldots, k, \ell = 1, \ldots, d$, from the training data, where D is the number of features, d is the number of discretization levels;

Estimate the posterior probabilities $P(C_j | A)$ for each test example x represented by a feature vector A;
Assign x to the class C^* such that $C^* = arg \max_{j=1,2} P(C_j | A)$;

This algorithm, implemented in *feature_discretization*, converts each feature individually. First, training data are processed, followed by test data. For this, the mean μ_i and standard deviation σ_i are first computed for each feature. If a gene expression level is below $\mu_i - \sigma_i / 2$ (under-expression), this level is replaced with -1; if a gene expression level is above $\mu_i + \sigma_i / 2$ (over-expression), it is set to +1; if a gene expression level belongs to $\left[\mu_i - \sigma_i / 2, \mu_i + \sigma_i / 2\right]$ (baseline), then its value is changed to 0. Thus, there are three categorical levels: -1, 0, and +1. It should be noticed that after the means and standard deviations of genes were computed based on the training data, they are used for the test data without re-computing as it is assumed that both training and test data came from the same distribution, which eliminates the necessity of separate computations for test data.

Once features have been made discrete, all the probabilities can be readily estimated for each class of data[2]. Univariate conditional probabilities are estimated by means of multinomial histogram-type technique using MATLAB® function *histc*. These probabilities are stored in a 2D array (look-up table), having $D \times d$ rows and k columns. The jth column of this table stores the probabilities in the following order:

$$P\left(A_1 = -1 | C_j\right), P\left(A_1 = 0 | C_j\right), P\left(A_1 = +1 | C_j\right),$$

$$P\big(A_2 = -1|C_j\big), P\big(A_2 = 0|C_j\big), P\big(A_2 = +1|C_j\big),$$

$$P\big(A_3 = -1|C_j\big), P\big(A_3 = 0|C_j\big), P\big(A_3 = +1|C_j\big),$$

$$\cdots\cdots\cdots\cdots\cdots\cdots\cdots\cdots\cdots\cdots\cdots\cdots$$

$$P\big(A_{D-1} = -1|C_j\big), P\big(A_{D-1} = 0|C_j\big), P\big(A_{D-1} = +1|C_j\big),$$

$$P\big(A_D = -1|C_j\big), P\big(A_D = 0|C_j\big), P\big(A_D = +1|C_j\big),$$

that is, three probabilities related to A_1 are first stored, followed by three probabilities for A_2, ..., followed by three probabilities associated with A_D.

To retrieve probabilities from this table in the test phase, an indexing mechanism, which takes into account the record sequence above, is implemented with the help of MATLAB® function *interp1*. The entire D dimensional test feature vector is discretized at once. Each continuous value is mapped into its discrete equivalent through 1D interpolation. MATLAB® function *prod* computes the product of all conditional probabilities specified by a given index vector.

Naïve Bayes is also implemented as an option of MATLAB® function *classify* from Statistics Toolbox™. The latest versions of this toolbox also include the object-oriented implementation of Naïve Bayes, which is used in our function *NaiveBayes_moo* below.

All functions implemented below return either true or uncalibrated posterior probabilities. The 'uncalibrated' implies that these are rather scores than true probabilities as they do not sum up to 1. Independently of the output type, these quantities are suitable for the Receiver Operating Characteristics (ROC) curve generation used in the classification performance evaluation (Fawcett, 2006), (Pepe, 2004). And one can always convert uncalibrated scores to the true probabilities by summing all scores for each instance and then dividing each score by this sum.

```
function [test_targets, varargout] = NaiveBayes(train_patterns,
train_targets, test_patterns)

% Naive Bayes classifier
% Original sources:
% Max Bramer, "Principles of data mining", Springer-Verlag,
```

```
% Berlin/Heidelberg, 2007
% David J. Hand, "Naive Bayes", in Xindong Wu and Vipin Kumar
(eds.),
% "The top ten algorithms in data mining", pp.163-178, Chapman
&
% Hall/CRC
% Press, Boca Raton, FL, 2009
% Pedro Domingos and Michael Pazzani, "On the optimality of the
simple
% Bayes classifier under zero-one loss", Machine Learning
29(2-3): 103-% 130,1997
%
% Inputs:
%    train_patterns  - Training patterns (DxN matrix, where D
- data
%                      dimensionality, N - number of patterns)
%    train_targets   - Training targets (1xN vector)
%    test_patterns   - Test patterns (DxNt matrix, where D -
data
%                      dimensionality, Nt - number of patterns)
%
% Outputs:
%    test_targets    - Test targets (1xNt vector)
%    prob            - Uncalibrated posterior probabilities
(CxNt
%                      matrix, where C - the number of classes)



% Get the number N of training patterns and a data dimensional-
ity D
[D,N] = size(train_patterns);

% Feature discretization
[train_patterns, test_patterns, cat_val] = feature_
discretization(train_patterns, test_patterns);

% Get unique labels and the number of classes
u = unique(train_targets);
nc = length(u);
```

```matlab
% Get the number of categories
mc = length(cat_val);

% Pre-allocate memory for probabilities
priorprob = zeros(1,nc);
condprob = zeros(mc*D,nc);

% Compute the probabilities from the training data
% (Naive Bayes classifier training)
for i = 1:nc
    % Get indices of patterns belonging to the ith class
    idx = find(train_targets == u(i));
    % Get the number of patterns of the ith class
    n = length(idx);
    % Compute the prior probability of the ith class
    priorprob(i) = n/N;

    % Compute the number of occurrences of each categorical
value for
    % each feature, given the ith class
    h = histc(train_patterns(:,idx), cat_val, 2)';
    % Convert a matrix into a column-vector
    h = h(:);
    % Compute conditional probabilities for the ith class
    condprob(:,i) = h/n;
end

% Pre-allocate memory for the output
test_targets = zeros(1,size(test_patterns,2));
% Pre-allocate memory for posterior probabilities
p = zeros(1,nc);

% Fixed increment to index a 2D look-up table of conditional
% probabilities
fix_inc = 1:mc:mc*D;

% Look-up table for converting a feature value to a variable
increment % for indexing a 2D look-up table of conditional
probabilities
lut = 0:mc-1;
```

```
% Classify test data with the trained Naive Bayes classifier
for j = 1:size(test_patterns, 2)
    % Variable increment to index a 2D look-up table of condi-
tional
    % probabilities
    % Use 1D interpolation to get an index into a look-up table
    var_inc = interp1(cat_val, lut, test_patterns(:,j));
    % var_inc is a column-vector so we need to transpose it!
    var_inc = var_inc';
    % Form the final index
    idx = fix_inc + var_inc;
    for i = 1:nc
        p(i) = priorprob(i)*prod(condprob(idx,i));
    end
    [dummy,i] = max(p);
    test_targets(j) = u(i);
    if nargout == 2
        prob(:,j) = p';
    end
end

if nargout == 2
    varargout{1} = prob;
end

return


% Perform discretization of each feature values into three
categories
function [train_patterns, test_patterns, cat_val] = feature_
discretization(train_patterns, test_patterns)

% Discretization of continuous features into three categories
% Source: Chris Ding and Hanchuan Peng, "Minimum redundancy
feature
% selection from microarray gene expression data", Journal of
% Bioinformatics and Computational Biology 3(2): 185-205, 2005
%
```

```
% Inputs:
%    train_patterns  - Training patterns (DxN matrix, where D
- data
%                      dimensionality, N - number of patterns)
%    test_patterns   - Test patterns (DxNt matrix, where D -
data
%                      dimensionality, Nt - number of patterns)
%
% Outputs:
%    train_patterns  - Training patterns (DxN matrix, where D
- data
%                      dimensionality, N - number of patterns)
%    test_patterns   - Test patterns (DxNt matrix, where D -
data
%                      dimensionality, Nt - number of patterns)
%    cat_val         - Categorical values (1x3 vector)



% Start from the training data

% Compute the mean and standard deviation for each feature
mu = mean(train_patterns, 2);
sigma = std(train_patterns, [], 2);

% Create a copy of train_patterns
X = zeros(size(train_patterns));

% Do discretization into categories
i = find(train_patterns < repmat(mu - sigma/2, [1 size(train_
patterns, 2)]));
X(i) = -1;
i = find(train_patterns > repmat(mu + sigma/2, [1 size(train_
patterns, 2)]));
X(i) = 1;
% Now X contains only three values: -1, 0, and +1

% Write the output
train_patterns = X;
```

```
clear X;

% Now process the test data in a similar fashion
% Note that mu and sigma are not re-computed since it was
assumed that
% both training and test data com from the same distribution

% Create a copy of test_patterns
X = zeros(size(test_patterns));

% Do discretization into categories
i = find(test_patterns < repmat(mu - sigma/2, [1 size(test_pat-
terns, 2)]));
X(i) = -1;
i = find(test_patterns > repmat(mu + sigma/2, [1 size(test_pat-
terns, 2)]));
X(i) = 1;
% Now X contains only three values: -1, 0, and +1

% Write the outputs
test_patterns = X;
cat_val = [-1 0 1];

return
function [test_targets, varargout] = NaiveBayes_moo(train_pat-
terns, train_targets, test_patterns)

% Naive Bayes classifier built on MATLAB Statistics Toolbox,
% object-oriented functionality
% Original sources:
% Max Bramer, "Principles of data mining", Springer-Verlag,
% Berlin/Heidelberg, 2007
% David J. Hand, "Naive Bayes", in Xindong Wu and Vipin Kumar
(eds.),
% "The top ten algorithms in data mining", pp.163-178, Chapman
&
% Hall/CRC Press, Boca Raton, FL, 2009
% Pedro Domingos and Michael Pazzani, "On the optimality of the
simple
% Bayes classifier under zero-one loss", Machine Learning
```

```
29(2-3): 103-% 130, 1997
%
% Inputs:
%    train_patterns   - Training patterns (DxN matrix, where D
- data
%                       dimensionality, N - number of patterns)
%    train_targets    - Training targets (1xN vector)
%    test_patterns    - Test patterns (DxNt matrix, where D -
data
%                       dimensionality, Nt - number of patterns)
%
% Outputs:
%    test_targets     - Test targets (1xNt vector)
%    prob             - Posterior probabilities (CxNt matrix,
where C -
%                       the number of classes)
%
% Note: it requires Statistics Toolbox 7, released in 2009



% Get a version number for Statistics Toolbox
v = ver('stats');
% If MATLAB release is before 2009, generate an error message
if isempty(v)
    error('No Statistics toolbox was found!');
else
    if str2num(v.Version) < 7.1
        error('Version 7.1 (or higher) of Statistics Toolbox is
needed to run this function!');
    end
end

% Feature discretization
[train_patterns, test_patterns, dummy] = feature_
discretization(train_patterns, test_patterns);

% Create Naive Bayes classifier object nb by fitting training
data
% Calling the constructor NaiveBayes for the same purpose won't
```

```
work!
% Pay attention to the necessity to transpose input arrays
since they
% should be of NxD rather than DxN in size
% Multivariate multinomial distribution for categorical fea-
tures is
% assumed with the priors computed from the training data
nb = NaiveBayes.fit(train_patterns', train_targets', 'Distribu-
tion', 'mvmn', 'Prior', 'empirical');
% Predict class labels for test data
test_targets = predict(nb, test_patterns', 'HandleMissing',
'on');
% Don't forget to convert test_targets into the default repre-
sentation
% of targets vectors
test_targets = test_targets';

if nargout == 2
    % Now get posterior probabilities of each class
    prob = posterior(nb, test_patterns', 'HandleMissing',
'on');
    % Covert probabilities to the default representation
    prob = prob';
    varargout{1} = prob;
end

return
```

REFERENCES

Blanco, R., Larrañaga, P., Inza, I., & Sierra, B. (2004). Gene selection for cancer classification using wrapper approaches. *International Journal of Pattern Recognition and Artificial Intelligence, 18*(8), 1373–1390. doi:10.1142/S0218001404003800

Bramer, M. (2007). *Principles of data mining*. London: Springer-Verlag.

De Ferrari, L., & Aitken, S. (2006). Mining housekeeping genes with a Naive Bayes classifier. *BMC Genomics, 7*, 277. doi:10.1186/1471-2164-7-277

Ding, C., & Peng, H. (2005). Minimum redundancy feature selection from microarray gene expression data. *Journal of Bioinformatics and Computational Biology*, *3*(2), 185–205. doi:10.1142/S0219720005001004

Domingos, P., & Pazzani, M. (1997). On the optimality of the simple Bayesian classifier under zero-one loss. *Machine Learning*, *29*(2-3), 103–130. doi:10.1023/A:1007413511361

Dougherty, J., Kohavi, R., & Sahami, M. (1995). Supervised and unsupervised discretization of continuous features. In A. Prieditis, & S. J. Russell (Ed.), *Proceedings of the 12th International Conference on Machine Learning, Tahoe City, CA* (pp. 194-202). San Francisco, CA: Morgan Kaufmann.

Everitt, B. (2006). *The Cambridge dictionary of statistics* (3rd ed.). Cambridge, UK: Cambridge University Press.

Fawcett, T. (2006). An introduction to ROC analysis. *Pattern Recognition Letters*, *27*(8), 861–874. doi:10.1016/j.patrec.2005.10.010

Fayyad, U. M., & Irani, K. B. (1993). Multi-interval discretization of continuous-valued attributes for classification learning. In R. Bajcsy (Ed.), *Proceedings of the 13th International Joint Conference on Artificial Intelligence, Chambéry, France* (pp. 1022-1027). San Francisco, CA: Morgan Kaufmann.

Finak, G., Bertos, N., Pepin, F., Sadekova, S., Souleimanova, M., & Zhao, H. (2008). Stromal gene expression predicts clinical outcome in breast cancer. *Nature Medicine*, *14*(5), 518–527. doi:10.1038/nm1764

Hand, D. J. (2009). Naïve Bayes. In Wu, X., & Kumar, V. (Eds.), *The top ten algorithms in data mining* (pp. 163–178). Boca Raton, FL: Chapman & Hall/CRC Press. doi:10.1201/9781420089653.ch9

Ho, K. M., & Scott, P. D. (1997). Zeta: a global method for discretization of continuous variables. In D. Heckerman, H. Mannila, & D. Pregibon (Ed.), *Proceedings of the 3rd International Conference on Knowledge Discovery and Data Mining, Newport Beach, CA* (pp. 191-194). Menlo Park, CA: AAAI Press.

Keller, A. D., Schummer, M., Hood, L., & Ruzzo, W. L. (2000). *Bayesian classification of DNA array expression data. Department of Computer Science and Engineering*. Seattle, WA: University of Washington.

Liu, H., Hussain, F., Tan, C. L., & Dash, M. (2002). Discretization: an enabling technique. *Data Mining and Knowledge Discovery, 6*(4), 393–423. doi:10.1023/A:1016304305535

Lu, Y., & Han, J. (2003). Cancer classification using gene expression data. *Information Systems, 28*(4), 243–268. doi:10.1016/S0306-4379(02)00072-8

Paul, T. K., & Iba, H. (2004). Selection of the most useful subset of genes for gene expression-based classification. *Proceedings of the 2004 Congress of Evolutionary Computation, Portland, OR* (pp. 2076-2083). Los Alamitos, CA: IEEE Computer Society Press.

Pepe, M. S. (2004). *The statistical evaluation of medical tests for classification and prediction* (1ˢᵗ paperback ed.). Oxford, UK: Oxford University Press.

Russell, S., & Norvig, P. (2003). *Artificial intelligence: a modern approach* (2nd ed.). Upper Saddle River, NJ: Pearson Education.

Wang, Y., Makedon, F. S., Ford, J. C., & Pearlman, J. (2005). HykGene: a hybrid approach for selecting marker genes for phenotype classification using microarray gene expression data. *Bioinformatics (Oxford, England), 21*(8), 1530–1537. doi:10.1093/bioinformatics/bti192

Zhou, X., Wang, X., & Dougherty, E. R. (2003). Binarization of microarray data on the basis of a mixture model. *Molecular Cancer Therapeutics, 2*(7), 679–684.

ENDNOTES

[1]   Let us consider a simple example from medical diagnosis, borrowed from (Russell & Norvig, 2003) and showing how to calculate $P(B|A)$. A doctor knows a priori that the disease meningitis causes the patient to have a stiff neck, say 50% of the time. The doctor also knows that the prior probability that a patient has meningitis is 1/50000, and the prior probability that any patient has a stiff neck is 1/20. Letting A be the proposition that the patient has a stiff neck and B be the proposition that the patient has meningitis, we have $P(A|B) = 0.5, P(B) = \dfrac{1}{50000}, P(A) = 1/20$ and hence, $P(B|A) = \dfrac{0.5 \times 1/50000}{1/20} = 0.0002$. That is, we expect only 1 in 5000 patients with a stiff neck to have meningitis.

2   Laplace correction can be used in order to avoid zero probability, which will make the posterior probability zero, too. The correction (see, e.g., (Hand, 2009)) will transform zero probability to a small nonzero value so that the product of the probabilities won't be zero anymore.

Chapter 5
Nearest Neighbor

NEAREST NEIGHBOR CLASSIFICATION

As follows from its name, a nearest neighbor (NN) classifier assigns a class label according to proximity or distance. The basic principle of this classifier is probably one of the easiest to comprehend: the closest neighbor of a given test instance (feature vector of gene expression levels) in the training set determines class membership of this test instance. So, basically, all that someone needs is distance computation (according to chosen distance metric) to a given test instance from every training instance. The shortest distance points to the neighbor whose class is assigned to the test instance. As one can see, the training phase is absent, which makes things even easier. Since only one neighbor decides on class membership, such a classifier is often called a 1-NN.

A natural extension of this idea is k-nearest neighbor (k-NN) classification, where k shortest distances are found and the label of a majority class is assigned to a test instance. In other words, the most common class among k nearest neighbors of an instance determines its class. The value of k is of importance as it influences

DOI: 10.4018/978-1-60960-557-5.ch005

on the classification results. Usually, it is chosen to be 1, 3, 5..., i.e. an odd number in order to avoid ties and decision uncertainties related to them[1].

Both 1-NN and k-NN classifications are graphically depicted in Figure 1, where a 5-point star denotes a test instance to be classified and black and empty circles correspond to the training instances of two different classes. In Figure 1 (left) the class of a test point is 'black circles' as one of these circles lies at the shortest distance from the test point, whereas in Figure 1 (right) the class of the test point changes to 'empty circles' in 5-NN classification.

When dealing with high dimensional data such as gene expressions, a word of caution should be taken into account: in high dimensional spaces, the data points "live" close to the edges of the data set (see (Hastie, Tibshirani, & Friedman, 2001), pp. 22-23), which means that as a data dimensionality (the number of features) grows, the distance to the nearest data point approaches to the distance to the farthest data point (Beyer, Goldstein, Ramakrishnan, & Shaft, 1999), (Steinbach & Tan, 2009). This fact does not make NN classifiers meaningless but simply implies that dimensionality reduction has to take place prior to NN classification as irrelevant or noisy features can severely affect the classification accuracy of NN algorithms. The dimensionality reduction methods that are described in this book are considered to be useful for reducing the harmful effect of noise and irrelevant/redundant features.

Many NN algorithms have been proposed in the literature. For the state-of-the-art overview in early days, the book (Dasarathy, 1990) provides a good introduction. Although microarray data sets are rather small in the number of samples, the recent book (Shakhnarovich, Darrell, & Indyk, 2006) can serve as a reference to how fast NN search can be done for large and huge data sets.

The 1-NN has a strong theoretical property: as the data set size increases[2], the 1-NN algorithm is guaranteed to yield the upper bounded error rate no more than twice the achievable optimum (Cover & Hart, 1967). This optimum is called the

Figure 1. 1-NN classification (left) and 5-NN classification (right)

Bayes error rate (Duda, Hart, & Stork, 2000) and it is the minimum achievable error rate given the distribution of the data[3].

In other words, if E_{1NN} and E^* are the 1-NN error rate and the Bayes error rate, respectively, then the following inequality holds as $N \to \infty$ (N is the training set size): $E_{1NN} \leq 2E^*$. For the proof of this statement and a more formal definition of the Bayes error, see, for example, (Duda, Hart, & Stork, 2000) or http://cgm.cs.mcgill.ca/~soss/cs644/projects/simard/nn_prob_err.html.

Alternatively, it is possible to say that no classifier can do better than half the error rate of a 1-NN classifier. Hence, an estimated lower bound for the Bayes error can be obtained by estimating the 1-NN error rate and dividing it into 2 ($E_{1NN} / 2 \leq E^*$).

For some values of k, the k-NN algorithm is guaranteed to approach to the Bayes error rate. If $k \to \infty$ while $k / N \to 0$, the k-NN error rate converges to the Bayes error rate as $N \to \infty$. However, in practice, one always has finite samples and the result depends on the distance metric used. Besides, the exact calculations of the Bayes error are impossible and instead lower and upper bounds for the Bayes error are often used (Ripley, 2007).

Metric Definition and Most Common Metrics

Let us first define a metric. Let $\mathcal{X}$ be any nonempty set representing microarray data. A function $\rho : \mathcal{X} \times \mathcal{X} \to [0, +\infty)$ (this can be read as a mapping from the $\mathcal{X} \times \mathcal{X}$ space to the half-open interval of nonnegative numbers) is a metric if for all $x, y \in \mathcal{X}$, the following properties hold (Kohane, Kho, & Butte, 2003):

$\rho(x, y) = 0$ if and only if $x = y$.

$\rho(x, y) = \rho(y, x)$.

$\rho(x, y) \leq \rho(x, z) + \rho(z, y)$, for all $z \in \mathcal{X}$.

The examples of common metrics include:
Let $\mathcal{X} = \mathbb{R}^D, D \in \mathbb{N}$. The classical Euclidean distance,

$$\rho(x, y) = \left[\sum_{i=1}^{D} (x_i - y_i)^2 \right]^{\frac{1}{2}}.$$

Let $\mathcal{X} = \mathbb{R}^D, D \in \mathbb{N}$. The city block (sum of absolute differences) distance,

$$\rho(x, y) = \sum_{i=1}^{D} |x_i - y_i|.$$

Let $\mathcal{X} = \mathbb{R}^D, D \in \mathbb{N}$. The cosine (1 minus the cosine of the included angle between vectors) distance,

$$\rho(x, y) = 1 - \frac{\langle x, y \rangle}{\langle x, x \rangle^{\frac{1}{2}} \langle y, y \rangle^{\frac{1}{2}}},$$

where $\langle x, y \rangle = \sum_{i=1}^{D} x_i y_i$ is the dot or scalar product between two vectors x and y.

Other metrics can be found in (Dudoit & Fridlyand, 2003), (Drăghici, 2003) and in MATLAB® documentation (functions *classify, pdist* and *knnclassify*).

K-Local Hyperplane Distance Nearest Neighbor Algorithm (HKNN)

As we mentioned before, there are many algorithms based on the nearest neighbor principle. Out of this variety, we would like to describe one of them, called the K-Local Hyperplane Distance Nearest Neighbor (HKNN) (Vincent & Bengio, 2002). In HKNN, distances are computed not between points but between a point and a hyperplane formed by *k* nearest neighbors of the point, all belonging to the same class. Below, the description of HKNN is adopted from (Vincent & Bengio, 2002) and (Okun, 2004).

HKNN is a modified k-NN algorithm intended to improve the classification performance of the conventional k-NN to the level of Support Vector Machine[4] (SVM), which is considered by many as the state-of-the-art in pattern recognition. Unlike the SVM, which builds a (non-linear) decision surface, separating different classes of the data, in a high (often infinite) dimensional feature space, HKNN tries to find this surface directly in input space. If one thinks of each class as a low dimensional manifold[5] embedded in a high dimensional space, it is quite reasonable to assume that this manifold is locally linear. Since the data are usually sparse, they leave "holes" in the manifold. These "holes" introduce artifacts in the decision surface generated by the conventional k-NN, thus negatively affecting the generalization ability of this method in classifying the unseen examples. To remedy this deficiency, the idea of HKNN is to fantasize the missing points[6], based on a local linear approximation of the manifold of each class.

The main steps of HKNN are shown in pseudo code in Figure 2 and graphically in Figure 3.

From these figures, it can be seen that the Euclidean distance to a hyperplane now dominates computations and the number of hyperplanes is equal to that of different classes of the data. The value of k does not have to be an odd number anymore: it can be a sufficiently large even number, e.g., 10.

To reach its objective, HKNN computes distances of each test point x to C local hyperplanes, where C is the number of different classes. The cth hyperplane is composed of k nearest neighbors of x in the training set, belonging to cth class. A test point x is associated with the class whose hyperplane is closest to x.

The local hyperplane can be expressed as

$$LH_c^k\left(x\right) = \left\{ p \middle| p = \sum_{i=1}^{k} \alpha_i N_i, \alpha \in \mathbb{R}^k, \sum_{i=1}^{k} \alpha_i = 1 \right\},$$

where $\left\{N_1, \dots, N_k\right\}$ are k nearest neighbors of x from class c and $\alpha = \left(\alpha_1, \dots, \alpha_k\right)'$. This equation implies that x's nearest neighbors form the locally linear hyperplane (standard definition of a linear hyperplane from linear algebra).

In order to get rid of the constraints $\sum_{i=1}^{k} \alpha_i = 1$, a reference point within the hyperplane is taken as an origin. One of the candidates to the reference point is the centroid of $\left\{N_1, \dots, N_k\right\}$:

Figure 2. Pseudo code of HKNNN

K-Local Hyperplane Distance Nearest Neighbor (HKNN) Algorithm
For each class c_i ($i = 1, \dots, C$) and a given test point x 　Find k nearest neighbors of x in c_i; 　Build a hyperplane formed by these neighbors by solving a small linear system; 　Compute a distance of x to the hyperplane; **End For** Assign x to the class associated with the shortest distance.

Figure 3. Graphical explanation of how HKNN works. The star denotes a test point; two hyperplanes formed by 5 examples from each class (i.e., k = 5) are represented by rectangles. Point-to-hyperplane distances are shown by arrows. For the case shown in this figure, the test point is assigned to the class of "black circles" according to HKNN.

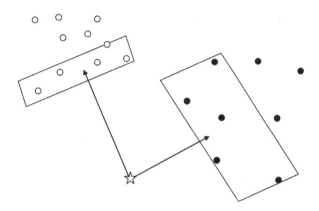

$$\overline{N} = \frac{1}{k}\sum_{i=1}^{k} N_i.$$

Then the same hyperplane can be expressed as

$$LH_c^k\left(x\right) = \left\{ p | p = \overline{N} + \sum_{i=1}^{k}\alpha_i V_i, \alpha \in \mathbb{R}^k \right\},$$

where $V_i = N_i - \overline{N}$.

The distance of x to the c th hyperplane is defined as

$$d\left(x, LH_c^k\left(x\right)\right) = \min_{p \in LH_c^k(x)} x - p =$$
$$\min_{\alpha \in \mathbb{R}^k} x - \overline{N} - \sum_{i=1}^{k}\alpha_i V_i.$$

To treat the problem of large k ($k > D$, where D is a data dimensionality)[7], a penalty term[8] is incorporated into the previous equation, punishing large values of α ("weight decay" penalty):

$$d\left(x, LH_c^k\left(x\right)\right) = \min_{p \in LH_c^k(x)} \left\| x - p \right\| =$$

$$\min_{\alpha \in \mathbb{R}^k} \left\| x - \overline{N} - \sum_{i=1}^{k} \alpha_i V_i \right\|^2 + \lambda \sum_{i=1}^{k} \alpha_i^2 .$$

This equation amounts to solving a linear system in α, which is expressed in the following matrix form:

$$\left(V'V + \lambda I\right)\alpha = V'\left(x - \overline{N}\right),$$

where x and $\overline{N}$ are D-dimensional column vectors, V is a $D \times k$ matrix whose columns are the vectors V_i, defined earlier, and I is the $D \times D$ identity matrix.

Thus, HKNN has two parameters to be pre-set: k and λ.

As α is unknown, one therefore needs to find it from the last equation, given a test point x and its k nearest neighbors in each class. For each class, there is its own vector α. If one looks at the last equation, it is easy to observe that it can be rewritten as $A\alpha = b$, where $A = V'V + \lambda I$ and $b = V'\left(x - \overline{N}\right)$ are both known (they can be computed). Hence, the solution of such a linear equation is well-known and it is $\alpha = A^{-1}b$, where A^{-1} is the matrix inverse of A. Once α is determined, it is used to compute $d\left(x, LH_c^k\left(x\right)\right)$ distance for each class according to the formula with the penalty term. Finally, the class of x is determined by $arg \min_{c=1,...,C} d\left(x, LH_c^k\left(x\right)\right)$, i.e. by the shortest distance among C distances.

The various modifications of HKNN were proposed in the literature (Cevikalp, Larlus, Douze, & Jurie, 2007), (Cevikalp, Triggs, & Polikar, Nearest hyperdisk methods for high-dimensional classification, 2008), (Yang & Kecman, 2008), (Chen, Warren, Yang, & Kecman, 2008), including kernel extensions. For bioinformatics applications of HKNN, see (Okun, 2004), (Niijima & Kuhara, 2005), (Nanni & Lumini, 2006). The Nearest Feature Line method (Li & Lu, 1999) is similar in spirit to HKNN but searches among all possible lines, formed by pairs of points rather than among C hyperplanes. Gene expression cancer classification using nearest neighbors can be found in many articles and book chapters, e.g., (Singh, et al., 2002), (Dudoit & Fridlyand, 2003), (Nutt, et al., 2003), (Kohane, Kho, & Butte, 2003), (Lu & Han, 2003), (Wang, Makedon, Ford, & Pearlman, 2005), (Okun, Valentini, & Priisalu, Exploring the link between bolstered classification error and dataset complexity for gene expression-based cancer classification, 2008).

MATLAB Code

The *k*-NN algorithm is implemented in MATLAB® Bioinformatics Toolbox™ in the function *knnclassify*. Our function *NearestNeighbor_m* demonstrates how to use it. The default metric is Euclidean, though other metrics can be specified as well (input parameter *distance*). If there are missing gene expression values, they can be imputed with *knnimpute* – another useful function from MATLAB® Bioinformatics Toolbox™ – built on the nearest neighbor principle from (Troyanskaya, et al., 2001).

For educational purposes, I also implemented *k*-NN in *NearestNeighbor*. For each test point, this function computes the distance to all the training points at once. After sorting these distances, only the class labels associated with the *k* smallest distances are further analyzed with MATLAB® functions *hist* and *max* in order to assign a class label to a given test point[9] (pay attention to the safe technique of enumerating of all unique class labels, provided by MATLAB® function *unique*, which I always recommend to follow). Besides the class labels, the matrix *prob* of probability estimates associated with each test point is returned by *NearestNeighbor*. The *j*th column of this matrix is a vector of probabilities assigned to the *j*th test point. The *i*th element of the *j*th vector is the probability of the *i*th class assigned to the *j*th test point. The sum of all elements (probabilities) of any column is equal to 1. The probability estimates are the optional output parameter, which implies that they are not returned, if the calling function specified only one output.

A faster but more memory-consuming variant of *NearestNeighbor* is implemented in *FastNearestNeighbor*. It merges training and test data together into a single matrix and all interpoint distances are computed. The distances of interest are between test and training points, i.e. only a part of the entire distance matrix as schematically shown in the code of *FastNearestNeighbor*. By extracting the desired distances, we then proceed as in *NearestNeighbor*.

Finally, let me present HKNN code in *HyperplaneDistanceNearestNeighbor*. My version assumes that the point-to-hyperplane distance is Euclidean but can be easily modified if other metric is desired. As HKNN has two parameters to define (k and λ), these are packed together in a vector *params* whose first element is k and the second element is λ. The rest of code follows precisely the HKNN description provided above. When solving a linear equation $A\alpha = b$, the MATLAB® operator \ (backslash or matrix left division) is used. That is, the solution is found as $\alpha = A \setminus b$ by means of Gaussian elimination with partial pivoting implemented in LAPACK (Linear Algebra Package) routines (http://www.netlib.org/lapack/lug/ and (Anderson, et al., 1999)).

In addition to the class labels of test data, *HyperplaneDistanceNearestNeighbor* can also return a vector of confidences linked to these labels. The confidence should not be confused with probability as the former does not sum up to 1 for all classes.

The confidence in class c is inversely proportional to the distance of a given test point x to the local hyperplane formed by k nearest neighbors of x in class c. Despite the fact that confidences are rather scores than proper probabilities, they can be used in classifier performance evaluation by means of the Receiver Operating Characteristics (ROC) Curve (Fawcett, 2006), (Pepe, 2004). To make scores look more like probabilities, one can define the confidence for each test instance as a vector

$\left[\dfrac{d_1}{d_1 + d_2}, \dfrac{d_2}{d_1 + d_2} \right]'$, where d_1, d_2 are the distances to the hyperplanes of two

classes. Combined together, all such vectors will form a $C \times N_t$ matrix akin to that produced by my implementations of the conventional k-nearest neighbors.

As both *NearestNeighbor* and *FastNearestNeighbor* output probability estimates, these are readily suitable for the ROC generation. *NearestNeighbor_m* cannot, however, produce either scores or probabilities, because MATLAB® function *knnclassify* is not equipped with this option. Therefore, if you really need a ROC curve, do not run *NearestNeighbor_m*.[10]

```
function test_targets = NearestNeighbor_m(train_patterns,
train_targets, test_patterns, k, distance)

% K-Nearest Neighbor classifier built on knnclassify from
MATLAB
% Bioinformatics Toolbox
%
% Inputs:
%    train_patterns  - Training patterns (DxN matrix, where D
- data
%                      dimensionality, N - number of patterns)
%    train_targets   - Training targets (1xN vector)
%    test_patterns   - Test patterns (DxNt matrix, where D -
data
%                      dimensionality, Nt - number of patterns)
%    k               - The number of nearest neighbors
%    distance        - Distance metric (string)
%
% Outputs:
%    test_targets    - Test targets (1xNt vector)
%
% Note: it requires Bioinformatis Toolbox 3, released in 2008
```

```
% Get a version number for Bioinformatics Toolbox
v = ver('bioinfo');
% If MATLAB release is before 2008, generate an error message
if isempty(v)
    error('No Bioinformatics toolbox was found!');
else
    if str2num(v.Version) < 3.1
        error('Version 3.1 (or higher) of Bioinformatics Tool-
box is needed to run this function!');
    end
end


% Default values
if nargin < 5
    distance = 'euclidean';
end
if nargin < 4
    k = 1;
end


% Classify test data
test_targets = knnclassify(test_patterns', train_patterns',
train_targets', k, distance);


return
function [test_targets, varargout] = NearestNeighbor(train_pat-
terns, train_targets, test_patterns, k)

% K-Nearest Neighbor classifier
%
% Inputs:
%    train_patterns   - Training patterns (DxN matrix, where D
- data
%                       dimensionality, N - number of patterns)
%    train_targets    - Training targets (1xN vector)
%    test_patterns    - Test patterns (DxNt matrix, where D -
data
%                       dimensionality, Nt - number of patterns)
%    k                - The number of nearest neighbors
%
```

```
% Outputs:
%    test_targets    - Test targets (1xNt vector)
%    prob            - Probability estimates (CxNt matrix, where
C -
%                       the number of classes)


% Get the number of training patterns
L           = length(train_targets);

% Get the number of classes
Uc          = unique(train_targets);
c           = length(Uc);

if (L < k),
   error('You specified more neighbors than there are points.')
end

% Get the number of test patterns
N               = size(test_patterns, 2);

% Pre-allocate memory for the outputs
test_targets    = zeros(1,N);
prob            = zeros(c,N);

% Classify test data
for i = 1:N
    dist                = sum((train_patterns - test_
patterns(:,i)*ones(1,L)).^2,1);

    [m, indices]    = sort(dist);

    n                   = hist(train_targets(indices(1:k)), Uc);

    [m, best]       = max(n);

    test_targets(i) = Uc(best);

    % Laplace correction of probability estimates
    prob(:,i) = (n + 1)/(k + c);
```

```
end

% If probability estimates are required in the calling function
if nargout == 2
    varargout{1} = prob;
end

return
function [test_targets, varargout] = FastNearestNeighbor(train_
patterns, train_targets, test_patterns, k)

% Fast K-Nearest Neighbor classifier
%
% Inputs:
%    train_patterns   - Training patterns (DxN matrix, where D
- data
%                       dimensionality, N - number of patterns)
%    train_targets    - Training targets (1xN vector)
%    test_patterns    - Test patterns (DxNt matrix, where D -
data
%                       dimensionality, Nt - number of patterns)
%    k                - The number of nearest neighbors
%
% Outputs:
%    test_targets     - Test targets (1xNt vector)
%    prob             - Probability estimates (CxNt matrix, where
C -
%                       the number of classes)
%
% It is memory-intensive code so that it can cause 'Out of
memory'
% message for large datasets, e.g., as large as 10^4 patterns!


% Get the number of training patterns
L          = length(train_targets);

% Get the number of classes
Uc         = unique(train_targets);
c          = length(Uc);
```

```
if (L < k),
   error('You specified more neighbors than there are points!')
end

% Combine training and test patterns into a single set
X = [train_patterns, test_patterns];
% Get the number of patterns in the combined set
Nc = size(X,2);

% Delete from memory
clear train_patterns test_patterns;

% Compute distances
X2 = sum(X.^2,1);
distance = repmat(X2,Nc,1)+repmat(X2',1,Nc)-2*X'*X;

% Delete from memory
clear X X2;

% Extract only training-to-test patterns distances from the
matrix show
% below:
%
%                 L                 N
%    *******************************
%    *                *                *
%    *                *                *
%    *                *                *  L
%    *        A       *       B        *
%    *                *                *
%    *                *                *
%    *******************************
%    *                *                *
%    *                *                *
%    *        C       *       D        *  N
%    *                *                *
%    *                *                *
%    *******************************
%
```

```
% Distances of interest lie in rectangles B or C.
% Extract distances from rectangle B
distance = distance(1:L,L+1:Nc);

% Sort distances along columns in ascending order
[sorted,indices] = sort(distance);

% Delete from memory
clear distance;

% Find class labels of test data
n               = hist(train_targets(indices(1:k,:)), Uc);

[m, best]       = max(n);

test_targets    = Uc(best);

% Laplace correction of probability estimates
prob = (n + 1)/(k + c);

% If probability estimates are required in the calling function
if nargout == 2
    varargout{1} = prob;
end

return
function [test_targets, varargout] = HyperplaneDistanceNearestN
eighbor(train_patterns, train_targets, test_patterns, params)

% K-Local Hyperplane Distance Nearest Neighbor classifier
% Original source: Pascal Vincent, Yoshua Bengio, "K-local
hyperplane
% and convex distance nearest neighbor algorithms", In Thomas
Glen
% Dietterich, Suzanna Becker, Zoubin Ghahramani (eds.), Advanc-
es in
% Neural Information Processing Systems 14, pp.985-992, MIT
Press,
% Cambridge, MA, 2002
%
```

```
% Inputs:
%     train_patterns   - Training patterns (DxN matrix, where D
- data
%                        dimensionality, N - number of patterns)
%     train_targets    - Training targets (1xN vector)
%     test_patterns    - Test patterns (DxNt matrix, where D -
data
%                        dimensionality, Nt - number of patterns)
%     params           - [k, lambda]
%                        k      - Number of nearest neighbors
%                        lambda - Penalty parameter
%
% Outputs:
%     test_targets     - Test targets (1xNt vector)
%     confidence       - Confidence in prediction of a test target
(1xNt %                              vector)



% The number of classes
Uc = unique(train_targets);
c = length(Uc);

% Get the number of test patterns
N               = size(test_patterns, 2);

% Pre-allocate memory for the outputs
test_targets    = zeros(1,N);
if nargout == 2
    confidence = zeros(1,N);
end

% Get parameters of a classifier
k = params(1);
lambda = params(2);

% Pre-allocate memory
T = cell(1,c);
```

```
% Extract training data of each class in a separate array
for i = 1:c
    T{i} = train_patterns(:,train_targets == Uc(i));
end

I = eye(k);

% Classify test patterns
for i = 1:N
    % Get a test pattern
    p = test_patterns(:,i);
    if ~all(isnan(p))
        dmin = bitmax;
        for j = 1:c
            % Computing interpoint distances from p to every
point in
% the training set belonging to the same class as class
% of p
            distance = sum((T{j} - repmat(p,1,size(T
{j},2))).^2,1);
            % Find k-nearest neighbors of p which belong to
class j
            [sorted,index] = sort(distance);
            neighborhood = index(1:k);
            % Compute the centroid of k nearest neighbors
            xc = mean(T{j}(:,neighborhood),2);
            % Compute a hyperplane formed by k nearest neigh-
bors
            V = T{j}(:,neighborhood) - repmat(xc,1,k);
            alpha = (V'*V + lambda*I)\(V'*(p - xc));
            % Compute a distance from p to the hyperplane
            d = norm(p - xc - sum(repmat(alpha',size(V,1),1).
*V,2))^2 + lambda*sum(alpha.^2),;
            % Compare the distance with the minimum obtained so
far
            if d < dmin
                dmin = d; label = Uc(j);
            end
        end
```

```
        % p is from class label
        test_targets(i) = label;
        if nargout == 2
            confidence(i) = 1./(eps+dmin);
        end
    else % end if ~isnan(p)
        test_targets(i) = NaN;
        if nargout == 2
            confidence(i) = NaN;
        end
    end
end

if nargout == 2
    varargout{1} = confidence;
end
return
```

REFERENCES

Anderson, E., Bai, Z., Bischof, C., Blackford, S., Demmel, J., & Dongarra, J. (1999). *LA-PACK user's guide* (3rd ed.). Philadelphia, PA: SIAM. doi:10.1137/1.9780898719604

Berger, M. (2003). *A panoramic view of Riemannian geometry*. Berlin, Heidelberg: Springer-Verlag.

Beyer, K., Goldstein, J., Ramakrishnan, R., & Shaft, U. (1999). When is "nearest neighbor" meaningful? In C. Beeri, & P. Buneman (Ed.), *Proceedings of the 7th International Conference on Database Theory, Jerusalem, Israel* (LNCS 1540, pp. 217-235). Berlin/Heidelberg: Springer-Verlag.

Cevikalp, H., Larlus, D., Douze, M., & Jurie, F. (2007). Local subspace classifiers: linear and nonlinear approaches. In *Proceedings of the 2007 IEEE Workshop on Machine Learning for Signal Processing, Thessaloniki, Greece* (pp. 57-62). Los Alamitos, CA: IEEE Computer Society Press.

Cevikalp, H., Triggs, B., & Polikar, R. (2008). Nearest hyperdisk methods for high-dimensional classification. In W. W. Cohen, A. McCallum, & S. T. Roweis (Ed.), *Proceedings of the 25th International Conference on Machine Learning, Helsinki, Finland* (pp. 120-127). New York, NY: ACM Press.

Chen, G., Warren, J., Yang, T., & Kecman, V. (2008). Adaptive K-local hyperplane (AKLH) classifiers on semantic spaces to determine health consumer web page metadata. In *Proceedings of the 21th IEEE International Symposium on Computer-Based Medical Systems, Jyväskylä, Finland* (pp. 287-289). Los Alamitos, CA: IEEE Computer Society Press.

Cover, T. M., & Hart, P. E. (1967). Nearest neighbor pattern classification. *IEEE Transactions on Information Theory, 13*(1), 21–27. doi:10.1109/TIT.1967.1053964

Dasarathy, B. V. (Ed.). (1990). *Nearest neighbor: pattern classification techniques.* Los Alamitos, CA: IEEE Computer Society Press.

Drăghici, S. (2003). *Data analysis tools for DNA microarrays.* Boca Raton, FL: Chapman & Hall/CRC Press.

Duda, R. O., Hart, P. E., & Stork, D. G. (2000). *Pattern classification* (2nd ed.). New York, NY: John Wiley & Sons.

Dudoit, S., & Fridlyand, J. (2003). Classification in microarray experiments. In Speed, T. (Ed.), *Statistical analysis of gene expression microarray data* (pp. 93–158). Boca Raton, FL: Chapman & Hall/CRC Press. doi:10.1201/9780203011232.ch3

Fawcett, T. (2006). An introduction to ROC analysis. *Pattern Recognition Letters, 27*(8), 861–874. doi:10.1016/j.patrec.2005.10.010

Hastie, T., Tibshirani, R., & Friedman, J. (2001). *The elements of statistical learning.* New York, NY: Springer-Verlag.

Kohane, I. S., Kho, A. T., & Butte, A. J. (2003). *Microarrays for an integrative genomics.* Cambridge, MA: MIT Press.

Li, S. Z., & Lu, J. (1999). Face recognition using the nearest feature line method. *IEEE Transactions on Neural Networks, 10*(2), 439–443. doi:10.1109/72.750575

Lu, Y., & Han, J. (2003). Cancer classification using gene expression data. *Information Systems, 28*(4), 243–268. doi:10.1016/S0306-4379(02)00072-8

Nanni, L., & Lumini, A. (2006). An ensemble of K-local hyperplanes for predicting protein-protein interactions. *Bioinformatics (Oxford, England), 22*(10), 1207–1210. doi:10.1093/bioinformatics/btl055

Niijima, S., & Kuhara, S. (2005). Effective nearest neighbor methods for multiclass cancer classification using microarray data. *Proceedings of the 16th International Conference on Genome Informatics, Yokohama, Japan,* (p. P051).

Nutt, C. L., Mani, D., Betensky, R. A., Tamayo, P., Cairncross, J. G., & Ladd, C. (2003). Gene expression-based classification of malignant gliomas correlates better with survival than histological classification. *Cancer Research, 63,* 1602–1607.

Okun, O. (2004). Protein fold recognition with k-local hyperplane distance nearest neighbor algorithm. *Proceedings of the 2nd European Workshop on Data Mining and Text Mining for Bioinformatics, Pisa, Italy* (pp. 51-57).

Okun, O., Valentini, G., & Priisalu, H. (2008). Exploring the link between bolstered classification error and dataset complexity for gene expression-based cancer classification. In Maeda, T. (Ed.), *New signal processing research* (pp. 249–278). New York, NY: Nova Science Publishers.

Pepe, M. S. (2004). *The statistical evaluation of medical tests for classification and prediction* (1st paperback ed.). Oxford, UK: Oxford University Press.

Ripley, B. (2007). *Pattern recognition and neural networks* (1st paperback ed.). Cambridge, UK: Cambridge University Press.

Shakhnarovich, G., Darrell, T., & Indyk, P. (Eds.). (2006). *Nearest-neighbor methods in learning and vision: theory and practice.* Cambridge, MA: MIT Press.

Singh, D., Febbo, P., Ross, K., Jackson, D., Manola, J., & Ladd, C. (2002). Gene expression correlates of clinical prostate cancer behavior. *Cancer Cell, 1*(2), 203–209. doi:10.1016/S1535-6108(02)00030-2

Steinbach, M., & Tan, P.-N. (2009). kNN: k-nearest neighbors. In X. Wu, & V. Kumar (Eds.), *The top ten algorithms in data mining* (pp. 151-162). Boca Raton, FL: Chapman & Hall/CRC Press.

Troyanskaya, O., Cantor, M., Sherlock, G., Brown, P., Hastie, T., & Tibshirani, R. (2001). Missing value estimation methods for DNA microarrays. *Bioinformatics (Oxford, England), 17*(6), 520–525. doi:10.1093/bioinformatics/17.6.520

Vincent, P., & Bengio, Y. (2002). K-local hyperplane and convex distance nearest neighbor algorithms. In Dietterich, T. G., Becker, S., & Ghahramani, Z. (Eds.), *Advances in Neural Information Processing Systems, 14* (pp. 985–992). Cambridge, MA: MIT Press.

Wang, Y., Makedon, F. S., Ford, J. C., & Pearlman, J. (2005). HykGene: a hybrid approach for selecting marker genes for phenotype classification using microarray gene expression data. *Bioinformatics (Oxford, England), 21*(8), 1530–1537. doi:10.1093/bioinformatics/bti192

Yang, T., & Kecman, V. (2008). Adaptive local hyperplane classification. *Neuro-computing, 71*(13-15), 3001–3004. doi:10.1016/j.neucom.2008.01.014

ENDNOTES

[1] For example, if one got 2 neighbors from class 1 and 2 neighbors from class 2 when doing 4-NN classification, what is the outcome? Class 1 or class 2? Of course, such ties could be randomly broken, but it would be a random k-NN classification. Setting k to an odd number effectively resolves this kind of ambiguity in two-class tasks.

[2] Often it is assumed going to infinity (infinity is denoted by the mathematical symbol ∞).

[3] This assumes that the form of the data distribution is known, which is quite rear. Often the data distributions for each class are considered to be Gaussian in order to obtain the analytical expression for the lower and upper bound of the Bayes error (Ripley, 2007).

[4] Support Vector Machine classifiers will be considered later in the book.

[5] A manifold is a mathematical term frequently used in topology and differential calculus. A D-dimensional manifold is a topological space that on a local scale is equivalent to the D-dimensional Euclidean space. For instance, a line is a 1-dimensional manifold, while a plane is a 2-dimensional manifold. While locally, manifolds can be said to be composed of Euclidean spaces, the global manifold structure does not have to resemble the Euclidean space. If readers want to dive into the sea of manifolds, I can recommend the following book: (Berger, 2003).

[6] This implies that such points are never explicitly generated.

[7] As soon as $k > D$, any instance in a D-dimensional space, including nonsensical instances, can be produced by a linear combination of k neighbors. So, a suitable penalty should prevent "fantasy" going too far from reality.

[8] This term penalizes moving away from the centroid, especially when the chosen value of k does not match well the locally linear approximation of the class manifold.

[9] Given a vector $v = [01001]$, *hist*$(v, [01])$ will return another vector $[32]$, meaning that there are three 0-elements and two 1-elements. That is, *hist* counts the frequency of occurrence of each unique value.

[10] Technically speaking, even if a classifier generates discrete labels, one can still build a ROC curve. However, in this case, it will be a single point instead of a curve.

Chapter 6
Classification Tree

TREE-LIKE CLASSIFIER

Classification tree is a part of tree-like classification algorithms broadly used in machine learning, data mining, statistics and their applications. The book (Breiman, Friedman, Olshen, & Stone, 1984), covering not only classification but also regression trees, describes this type of classifier in detail. In fact, the full name – CART – which stands for 'Classification and Regression Trees', is frequently used to denote this algorithm. Brief history, preceding the emergence of CART, is presented in (Steinberg, 2009). Due to its popularity and simple interpretation of how classification was done, CART is described in many textbooks (Martinez & Martinez, 2002), (Hastie, Tibshirani, & Friedman, 2003), (Russell & Norvig, 2003), (Kuncheva, 2004), (Bishop, 2006), (Ripley, 2007), (Rokach & Maimon, 2008), (Izenman, 2008). The related algorithms to CART are ID3, C4.5 and C5.0 (Quinlan, 1993), which are not discussed in this book but their description can be found, for example, in (Quinlan, 1993), (Wu & Kumar, 2009).

The CART decision tree is a *binary* multi-stage decision procedure. The tree is made up of nodes and branches. Nodes can be either internal or terminal. The

DOI: 10.4018/978-1-60960-557-5.ch006

terminal nodes are also called leaves. Internal nodes split into two children. These children can be 1) both internal nodes, 2) both terminal nodes or 3) one child can be terminal while the other can be internal. Terminal nodes (leaves) do not have children. A terminal node has a class label associated with it, such that all instances falling into this node are assigned to that class. A single-split tree with only two terminal nodes is called a stump. That is, the stump divides the data into two classes based on a single feature value.

Why binary splits? Rather than splitting each node into two nodes, we could consider multiway splits into more than two groups. However, multiway splits are not very good strategy as they fragment the data too quickly, leaving insufficient data for the next level. Moreover, any multiway split can be represented as a sequence of binary splits so there is no reason to prefer multiway splits over binary ones.

In tree construction the data are handled in their raw form as feature vectors. In the root node the entire data set is split into two children nodes. Each of the children is then further split into grandchildren nodes that are in turn split to obtain great-grandchildren, etc., i.e. the original data set is recursively split between tree nodes as the tree grows.

The tree is usually grown to a maximal size (the tree-growing process stops when no further splits are possible because of the lack of data). Such a tree can easily lead to overfitting when minute details are learned at the expense of generalization.

The fully-grown tree can be then pruned back towards the root in order to remove splits that contribute least to the overall performance.

The CART mechanism includes (optional) automatic class balancing and automatic missing value handling, and allows for cost-sensitive learning (where different classes have different costs of misclassification) and probability estimation. CART can handle both continuous and categorical features.

CART is an unstable classifier whose decision splits and predictions can change due to small changes in the training data (this is called high variance). This is in the contrast to Naïve Bayes and k-nearest neighbors, considered earlier in this book, that are relatively stable with respect to small changes of the training data (it is said that these both classifiers have low variance). Due to its instability, CART is applied for bioinformatics tasks in classifier ensembles ((Díaz-Uriarte & Alvarez de Andrés, 2006), (Statnikov, Wang, & Aliferis, 2008)) rather than as a stand-alone classifier.

The major reason for instability of trees is the hierarchical nature of the process (Hastie, Tibshirani, & Friedman, 2003), leading to the effect of error propagation from the place it was made down to all the splits below it.

Hence, the practical usefulness of a single tree in making predictions, e.g., regarding cancer outcome or class, is rather limited (who needs an oracle that answers each time differently to the same question?). Instead, an ensemble of full-grown (i.e. non-pruned) classification trees provides better (both stable and more accurate) results. Due to these reasons, I omit the detailed discussion of tree pruning procedures and send an interested reader to (Breiman, Friedman, Olshen, & Stone, 1984) or any other textbooks cited in the beginning of this section. The description of ensembles of trees is given later in this book.

CART asks a sequence of hierarchical Boolean questions (is $x_j \leq \theta_j$?, where θ_j is a threshold value). Therefore it is relatively simple to understand and interpret the rules generated by CART. This is a key property that makes tree-based models popular in fields such as medical diagnosis.

The example of the classification tree is presented in Figure 1, which uses the famous Fisher's iris flower data (see http://en.wikipedia.org/wiki/Iris_flower_data_set). The task is to predict three species of iris flower (setosa, virginica, and versicolor) based on four measured features (sepal length (SL), sepal width (SW), petal length (PL), petal width (PW)). As you can see, features are continuous (lengths and widths) whereas targets (class labels) are categorical.

The built decision tree in Figure 1 can be expressed by the following rules:

```
if PL<2.45 then class = setosa;
else
              if PW<1.75 then
```

Figure 1. Classification tree for Fisher's iris flower data (MATLAB® screenshot)

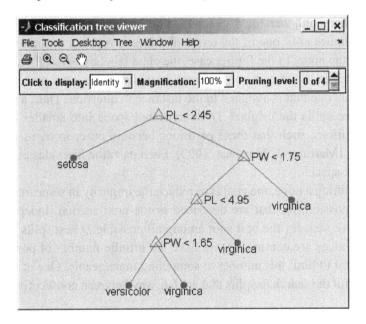

```
                              if PL<4.95 then
                                       if PW<1.65 then class = ver-
sicolor; else class = virginica; end
                             else class = virginica;
                          end
               else class = virginica;
               end
end
```

The tree predicts class labels at the circular leaf nodes based on a series of questions about the iris at the triangular branching nodes. A true ('yes') answer to any question follows the branch to the left; a false ('no') follows the branch to the right. It can be noticed that sepal length and width do not influence a decision about class labels measurements for predicting species, which implies that these features can go unmeasured in new data.

Algorithm Description

The simplified[1] algorithm for the maximal tree growing adopted from (Martinez & Martinez, 2002) is given in Figure 2. We assume that the data are collected in the $D \times N$ matrix X whose rows are features and columns are instances (observations).

By default, all DD features are considered for each decision split. At any stage, if the continuous value for a feature is less than or equal to some number (threshold), the decision is to move to the left child. If the continuous value for a feature is larger than threshold, then the decision is to move to the right child. This splitting process continues until one of the terminal nodes is reached or there are too few instances in the node. In the former case, the class label associated with the terminal node is given to the instance in question, while in the latter case, the class of majority is the one that is assigned to the instance in question. Thus, a binary classification tree splits the original D dimensional space into smaller and smaller cuboids partitions, such that these partitions become purer in terms of the class membership (Martinez & Martinez, 2002). Each partition has sides parallel to the axes of input space[2].

When splitting a node, the goal is to reduce the impurity in some manner. There are several possibilities that are described in the next section. Independently of the choice, we seek for the best split among all possible D best splits. Since gene expression values are continuous, we have the infinite number of possible splits. Thus, we need to limit this number to something manageable. One of the conventions is to limit the search to splits that are halfway between consecutive values for

Figure 2. Pseudo code of the maximal tree-growing algorithm

Maximal tree-growing algorithm

Assign all training data to the root node;
Set the maximum number of observations n_{max} that will be allowed in a terminal node;
Estimate the prior probabilities for different data classes;
Define the root node as a nonterminal node;

For every nonterminal node do
    If the number of training instances is smaller than or equal to n_{max} or
        all instances belong to the same class, then mark the node as terminal and assign
        the class label to it;
    Else /* search for the best split */
        For each feature do
Sort its values in ascending order of magnitude. Let $x_{i(j)}$ be these ordered values;

Determine all splits $s_{i(j)}$ as $x_{i(j)} + \left(x_{i(j)} - x_{i(j+1)} \right) / 2$;

            a. For each proposed split, evaluate the impurity and its decrease;
            b. Pick the best split, which is the one yielding the largest decrease in
                impurity;
        End
        Out of D best splits found in the previous step, split the current node on the
        feature that yields the best overall split;
        For that split, determine the training data partitioning between the left and
        right children;
    End
End

---

that feature (see Figure 2). For each proposed split, the impurity is evaluated and the split yielding the largest decrease in impurity is selected.

It should be noted that a greedy optimization used in the tree construction significantly relaxes the computational demands on otherwise infeasible task of finding the optimal tree structure (optimal split features, optimal split thresholds).

Splitting Rules

These rules are based on measuring the impurity $i(t)$ for a node t. The impurity is then used to calculate the decrease in impurity due to node split into left and right children. The decrease $\Delta i(s, t)$ indicates the goodness of the split s at node t and it is defined as

$$\Delta i\left(s,t\right) = i\left(t\right) - p_R i\left(t_R\right) - p_L i\left(t_L\right),$$

where p_L and p_R are the proportion of data that are sent to the left t_L and right t_R child nodes by the split s.

Breiman et al. (Breiman, Friedman, Olshen, & Stone, 1984) advocated employing the Gini's diversity index defined as

$$i\left(t\right) = \sum_{C_i \neq C_j} P\left(C_i|t\right) P\left(C_j|t\right) = 1 - \sum_{j=1}^{k} P^2\left(C_j|t\right),$$

where $P\left(C_j|t\right)$ is the probability that an instance is in class C_j, given it is in node t, estimated as

$$P\left(C_j|t\right) = \frac{P\left(C_j,t\right)}{\sum_{i=1}^{k} P\left(C_i,t\right)}, \; P\left(C_j,t\right) = \frac{\pi_j n_j\left(t\right)}{n_j},$$

where $P\left(C_j,t\right)$ is the joint probability than an instance will be in node t and it will belong to class C_j; π_j is the prior probability that an instance belongs to class C_j (π_j can be estimated from the data as N_j / N, where N is the total number of training instances and N_j is the number of training instances belonging to class C_j); $n_j\left(t\right)$ is the number of training instances that fall into node t and belonging to class C_j.

In the two-class case, the Gini index is reduced to

$$i\left(t\right) = 2P\left(C_2|t\right)\left(1 - P\left(C_2|t\right)\right),$$

and it yields results equivalent to the twoing rule.

Another popular splitting rule is cross-entropy or deviance, where the impurity is computed according to

$$i\left(t\right) = -\sum_{j=1}^{k} P\left(C_j|t\right) \log P\left(C_j|t\right),$$

which becomes

$$i(t) = -P(C_2|t)\log P(C_2|t) - \left(1 - P(C_2|t)\right)\log\left(1 - P(C_2|t)\right)$$

in the two-class case.

Prior Probabilities in Class Balancing

It is well known that many algorithms do not perform well if the training data are highly imbalanced. In other words, there are many more instances of one class than those of another class. For microarray gene expression data, this is not common so that class imbalance does not seem to be an issue. Nevertheless, it is useful to know how CART handles this.

In fact, there is nothing to worry about since CART does not require any intervention from a user to handle imbalance. This is because CART makes use of the prior probabilities in the calculations of the node impurity (see the previous section). This is equivalent to data reweighting in order to achieve class rebalancing.

Missing Value Handling

CART is one of few algorithms that can automatically handle missing values. It does so via the mechanism of "surrogate" or substitute splitters.

When considering a feature for a split, only the instances for which that feature is not missing are used to find the best primary split point. The first surrogate is the feature and its value that best mimic the split of the training data achieved by the primary split. The second surrogate is the feature and its value that does second best, etc.

When sending instances down the tree in either training or test phase, the first surrogate is used if the primary feature value is missing. If the first surrogate is missing, too, then the second one is taken for the node split, and so on. If all surrogates are missing, the default rule assigns the instance to the larger child node (after adjusting for priors). Ties are broken by moving an instance to the left.

Surrogate splits exploit correlations between the missing feature and other features chosen to substitute it. The higher the correlation is, the smaller the loss of information due to the missing value.

MATLAB Code

Contrary to usual habit, I decided not to implement the classification tree but utilize the MATLAB® implementation of CART – function *classregtree*, which is based on object-oriented paradigm. This function provides a lot of *options* described in (Breiman, Friedman, Olshen, & Stone, 1984) so that my own contribution would be insignificant. Those readers who want to see how CART can be implemented in MATLAB® are advised to look in (Martinez & Martinez, 2002), (Stork & Yom-Tov, 2004)[3] or (Kuncheva, 2004).

Nevertheless I call *classregtree* from the function *classification_tree_moo*. Except for customary inputs specifying the training and test data, *classification_tree_moo* expects five more parameters to set up: *viewflag*, *pruneflag*, *k*, *split_crit*, and *cost*. The parameter *viewflag* can take two values: 'on' (display a tree – default) and 'off' (don't display). The parameter *pruneflag* can take two values: 'on' (perform tree pruning) and 'off' (leave a tree without pruning – default). The parameter *k* is the minimum number of instances any impure node shall contain in order to be split (default is 10). The parameter *split_crit* is the node splitting criterion: Gini's diversity index (default), twoing rule, or maximum deviance reduction. Finally, the parameter *cost* is the cost matrix for cost-sensitive learning (default is the matrix with all ones except for the elements on the main diagonal, which are zeroes; this implies no cost-sensitive learning, i.e. the misclassification cost for all classes is considered to be the same).

By default, all features are considered for any node split, but there is an option to specify the number of randomly selected (out of all original features) features for each split[4] (this option is only available in the newest release of MATLAB® Statistics Toolbox™). Prior probabilities for data classes if necessary are set up by the parameter *priorprob*. As was said earlier, prior probability estimation is embedded into CART computations so that one does not have much to worry.

Since *classregtree* was designed with categorical target values in view, if the training targets are continuous, they are first transformed into categorical by using function *num2cat*. After that, the classification tree is built on the training data according to the parameters specified, followed by the classification of the test data by the constructed tree and assigning class labels to the test instances. Since these labels are categorical, they are converted to continuous values by another function *cat2num* (pay attention that the second argument of this function is a vector of unique class labels). Besides discrete class labels, *classification_tree_moo* can also return class probability estimates for each test instance. These estimates may be useful in performance evaluation when plotting Receiver Operating Characteristics (ROC) curves (Fawcett, 2006), (Pepe, 2004).

After such a description of coding efforts, it is advantageous to look deeper into the rich functionality of *classregtree*. The call to *classregtree* creates the object *t* of class *classregtree*. Using the created object, it is now possible to invoke numerous useful functions belonging to this class. The complete list of these functions includes:

catsplit (categorical splits used for branches in a tree),
children (child nodes),
classcount (class counts),
classprob (class probabilities),
cutcategories (cut categories),
cutpoint (cut point values),
cuttype (cut types: 'continuous', 'categorical', ''),
cutvar (cut variable names),
eval (predict test targets),
isbranch (test a node for branch),
nodeerr (node resubstitution errors),
nodeprob (node probabilities),
nodesize (node size – the number of instances in each node),
numnodes (the number of nodes in a tree),
parent (parent nodes),
prune (prune a tree),
risk (node risks – result of *nodeerr* weighed by the result of *nodeprob* ; the element-wise product of the outputs of *nodeerr* and *nodeprob*),
test (misclassification cost attained through resubstitution, cross-validation or independent test set),
type (tree type: 'regression' or 'classification'),
varimportance (feature importance obtained by summing changes in the risk (differences between the risk for the parent node and the total risk for two children) due to splits on every feature),
view (plot a tree).

```
function [test_targets, varargout] = classification_tree_
moo(train_patterns, train_targets, test_patterns, viewflag,
pruneflag, k, split_crit, cost)

% Classification tree built on MATLAB Statistics Toolbox ob-
ject-
% oriented functionality
% Original source:
```

```
% Leo Breiman, Jerome H. Friedman, Richard A. Olshen, Charles
J. Stone,
% "Classification and regression trees", Wadsworth, New York,
NY, 1984
%
% Inputs:
%   train_patterns  - Training patterns (DxN matrix, where D -
data
%                     dimensionality, N - number of patterns)
%   train_targets   - Training targets (1xN vector)
%.  test_patterns   - Test patterns (DxNt matrix, where D -
data
%                     dimensionality, Nt - number of patterns)
%   viewflag        - View flag ('on' - display a tree, 'off' -
don't
%                     display)
%   pruneflag       - Pruning flag ('on' - with tree pruning,
'off' -
%                     without it)
%   k               - The minimum number of patterns impure
nodes shall
%                      contain in order to be split
%   split_crit      - Node splitting criterion (character
string)
%                        Legitimate values:
%                        'gdi' - Gini's diversity index,
%                        'twoing' - twoing rule,
%                        'deviance' - maximum deviance reduction
%   cost            - Cost cxc matrix (c - the number of class-
es) used
%                     in cost-sensitive learning where different
%                     classes have different costs of
%                     misclassification
%
% Outputs:
%   test_targets    - Test targets (1xNt vector)
%   prob            - Probability estimates (CxNt matrix, where
C -
%                     the number of classes)
%
```

```
% Note: this function is built on MATLAB functionality. In
other words,
% MATLAB implementation of a classification tree was taken
rather than % our own implementation



% Get the number of classes
c = length(unique(train_targets));

% Set default values
if nargin < 8
    % All save diagonal elements are ones
    % Diagonal elements are zeroes
    cost = tril(ones(c,c), -1) + triu(ones(c,c), 1);
end
if nargin < 7
    split_crit = 'gdi';
end
if nargin < 6
    k = 10;
end
if nargin < 5
    pruneflag = 'off';
end
if nargin < 4
    viewflag = 'on';
end

% Check the name of a split criterion
switch split_crit
    case {'gdi'; 'twoing'; 'deviance'},
        fprintf(1, 'Split criterions is %s\n', split_crit);
    otherwise
        error('Unknown split criterion!');
end

% Convert numerical labels into categorical variables
train_targets = num2cat(train_targets);
```

```matlab
% Create a tree from training data
t = classregtree(train_patterns', train_targets', 'method',
'classification',...
    'prune', pruneflag, 'splitmin', k, 'splitcriterion', split_
crit, 'cost', cost);

% Graphically display the obtained tree if the flag is on
if strcmpi(viewflag, 'on')
    view(t);
end

% Predict class labels for test data
[test_targets, nodes] = eval(t, test_patterns');
% Don't forget to convert test_targets into the default repre-
sentation
% of targets vectors
test_targets = test_targets';
% Convert categorical labels back into the numerical represen-
tation
test_targets = cat2num(test_targets, unique(train_targets));

if nargout == 2
    % Get class probability estimates for given nodes of the
tree
    prob = classprob(t, nodes);
    % Convert probabilities to the default representation
    prob = prob';
    varargout{1} = prob;
end

return
function y = cat2num(x, u)

% Categorical-to-numerical conversion of elements of a row vec-
tor

if isnumeric(x)
    % No conversion is needed
    y = x;
    return;
```

```matlab
end

% Pre-allocate memory
y = zeros(1,length(x));

% Perform conversion
for i = 1:length(x)
    % String '0' will be converted into 0
    % String '1' will be converted into 1
    y(i) = str2double(char(x(i)));
    % If a categorical value is like 'setosa', then the conver-
sion will
    % lead to NaN
    % Hence, we need to proceed differently by assigning to a
label
    % a real number corresponding to the index of a given cat-
egorical
    % value in vector u
    if isnan(y(i))
        y(i) = find(strcmp(u, x(i)));
    end
end

return
function y = num2cat(x)

% Numerical-to-categorical conversion of elements of a row-vec-
tor

if iscellstr(x)
    % No conversion is needed
    y = x;
    return;
end

% Pre-allocate memory
y = cell(1,length(x));

% Perform conversion
for i = 1:length(x)
```

```
    y(i) = cellstr(num2str(x(i)));
end


return
```

REFERENCES

Bishop, C. M. (2006). *Pattern recognition and machine learning*. New York, NY: Springer Science-Business Media.

Breiman, L., Friedman, J. H., Olshen, R. A., & Stone, C. J. (1984). *Classification and regression trees*. New York, NY: Wadsworth.

Díaz-Uriarte, R., & Alvarez de Andrés, S. (2006). Gene selection and classification of microarray data using random forest. *BMC Bioinformatics*, *7*(3).

Duda, R. O., Hart, P. E., & Stork, D. G. (2000). *Pattern classification* (2nd ed.). New York, NY: John Wiley & Sons.

Fawcett, T. (2006). An introduction to ROC analysis. *Pattern Recognition Letters*, *27*(8), 861–874. doi:10.1016/j.patrec.2005.10.010

Hastie, T., Tibshirani, R., & Friedman, J. (2003). *The elements of statistical learning: data mining, inference, and prediction*. New York, NY: Springer-Verlag.

Izenman, A. J. (2008). *Modern multivariate statistical techniques: regression, classification, and manifold learning*. New York, NY: Springer Science+Business Media.

Kuncheva, L. I. (2004). *Combining pattern classifiers: methods and algorithms*. Hoboken, NJ: John Wiley & Sons. doi:10.1002/0471660264

Martinez, W. L., & Martinez, A. R. (2002). *Computational statistics handbook with MATLAB*. Boca Raton, FL: Chapman & Hall/CRC Press.

Pepe, M. S. (2004). *The statistical evaluation of medical tests for classification and prediction* (1st paperback ed.). Oxford, UK: Oxford University Press.

Quinlan, J. R. (1993). *C4.5: programs for machine learning*. San Mateo, CA: Morgan Kaufmann.

Ripley, B. (2007). *Pattern recognition and neural networks* (1st paperback ed.). Cambridge, UK: Cambridge University Press.

Rokach, L., & Maimon, O. (2008). *Data mining with decision trees: theory and applications*. Singapore: World Scientific.

Russell, S., & Norvig, P. (2003). *Artificial intelligence: a modern approach* (2nd ed.). Upper Saddle River, NJ: Pearson Education.

Statnikov, A., Wang, L., & Aliferis, C. F. (2008). A comprehensive comparison of random forests and support vector machines for microarray-based cancer classification. *BMC Bioinformatics*, *9*(319).

Steinberg, D. (2009). CART: classification and regression trees. In Wu, X., & Kumar, V. (Eds.), *The top ten algorithms in data mining* (pp. 179–202). Boca Raton, FL: Chapman & Hall/CRC Press. doi:10.1201/9781420089653.ch10

Stork, D. G., & Yom-Tov, E. (2004). *Computer manual in MATLAB to accompany Pattern Classification* (2nd ed.). Hoboken, NJ: John Wiley & Sons.

Wu, X., & Kumar, V. (Eds.). (2009). *The top ten algorithms in data mining*. Boca Raton, FL: Chapman & Hall/CRC Press.

ENDNOTES

[1]   The complete algorithm description with all options can be found in (Breiman, Friedman, Olshen, & Stone, 1984). It would take too much space to reproduce this description here.

[2]   It can be a drawback if the optimal decision boundary between two classes runs at 45 degrees to the axes. In this case, one would need a large number of axis-parallel splits of the feature space in order to approximate the decision boundary.

[3]   This is the complement to (Duda, Hart, & Stork, 2000), describing Classification Toolbox – MATLAB® software implementing many algorithms mentioned in (Duda, Hart, & Stork, 2000).

[4]   A subset of randomly selected features varies from split to split; it is not the same for all tree splits.

Chapter 7
Support Vector Machines

SUPPORT VECTOR MACHINES

This chapter is based on the famous book of Nello Cristianini and John Shawe-Taylor (Cristianini & Shawe-Taylor, 2000). It was one of the first introductory books to Support Vector Machines (SVMs) – a new generation learning system based on the recent advances in statistical learning theory. Currently, SVM is one of the standard tools for machine learning and data mining and it found many applications in bioinformatics, too.

To date, there are quite many books on this topic. I included the following non-exhaustive list for you: (Vapnik, The nature of statistical learning theory, 1995), (Vapnik, Statistical learning theory, 1998), (Schölkopf, Burges, & Smola, Advances in kernel methods: support vector machine learning, 1999), (Smola, Bartlett, Schölkopf, & Schuurmans, 2000), (Cristianini & Shawe-Taylor, 2000), (Kecman, 2001), (Schölkopf & Smola, 2002), (Herbrich, 2002), (Suykens, van Gestel, De Brabanter, De Moor, & Vandewalle, 2002), (Shawe-Taylor & Cristianini, 2004), (Schölkopf, Tsuda, & Vert, Kernel methods in computational biology, 2004), (Abe, 2005), (Neuhaus & Bunke, 2007), (Steinwart & Christmann, 2008), (Hamel, 2009)

DOI: 10.4018/978-1-60960-557-5.ch007

Chapters on SVMs are also included in many textbooks, e.g., (Duda, Hart, & Stork, 2000), (Bishop, 2006), (Hastie, Tibshirani, & Friedman, 2003), (Izenman, 2008), (Wu & Kumar, 2009).

Gene expression based cancer classification using SVMs can be found, e.g., in (Furey, Cristianini, Duffy, Bednarski, Schummer, & Haussler, 2000), (Zhang & Ke, 2000), (Guyon, Weston, Barnhill, & Vapnik, 2002), (Wang, Makedon, Ford, & Pearlman, 2005), (Tan & Yang, 2008).

This chapter is the longest one in the book, so be ready for a long journey as I will guide you, dear readers, through jungles of mathematics threatening to seize you with boredom. I have carefully read the book of (Cristianini & Shawe-Taylor, 2000) and decoded difficult to understand text. Thus, your task is to trust me and be good followers. If Cristianini and Shawe-Taylor made the path in "SVM jungles", I am going to cut all branches of trees and lianas along this path so that you could comfortably walk along it as if you were on the main avenue of a big city (let us assume here that there are no big crowds).

Notations

Let X and Y denote the input space and output domain, respectively. Usually, $X \subseteq \mathbb{R}^n$, while for binary classification $Y \in \{-1, 1\}$, and for m-class classification $Y = \{1, 2, \ldots, m\}$. A training set is denoted by

$$S = \left(\left(x_1, y_1 \right), \ldots, \left(x_\ell, y_\ell \right) \right) \subseteq \left(X \times Y \right)^\ell,$$

where ℓ is the number of instances, and $x_i, i = 1, 2, \ldots, \ell$ is an n-dimensional column-vector (to produce a row vector, one can take the transpose x_i'.

I will often talk of vectors as points in some space of features. For example, any point on a 2D plot is characterized by two features – its coordinates.

In this chapter and in several other chapters, we will frequently utilize the following mathematical operator: $\langle a, b \rangle$, where a and b are vectors of features of the same length. This operator denotes inner product that takes two vectors of equal length as input and produces a scalar as output. In Euclidean spaces[1], the inner product is called the dot or scalar product. How the inner product is calculated is shown in the next section.

Linear Classification and its Deficiency

Let us start from discriminating between two classes of objects when classes can be linearly separated. This is a well-studied case in traditional statistics and neural networks, and linear classification forms a basis of Support Vector Machines.

Binary classification is frequently performed by using a real-valued function $f : X \subseteq \mathbb{R}^n \to \mathbb{R}$ (mathematical notation like $\mathbb{R}^n \to \mathbb{R}$ means the mapping or projection from n-dimensional space of real numbers to 1-dimensional space of real numbers) in the following way: the input vector x is assigned to the positive class, if $f(x) \geq 0$, otherwise to the negative class. We consider the case, where $f(x)$ is a linear function of x, so that it can be written as

$$f(x) = \langle w, x \rangle + b = \sum_{i=1}^{n} w_i x_i + b,$$

where $(w, b) \in \mathbb{R}^n \times \mathbb{R}$ are the parameters that control the function and the decision rule is given by the sign of $f(x)$, where the sign of $f(0)=1$. The learning methodology implies that *these parameters must be learned from the data*.

To better understand all peculiarities, let us turn to a geometric interpretation in Figure 1. In this figure, there are two classes of the data points marked by filled and empty circles. The think straight line which we call a hyperplane (the term borrowed from geometry and generalizing the concept of the conventional 2D plane into a different (larger) number of dimensions) separates the input space X into the positive region above the line and the negative region below it[2]. The hyperplane equation is $\langle w, x \rangle + b = 0$. The vector w determines a direction perpendicular to the hyperplane, while varying the value of b moves the hyperplane parallel to itself. By borrowing terms from the neural networks literature, w and b are called the *weight vector* and *bias*, respectively. Notice that w is an n-dimensional column vector whereas b is a scalar.

The problem of two-class discrimination can also be solved by defining w_i and b_i for each class. Each time a new instance has to be classified, the point x is assigned to class 1 if $\langle w_1, x \rangle + b_1 \geq \langle w_{-1}, x \rangle + b_{-1}$, to class -1 otherwise.

Linear classifiers are limited to the case when different classes are linearly separable, i.e. a hyperplane can well separate different classes. When classes are linearly non-separable, a linear classifier fails. To remedy this problem, kernel representations offer a solution by projecting the data into a high dimensional feature

Figure 1. A separating hyperplane for two-dimensional data

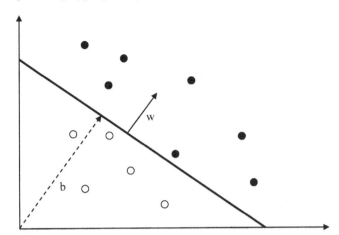

space to increase the computational power of the linear learning machines. The use of kernels provides one of the main building blocks of Support Vector Machines.

Learning in Feature Space

The complexity of the decision function to be learned depends on the way it is represented, and the difficulty of the learning task can vary accordingly. So one common preprocessing strategy in machine learning involves changing the representation of the data: $x \mapsto \phi(x)$. This step is equivalent to mapping the input space X into a new space, $F = \{\phi(x) | x \in X\}$. The space X is referred to as the input space, while F is called the feature space.

Figure 2 shows an example of a feature mapping from a two dimensional input space to a two dimensional feature space, where the data cannot be separated by a linear function in the input space, but can be in the feature space. Such mappings can be made into very high dimensional spaces where linear separation becomes much easier.

However, the larger dimensionality is, the more computational efforts are needed for learning and the more vulnerable a learning procedure to the phenomenon sometimes referred to as *the curse of dimensionality* when during training a classifier the overfitting occurs (i.e. the classifier learns every minor detail, often caused by noise, at the expense of generality), while performance on the test set rapidly degrades as the number of features grows. This implies that the growth in the number of feature not only does not improve the generalization to unseen data, but makes it even harder since it brings unnecessary redundancy.

Figure 2. A feature map can simplify the classification task

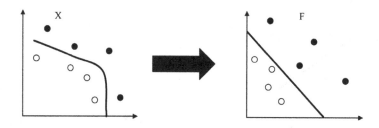

On the other hand, one can perform feature reduction in order to avoid the curse of dimensionality and to lower computational load. However, removing some features cannot always guarantee that these features are not essential for performing the classification task.

We came to a contradiction: on the one hand, good generalization requires dimensionality reduction; on the other hand, kernel representation used in SVMs projects the data into high (or even infinite) dimensional space, i.e. no dimensionality reduction occurs. This puzzle is resolved in the next section.

Implicit Mapping into Feature Space

In order to learn non-linear relations with a linear classification algorithm, i.e. the one expecting the data to be linearly separated into distinct classes, one needs to select a set of non-linear features and to convert the data into the new representation. This is equivalent to applying a non-linear mapping of the data to a feature space, in which the linear classifier can be used. Hence, the decision function is as follows

$$f(x) = \sum_{i=1}^{\ell} w_i \phi(x) + b,$$

where $\phi : X \mapsto F$ is a non-linear map from the input space to some feature space. This means that first a non-linear mapping transforms the data into a feature space, and then a linear classifier classifies the data in the feature space. These two steps are merged together by using a *kernel* function.

A kernel is a function K, such that $\forall x, z \in X$

$$K(x, z) = \langle \phi(x), \phi(z) \rangle,$$

where ϕ is a mapping from X to F.

Notice that when using the kernel representation, the inputs never appear isolated but always in the form of inner products between pairs of inputs[3]. This implies that ϕ is the implicit mapping from X to F, because it does not require to map each input individually. We do not need to know the underlying feature map in order to be able to learn in the feature space.

An important consequence of using kernels is that **the dimension of the feature space need not affect the computation**[4]. The use of kernels makes it possible to map the data **implicitly** into a feature space and to train a linear classifier in such a space. The only information to be necessary is the *kernel matrix* denoted by K.

However, not every function can be a kernel and the next section describes the necessary and sufficient conditions to be satisfied.

Conditions for a Function to Be a Kernel

Condition 1. The function must be symmetric:

$$K\left(x,z\right)=\left\langle\phi\left(x\right),\phi\left(z\right)\right\rangle=\left\langle\phi\left(z\right),\phi\left(x\right)\right\rangle=K\left(z,x\right).$$

Condition 2. The function must satisfy the inequalities that follow from the Cauchy-Schwarz inequality[5]:

$$K\left(x,z\right)^{2}=\left\langle\phi\left(x\right),\phi\left(z\right)\right\rangle^{2}\leq\left\|\phi\left(x\right)\right\|^{2}\left\|\phi\left(z\right)\right\|^{2}=\left\langle\phi\left(x\right),\phi\left(x\right)\right\rangle\left\langle\phi\left(z\right),\phi\left(z\right)\right\rangle=K\left(x,x\right)K\left(z,z\right).$$

Condition 3. (Mercer condition) The matrix

$$K=\left(K\left(x_{i},x_{j}\right)\right)_{i,j=1}^{n}$$

is positive semi-definite[6].

The first two conditions are necessary but not sufficient for the function to be a kernel. The last condition must be satisfied for this (sufficient condition). Thus, if one wishes to construct a new kernel, condition 3 must be first checked.

Let us verify the third condition for the following function: $K\left(Ax,Az\right)$, where A is a matrix. According to the definition of the inner product,

$$K\left(Ax,Az\right)=\left\langle Ax,Az\right\rangle=x^{'}A^{'}Az=x^{'}Bz,$$

where by construction $B = A'A$ is a square symmetric matrix. We need to prove that in addition to that, B is also positive semi-definite. By definition, if B is indeed positive semi-definite, then the following condition, resulting from the eigenvector equation ($Bx = \lambda x$), must be satisfied: $x'Bx = \lambda \geq 0$. Since $B = A'A$, by substituting this to the last inequality, we obtain

$$\lambda = x'Bx = x'A'Ax = \left(Ax\right)'\left(Ax\right) = \left\langle Ax, Ax \right\rangle = \left\|Ax\right\|^2 \geq 0,$$

which holds for any column-vector x.

That is, if K is a kernel function, then the linear transformation A of inputs x leads to another function, which nevertheless remains to be a kernel.

Typical Kernels

The following kernels are typical:

$K\left(x, z\right) = x'z$ (linear kernel),

$K\left(x, z\right) = \left(\tau + x'z\right)^d$ (polynomial kernel of degree d),

$K\left(x, z\right) = exp\left(-\sigma\left\|x - z\right\|^2\right)$ (RBF (Radial Basis Function) or Gaussian kernel),

$K\left(x, z\right) = \tanh\left(\kappa_1 x'z + \kappa_2\right)$ (MLP or multi-layer perceptron kernel).

In these definitions, only x and z are vectors while other symbols denote scalars. The Mercer condition holds for all σ values in the RBF kernel and positive τ values in the polynomial kernel, but not for all possible choices of κ_1, κ_2 in the MLP kernel case.

Optimization Problem: Basic Definitions

Using kernels provides one foundation for introducing support vector machines. Another foundation has its roots in optimization[7] theory (branch of mathematics). In the case of linear learning machines as those considered before, the task is to find a vector of parameters that optimizes a certain cost function, typically subject to some constraints. The general (*primal*) optimization problem can be therefore stated as follows:

Given functions f, g_i, $i = 1, 2, \ldots, k$ and h_i, $i = 1, 2, \ldots, m$, defined over a domain $\Omega \in \mathbb{R}^n$, minimize $f(w)$, $w \in \Omega$,

subject to $g_i(w) \leq 0$, $i = 1, \ldots, k$,

$h_i(w) = 0$, $i = 1, \ldots, m$,

where $f(w)$ is called the *objective function*, and the remaining relations are called, respectively, the *inequality* and *equality constraints*. The optimal value of the objective function is called the *value of the optimization problem*.

To simplify the notation I will write $g(w) \leq 0$ to indicate $g_i(w) \leq 0$, $i = 1, \ldots, k$. The expression $h(w)$ has a similar meaning for the equality constraints.

Since maximization problems can be converted to minimization ones by reversing the sign of $f(w)$, the choice of minimization does not represent a restriction. Similarly any constraints can be rewritten in the above form.

The region of the domain where the objective function is defined and where all the constraints are satisfied is called the *feasible region*, and it will be denoted by

$$R = \left\{ w \in \Omega : g(w) \leq 0, h(w) = 0 \right\}.$$

A solution of the optimization problem is a point $w^* \in R$ such that there exists no other point $w \in R$ for which $f(w) < f(w^*)$. Such a point is also known as a global minimum. A point $w^* \in \Omega$ is called a local minimum of $f(w)$ if there exists $\varepsilon > 0$ such that $f(w) \geq f(w^*)$, $\forall w \in \Omega$ such that $\|w - w^*\| < \varepsilon$.

An optimization problem in which the objective function, inequality and equality constraints are all linear functions is called a *linear programme*. If the objective function is quadratic while the constraints are all linear, the optimization problem is called a *quadratic programme*.

An inequality constraint $g_i(w) \leq 0$ is said to be *active* if the solution w^* satisfies $g_i(w^*) = 0$, otherwise it is said to be *inactive*. In this sense, equality constraints are always active. Further in the next sections, quantities called *slack variables* and denoted by ξ are introduced, to transform an inequality constraint into an equality one as follows:

$$g_i(w) \leq 0 \Leftrightarrow g_i(w) + \xi_{i,} = 0 \text{ with } \xi_i \geq 0.$$

We will consider restricted classes of convex optimization problems called convex quadratic programmes, because it is this class that proves adequate for the task of training SVMs.

A real-valued function $f(w)$ is called *convex* for $w \in \mathbb{R}^n$ if, $\forall w, u \in \mathbb{R}^n$, and for any $\theta \in (0,1)$,

$$f\left(\theta w + (1-\theta)u\right) \leq \theta f\left(w\right) + (1-\theta) f\left(u\right).$$

If a strict inequality holds, the function is said to be *strictly convex*. A function that is twice differentiable will be convex provided its Hessian matrix is positive semi-definite.

If a function f is convex, any local minimum w^* of the unconstrained (i.e., without constraints) optimization problem with objective function f is also a global minimum, since for any $u \neq w^*$, by definition of a local minimum there exists θ sufficiently close to 1 that

$$f\left(w^*\right) \leq f\left(\theta w^* + (1-\theta)u\right) \leq \\ \theta f\left(w^*\right) + (1-\theta) f\left(u\right).$$

The last inequality follows from the definition of a convex function. From the last inequality, it also follows that $f\left(w^*\right) < f\left(w\right)$. It is this property of convex functions that renders optimization problems tractable when the functions and sets involved are convex.

A set $\Omega \in \mathbb{R}^n$ is called *convex* if, $\forall w, u \in \Omega$, and for any $\theta \in (0,1)$, the point $\left(\theta w + (1-\theta)u\right) \in \Omega$.

An optimization problem in which the set Ω, the objective function and all of the constraints are convex is said to be *convex*.

For the purposes of the SVM training, we restrict ourselves to the case where the constraints are linear, the objective function is convex and quadratic and $\Omega = \mathbb{R}^n$, hence we consider convex quadratic programmes.

Quadratic Programming

In the previous section, we mentioned the quadratic programme. This type of mathematical optimization problem forms a core of SVMs. We would like to briefly describe its formulation here reproduced from http://encyclopedia.thefreedictionary. com/quadratic%20programming.

Let us assume that $x \in \mathbb{R}^n$ as before, the $n \times n$ matrix E is positive semi-definite and h is any $n \times 1$ vector. Then the quadratic programme can be formulated like this:

Minimize (with respect to x) $f(x) = 0.5x'Ex + h'x$ with at least one instance of the following type of constraints:

$Ax \leq b$ (inequality constraints),
$Cx = d$ (equality constraints).

If there are only equality constraints, the quadratic problem can be solved by a linear system. Otherwise, one of the most common methods of solving this problem is an inferior point method, such as LOQO (http://www.princeton.edu/~rvdb/).

Lagrangian Theory

In constrained problems, one needs to define a function, known as the Lagrangian, that incorporates information about both the objective function and the constraints, and whose stationarity can be used to find solutions. Precisely, the Lagrangian is defined as the objective function plus a linear combination of the constraints, where the coefficients of the combination are called the Lagrange multipliers.

Given an optimization problem with objective function $f(w)$, and equality constraints $h_i(w) = 0$, $i = 1, \ldots, m$, the *Lagrangian function* is defined as

$$L(w, \beta) = f(w) + \sum_{i=1}^{m} \beta_i h_i(w),$$

where the coefficients β_i are called the *Lagrange multipliers*.

A necessary condition (Lagrange theorem) for a normal point w^* to be a minimum of $f(w)$, subject to $h_i(w) = 0$, $i = 1, \ldots, m$ is

$$\frac{\partial L(w^*, \beta^*)}{\partial w} = 0,$$

$$\frac{\partial L(w^*, \beta^*)}{\partial \beta} = 0,$$

for some values β^*. The above conditions are also sufficient provided that $L\left(w, \beta^*\right)$ is a convex function of w.

The first of the two conditions gives a new system of equations, whereas the second returns the equality constraints. By jointly solving the two systems one obtains the solution.

We now consider the most general case where the optimization problem contains both equality and inequality constraints.

Given an optimization problem with domain $\Omega \subseteq \mathbb{R}^n$,

minimize $f\left(w\right)$, $w \in \Omega$,

subject to $g_i\left(w\right) \leq 0$, $i = 1, \dots, k$,

$h_i\left(w\right) = 0$, $i = 1, \dots, m$,

with f convex and g_i, h_i affine, we define the *generalized Lagrangian function* as

$$L\left(w, \alpha, \beta\right) = f\left(w\right) + \sum_{i=1}^{k} \alpha_i g_i\left(w\right) + \sum_{i=1}^{m} \beta_i h_i\left(w\right) = \\ f\left(w\right) + \alpha' g\left(w\right) + \beta' h\left(w\right).$$

Then necessary and sufficient conditions (Kuhn-Tucker theorem) for a normal point w^* to be an optimum are the existence of α_i^*, β_i^* such that

$$\frac{\partial L\left(w^*, \alpha^*, \beta^*\right)}{\partial w} = 0,$$

$$\frac{\partial L\left(w^*, \alpha^*, \beta^*\right)}{\partial \beta} = 0,$$

$$\alpha_i^* g_i\left(w^*\right) = 0, \ i = 1, \dots, k,$$

$$g_i\left(w^*\right) \leq 0, \ i = 1, \dots, k,$$

$$\alpha_i^* \geq 0, \ i = 1, \dots, k.$$

The third relation is known Karush-Kuhn-Tucker complementary condition. It implies that for active constraints, $\alpha_i^* \geq 0$, whereas for inactive constraints $\alpha_i^* = 0$.[8] In other words, a solution point can be in one of the two positions with respect to an inequality constraint: either in the interior of the feasible region with the constraint inactive, or on the boundary defined by that constraint with the constraint active.

The readers may wonder what Lagrange multipliers give in practice? The original problem included both an objective function and constraints. By incorporating constraints into the objective function, we converted the constrained problem into an unconstrained one, i.e. we made the task somewhat simpler, since an unconstrained problem can be solved, for example, by the conventional gradient method.

Duality

Lagrangian treatment of convex optimization problems leads to an alternative dual description, which often turns out to be easier to solve than the primal problem since handling inequality constraints directly is difficult. The dual problem is obtained by introducing Lagrange multipliers, also called the dual variables. Dual methods are based on the idea that the dual variables are the fundamental unknowns of the problem.

The *Lagrangian dual problem* of the primal problem:

maximize $\theta(\alpha, \beta)$

subject to $\alpha \geq 0$,

where $\theta(\alpha, \beta) = \inf_{w \in \Omega} L(w, \alpha, \beta)$.

We can transform the primal into a dual by simply setting to zero the derivatives of the Lagrangian with respect to the primal variables, and substituting the relations so obtained back into the Lagrangian, hence removing the dependence on the primal variables. This corresponds to explicitly computing the function

$$\theta(\alpha, \beta) = \inf_{w \in \Omega} L(w, \alpha, \beta).$$

The resulting function contains only dual variables and must be maximized under simpler constraints. This strategy is adopted for Support Vector Machines.

Functional and Geometrical Margin

In the subject studied in this chapter, the quantity called the *margin* plays a central role. The *(functional) margin of an example* (x_i, y_i) with respect to a hyperplane (w, b) to be the quantity

$$\gamma_i = y_i (\langle w, x_i \rangle + b).$$

Note that $\gamma_i > 0$ implies correct classification of x_i.[9] The *(functional) margin distribution of a hyperplane* (w, b) *with respect to a training set S* is the distribution of the margins of the examples in S. The minimum of the margin distribution is referred to as the *(functional) margin of a hyperplane* (w, b) *with respect to a training set S*. In both definitions if we replace functional margin by *geometric margin* we obtain the equivalent quantity for the normalized linear function $\left(\dfrac{1}{w} w, \dfrac{1}{w} b \right)$, which measures the Euclidean distance of the points from the decision boundary in the input space (Figure 3 shows the geometric margin at two points with respect to a hyperplane in two dimensions)[10]. Finally, the *margin of a training set S* is the maximum geometric margin over all hyperplanes (see Figure 4). A hyperplane realizing this maximum is a *maximal margin hyperplane*. The size of its margin will be positive for a linearly separable training set. The larger the margin is, the better for classification.

Assembling Things Together

Let us now combine individual components of SVMs together in order to get a general picture. We can say that SVMs are systems for training the linear learning machines in a high dimensional feature space by employing the optimization theory. An important feature of SVMs is that they produce sparse dual representations (thanks to the fact that some α 's are equal to zero due to the Karush-Kuhn-Tucker conditions), resulting in extremely efficient algorithms. Another important feature of SVM is that due to Mercer condition on the kernels the corresponding optimization problems are convex and hence have no local minima. These two facts clearly distinguish SVMs from many other pattern recognition algorithms.

The Maximal Margin Classifier

The simplest model of Support Vector Machine, which was also the first to be introduced, is the so-called maximal margin classifier. It works only for two-class data which are linearly separable in the feature space[11]. Nonetheless it is the easiest algorithm to understand, and it forms the main block for the more complex Support Vector Machines.

Let us derive the optimization problem to be solved for points belonging to either "positive" or "negative" class. Recall that a functional margin of 1 (or equivalently geometric margin) implies

Figure 3. The geometrical margin of two points

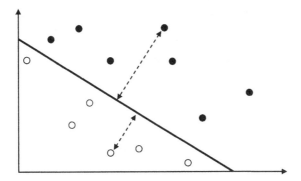

Figure 4. The margin of a set of points

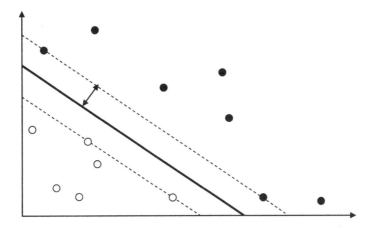

$$\left\langle w, x^+ \right\rangle + b = +1,$$
$$\left\langle w, x^- \right\rangle + b = -1,$$

while to compute the geometric margin we must normalized w by its norm. The geometric margin γ is then the functional margin of the resulting classifier

$$\gamma = \frac{1}{2}\left(\left\langle \frac{w}{\|w\|}, x^+ \right\rangle - \left\langle \frac{w}{\|w\|}, x^- \right\rangle \right) = \frac{1}{2\|w\|}\left(\left\langle w, x^+ \right\rangle - \left\langle w, x^- \right\rangle \right) = \frac{1}{\|w\|},$$

where the final result is obtained from subtracting the second formula above from the first one[12].

Hence, the resulting geometric margin will be equal to $1/\|w\|$ and we have to solve the following optimization problem under linear inequality constraints.

Given a linearly separable training sample $S = \left((x_1, y_1), \ldots, (x_\ell, y_\ell) \right)$, the hyperplane (w, b) that solves the optimization problem

minimize $\langle w, w \rangle$,
subject to $y_i \left(\langle w, x_i \rangle + b \right) \geq 1$, $i = 1, \ldots, \ell$,

realizes the maximal margin hyperplane with geometric margin $\gamma = 1 / \|w\|$. Minimization is done by both w and b.

An amused reader may wonder how we came to such an optimization problem and its constraints. First of all, we seek to maximize margin between classes in order to separate them as much as possible; hence, the norm of w needs to be minimized. Second, the condition $y_i \left(\langle w, x_i \rangle + b \right) = 1$ is fulfilled only for the points of both classes that are closest to the maximal margin hyperplane. In the interior of either class, this equality is turned to the "greater than or equal to" inequality[13]. Thus, the optimization task can be read as follows: maximize interclass separation by drawing such a hyperplane between classes that points of both classes are correctly classified.

We now consider how to transform this optimization problem into its corresponding dual problem. The Lagrangian is

$$L\left(w, b, \alpha\right) = \frac{1}{2}\langle w, w \rangle - \sum_{i=1}^{\ell} \alpha_i \left[y_i \left(\langle w, x_i \rangle + b \right) - 1 \right],$$

where $\alpha_i \geq 0$ are the Lagrange multipliers[14].

The corresponding dual is found by differentiating the Lagrangian with respect to w and b

$$\frac{\partial L\left(w, b, \alpha\right)}{\partial w} = w - \sum_{i=1}^{\ell} y_i \alpha_i x_i = 0,$$
$$\frac{\partial L\left(w, b, \alpha\right)}{\partial b} = \sum_{i=1}^{\ell} y_i \alpha_i = 0.$$

That is, we obtained

$$w = \sum_{i=1}^{\ell} y_i \alpha_i x_i,$$

$$0 = \sum_{i=1}^{\ell} y_i \alpha_i,$$

which when substituted into the primal results in

$$L(w, b, \alpha) = \frac{1}{2} \langle w, w \rangle - \sum_{i=1}^{\ell} \alpha_i \left[y_i \left(\langle w, x_i \rangle + b \right) - 1 \right]$$

$$= \frac{1}{2} \langle w, w \rangle - \sum_{i=1}^{\ell} \alpha_i y_i \langle w, x_i \rangle - b \sum_{i=1}^{\ell} \alpha_i y_i + \sum_{i=1}^{\ell} \alpha_i$$

$$= \frac{1}{2} \sum_{i,j=1}^{\ell} y_i y_j \alpha_i \alpha_i \langle x_i, x_j \rangle - \sum_{i,j=1}^{\ell} y_i y_j \alpha_i \alpha_j \langle x_i, x_j \rangle + \sum_{i=1}^{\ell} \alpha_i$$

$$= \sum_{i=1}^{\ell} \alpha_i - \frac{1}{2} \sum_{i,j=1}^{\ell} y_i y_j \alpha_i \alpha_j \langle x_i, x_j \rangle.$$

By summarizing results, we have the following dual problem instead of the primal. Consider a linearly separable training sample $S = \left((x_1, y_1), \ldots, (x_\ell, y_\ell) \right)$ and suppose the parameters α^* solve the following quadratic optimization problem:

maximize $W(\alpha) = \displaystyle\sum_{i=1}^{\ell} \alpha_i - \frac{1}{2} \sum_{i,j=1}^{\ell} y_i y_j \alpha_i \alpha_j \langle x_i, x_j \rangle,$

subject to $\displaystyle\sum_{i=1}^{\ell} y_i \alpha_i = 0, \ \alpha_i \geq 0, \ i = 1, \ldots, \ell.$

Then the weight vector $w^* = \displaystyle\sum_{i=1}^{\ell} y_i \alpha_i^* x_i$ realizes the maximal margin hyperplane with geometric margin $\gamma = 1 / \left\| w^* \right\|$.

The value b does not appear in the dual problem and so b^* must be found making use of the primal constraints:

$$b^* = - \frac{\displaystyle\max_{y_i = -1} \left(\langle w^*, x_i \rangle \right) + \min_{y_i = +1} \left(\langle w^*, x_i \rangle \right)}{2}.$$

This formula is derived from the two conditions:

$$\left\langle w^*, x^+ \right\rangle + b^* = +1,$$

$$\left\langle w^*, x^- \right\rangle + b^* = -1.$$

When adding them together we have

$$b^* = -\frac{\left\langle w^*, x^+ \right\rangle + \left\langle w^*, x^- \right\rangle}{2}.$$

Since we minimize w and b, it means

$$\min_{y_i=+1}\left(\left\langle w^*, x_i^+ \right\rangle\right) \text{ and } \max_{y_i=-1}\left(\left\langle w^*, x_i^- \right\rangle\right).$$

The Karush-Kuhn-Tucker complementarity conditions provide useful information about the structure of the solution. The conditions state that the optimal solutions α^*, $\left(w^*, b^*\right)$ must satisfy

$$\alpha_i^*\left[y_i\left(\left\langle w^*, x_i \right\rangle + b^*\right) - 1\right] = 0, \ i = 1, \dots, \ell.$$

This implies that only for inputs x_i for which the functional margin is one and therefore lie closest to the hyperplane are the corresponding α_i^* non-zero. All the other parameters α_i^* are zero. Hence, in the expression for the weight vector w^* only these points are involved. It is for this reason that they are called support vectors (see Figure 5). We will denote the set of indices of the support vectors with sv.

Furthermore the optimal hyperplane can be expressed in the dual representation in terms of this subset of the parameters:

$$f\left(x, \alpha^*, b^*\right) = \sum_{i=1}^{\ell} y_i \alpha_i^* \left\langle x_i, x \right\rangle + b^* = \sum_{i \in sv} y_i \alpha_i^* \left\langle x_i, x \right\rangle + b^*.$$

The Lagrange multipliers associated with each point become the dual variables, giving them an intuitive interpretation quantifying how important a given training

Figure 5. A maximal margin hyperplane with its support vectors highlighted as circles of larger radius

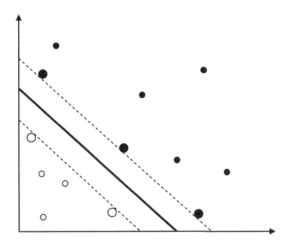

point is in forming the final solution. Points that are not support vectors have no influence.

Another important consequence of the Karush-Kuhn-Tucker complementarity conditions is that $j \in sv$,

$$y_j f\left(x_j, \alpha^*, b^*\right) = y_j \left(\sum_{i \in sv} y_i \alpha_i^* \left\langle x_i, x_j \right\rangle + b^*\right) = 1.$$

This follows from the fact that for $j \in sv$, i.e. for $\alpha_j \neq 0$,

$$y_j \left(\left\langle w^*, x_j \right\rangle + b^*\right) - 1 = 0.$$

Substituting $w^* = \sum_{i \in sv} y_i \alpha_i^* x_i$ into the last formula, we obtain

$$y_j \left(\sum_{i \in sv} y_i \alpha_i^* \left\langle x_i, x_j \right\rangle + b^*\right) = 1.$$

Therefore

$$y_j \sum_{i \in sv} y_i \alpha_i^* \left\langle x_i, x_j \right\rangle = 1 - y_j b^*.$$

As a result,

$$\left\langle w^*, w^* \right\rangle = \sum_{i,j=1}^{\ell} y_i y_j \alpha_i^* \alpha_j^* \left\langle x_i, x_j \right\rangle = \sum_{j\in sv} \alpha_j^* y_j \sum_{i\in sv} y_i \alpha_i^* \left\langle x_i, x_j \right\rangle = \sum_{j\in sv} \alpha_j^* \left(1 - y_j b^*\right) = \sum_{j\in sv} \alpha_j^* - b^* \sum_{j\in sv} y_j \alpha_j^* = \sum_{j\in sv} \alpha_j^*.$$

From the last equation, the geometric margin $\gamma = 1 / \|w\|$ is equal to $\left(\sum_{i\in sv} \alpha_i^*\right)^{-\frac{1}{2}}$.

Both the dual objective and the decision function have the property that the data only appear inside an inner product. This will make it possible to find and use optimal hyperplanes in the feature space through the use of kernels as stated below.

Consider a training sample $S = \left((x_1, y_1), \ldots, (x_\ell, y_\ell)\right)$ that is linearly separable in the feature space implicitly defined by the kernel $K(x, z)$ and suppose the parameters α^* and b^* solve the following quadratic optimization problem:

$$\text{maximize } W(\alpha) = \sum_{i=1}^{\ell} \alpha_i - \frac{1}{2}\sum_{i,j=1}^{\ell} y_i y_j \alpha_i \alpha_j K(x_i, x_j),$$

$$\text{subject to } \sum_{i=1}^{\ell} y_i \alpha_i = 0, \ \alpha_i \geq 0, \ i = 1, \ldots, \ell.$$

Then the decision rule given by $sgn(f(x))$, where $f(x) = \sum_{i=1}^{\ell} y_i \alpha_i^* K(x_i, x) + b^*$, is equivalent to the maximal margin hyperplane in the feature space implicitly defined by the kernel $K(x, z)$ and that hyperplane has geometric margin

$$\gamma = \left(\sum_{i\in sv} \alpha_i^*\right)^{-\frac{1}{2}}.$$

Note that the dual problem is solved in α, not in w. The matrix related to the quadratic term in α in this quadratic form (i.e. the matrix $\left(y_i y_j K(x_i, x_j)\right)_{i,j=1}^{\ell}$) is positive definite or positive semi-definite[15]. In the case that the matrix is positive definite (all eigenvalues strictly positive), the solution α to this quadratic optimization problem is *global and unique*. When the matrix is positive semi-definite (all eigenvalues positive but zero eigenvalues are possible as well) then the solution is global but not necessarily unique.

Slack Variables

When the data are not linearly separable, another measure called the *margin slack variable* is used to measure the amount of 'non-separability'.

Let us fix a value $\gamma > 0$. Then the margin slack variable of an instance (x_i, y_i) with respect to the hyperplane (w, b) and target margin γ_i is

$$\xi\Big((x_i, y_i), (w, b), \gamma\Big) = \xi_i = \max\Big(0, \gamma - y_i\big(w, x_i + b\big)\Big).$$

This quantity measures how much a point fails to have a margin of γ from the hyperplane. If $\xi_i > \gamma$, then x_i is misclassified by (w, b). The norm ξ measures the amount by which the training set fails to have margin γ, and takes into account any misclassifications of the training data.

Figure 6 shows the slack variables for two misclassified points. All the other points in the figure have their slack variable equal to zero, because they have a positive margin (signs of both y_i and $w, x_i + b$ are the same) that is larger than γ. For misclassified points the quantity $y_i\big(w, x_i + b\big)$ is negative; hence $\xi_i > 0$.

Soft Margin Optimization

The maximal margin classifier cannot be used in many real-world problems: if the data are noisy, there will in general be no linear separation in the feature space. To cope with this, slack variables were introduced in the previous section to allow the margin to be violated, i.e. to take into consideration more training points than just those closest to the decision boundary. With this in mind, there are the following two cases.

2-Norm Soft Margin

Let us consider the following primal problem:

$$\text{minimize } \langle w, w \rangle + C\sum_{i=1}^{\ell}\xi_i^2 \text{ over } \xi, w, b,$$
$$\text{subject to } y_i\Big(\big\langle w, x_i \big\rangle + b\Big) \geq 1 - \xi_i, \ i = 1, \dots, \ell.$$

Figure 6. Two slack variables

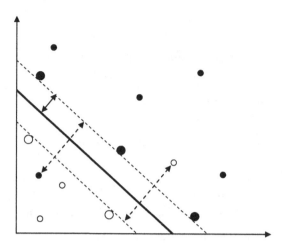

In practice the parameter C (positive scalar whose typical values can be 1, 10, 100, 1000) is varied through a wide range of values. In (Cristianini & Shawe-Taylor, 2000) on p.106 it is shown that

$$\left\langle w^{*}, w^{*} \right\rangle = \sum_{i \in sv} \alpha_{i}^{*} - \frac{1}{C} \left\langle \alpha^{*}, \alpha^{*} \right\rangle.$$

Hence, as the parameter C runs through a range of values, the norm $\|w\|$ varies smoothly through a corresponding range. As a result, choosing a particular value for C corresponds to choosing a value for $\|w\|$, and then minimizing $\|\xi\|$ for that $\|w\|$.

The primal Lagrangian for this problem is

$$L\left(w, b, \xi, \alpha\right) = \frac{1}{2}\left\langle w, w \right\rangle + \frac{C}{2}\sum_{i=1}^{\ell}\xi_{i}^{2} - \sum_{i=1}^{\ell}\alpha_{i}\left[y_{i}\left(\left\langle w, x_{i} \right\rangle + b\right) - 1 + \xi_{i}\right],$$

where $\alpha_{i} \geq 0$ are the Lagrange multipliers.

The corresponding dual is found by differentiating with respect to w, b, ξ:

$$\frac{\partial L\left(w, b, \xi, \alpha\right)}{\partial w} = w - \sum_{i=1}^{\ell} y_{i}\alpha_{i}x_{i} = 0,$$

$$\frac{\partial L\left(w,b,\xi,\alpha\right)}{\partial \xi} = C\xi - \alpha = 0,$$

$$\frac{\partial L\left(w,b,\xi,\alpha\right)}{\partial x} = \sum_{i=1}^{\ell} y_i \alpha_i = 0,$$

and substituting the relations obtained into the primal:

$$L\left(w,b,\xi,\alpha\right) = \sum_{i=1}^{\ell}\alpha_i - \frac{1}{2}\sum_{i,j=1}^{\ell} y_i y_j \alpha_i \alpha_j \left\langle x_i, x_j \right\rangle + \frac{1}{2C}\left\langle \alpha,\alpha \right\rangle - \frac{1}{C}\left\langle \alpha,\alpha \right\rangle = \sum_{i=1}^{\ell}\alpha_i - \frac{1}{2}\sum_{i,j=1}^{\ell} y_i y_j \alpha_i \alpha_j \left\langle x_i, x_j \right\rangle - \frac{1}{2C}\left\langle \alpha,\alpha \right\rangle.$$

Hence maximizing the above objective function over α is equivalent to maximizing

$$W\left(\alpha\right) = \sum_{i=1}^{\ell}\alpha_i - \frac{1}{2}\sum_{i,j=1}^{\ell} y_i y_j \alpha_i \alpha_j \left(\left\langle x_i, x_j \right\rangle + \frac{1}{C}\delta_{ij}\right),$$

where δ_{ij} is the Kronecker δ defined to be 1 if $i = j$ and 0 otherwise. The corresponding Karush-Kuhn-Tucker complementarity conditions are

$$\alpha_i \left[y_i \left(\left\langle w, x_i \right\rangle + b\right) - 1 + \xi_i \right] = 0, i = 1,\ldots,\ell.$$

By employing this result we can formulate a more general task as follows.

Consider a training sample $S = \left(\left(x_1, y_1\right),\ldots,\left(x_\ell, y_\ell\right)\right)$ and the feature space implicitly defined by the kernel $K\left(x,z\right)$ and suppose the parameters α^* solve the following quadratic optimization problem:

$$\text{maximize } W\left(\alpha\right) = \sum_{i=1}^{\ell}\alpha_i - \frac{1}{2}\sum_{ij=1}^{\ell} y_i y_j \alpha_i \alpha_j \left[K\left(x_i, x_j\right) + \frac{1}{C}\delta_{ij}\right],$$

$$\text{subject to } \sum_{i=1}^{\ell} y_i \alpha_i = 0,\ \alpha_i \geq 0,\ i = 1,\ldots,\ell.$$

Let $f(x) = \sum_{i=1}^{\ell} y_i \alpha_i^* K(x_i, x) + b^*$, where b^* is chosen so that $y_i f(x_i) = 1 - \alpha_i^* / C$ for any i with $\alpha_i^* \neq 0$. Then the decision rule given by $sgn(f(x))$ is equivalent to the hyperplane in the feature space implicitly defined by the kernel $K(x, z)$, which solves the abovementioned optimization problem, where the slack variables are defined relative to the geometric margin (for derivation of this formula see p.106 of (Cristianini & Shawe-Taylor, 2000)):

$$\gamma = \left(\sum_{i \in sv} \alpha_i^* - \frac{1}{C} \langle \alpha^*, \alpha^* \rangle \right)^{-\frac{1}{2}}.$$

The value of b^* is chosen using the relation $\alpha_i = C\xi_i$ (see the result of $\dfrac{\partial L(w, b, \xi, \alpha)}{\partial \xi}$ on the previous page) and the primal constraints that by the KKT complimentarity conditions

$$\alpha_i \left[y_i \left(\langle w, x_i \rangle + b \right) - 1 + \xi_i \right] = 0, i = 1, \ldots, \ell$$

must be equalities for non-zero α_i.[16] Therefore we have $y_i f(x_i) = 1 - \alpha_i^* / C$.

We still have a quadratic optimization problem and therefore the same methods can be used as those for the maximal margin hyperplane. The only change is the addition of $1/C$ to the diagonal of the inner product matrix associated with with the training set. This has the effect of adding $1/C$ to the eigenvalues of the matrix, making the problem better conditioned. We can therefore view the new problem as simply a change of kernel

$$K^\dagger(x, z) = K(x, z) + \frac{1}{C}\delta_{xz}.$$

1-Norm Soft Margin

Let us consider another primal problem:

$$\text{minimize } \langle w, w \rangle + C\sum_{i=1}^{\ell} \xi_i \text{ over } \xi, w, b,$$

subject to $y_i \left(\langle w, x_i \rangle + b \right) \geq 1 - \xi_i, \ \xi_i \geq 0, \ i = 1, \ldots, \ell.$

The parameter C (positive scalar) was introduced in the previous section. In contrast to the objective function of the 2-norm soft margin, the objective function of the 1-norm soft margin contains just ξ_i instead of ξ_i^2. As we will see later, this distinction is important.

So, let us write down the Lagrangian for this optimization problem is

$$L\left(w, b, \xi, \alpha, r\right) = \frac{1}{2}\langle w, w\rangle + C\sum_{i=1}^{\ell}\xi_i - \sum_{i=1}^{\ell}\alpha_i\left[y_i\left(\langle w, x_i\rangle + b\right) - 1 + \xi_i\right] - \sum_{i=1}^{\ell}r_i\xi_i,$$

with $\alpha_i \geq 0$ and $r_i \geq 0$ (notice that there are two sets of Lagrange multiplies!). The corresponding dual is found by differentiating with respect to w, ξ and b

$$\frac{\partial L\left(w, b, \xi, \alpha, r\right)}{\partial w} = w - \sum_{i=1}^{\ell}y_i\alpha_i x_i = 0,$$

$$\frac{\partial L\left(w, b, \xi, \alpha, r\right)}{\partial \xi_i} = C - \alpha_i - r_i = 0,$$

$$\frac{\partial L\left(w, b, \xi, \alpha, r\right)}{\partial b} = \sum_{i=1}^{\ell}y_i\alpha_i = 0,$$

and substituting the relations obtained back into the primal:

$$L\left(w, b, \xi, \alpha, r\right) = \sum_{i=1}^{\ell}\alpha_i - \frac{1}{2}\sum_{i,j=1}^{\ell}y_i y_j\alpha_i\alpha_j\langle x_i, x_j\rangle,$$

which is identical to that for the maximal margin. The only difference is that the constraint $C - \alpha_i - r_i = 0$ along with $r_i \geq 0$ leads to $\alpha_i \leq C$, while $\xi_i \neq 0$ only if $r_i = 0$ (ξ_i and r_i cannot be simultaneously equal to zero) and therefore $\alpha_i = C$. The Karush-Kuhn-Tucker complementarity conditions are therefore

$$\alpha_i\left[y_i\left(\langle w, x_i\rangle + b\right) - 1 + \xi_i\right] = 0, \ i = 1, \ldots, \ell,$$

$$\xi_i r_i = \xi_i\left(C - \alpha_i\right), \ i = 1, \ldots, \ell.$$

Notice that the KKT conditions implies that non-zero slack variables can only occur when $\alpha_i = C$. Therefore for such points their geometric margin is less than $1/w$ due to subtraction of the positive quantity $1 - \xi_i$ from γ_i in the first condition. In contrast, points for which $0 < \alpha_i < C$ lie at the target distance of $1/w$ from the separating hyperplane due to the fact that $\xi_i = 0$. Such points are support vectors. Summarizing, the KKT conditions for

support vectors: $r_i \neq 0$, $\xi_i = 0$,
non-support vectors: $r_i = 0$, $\xi_i \neq 0$.

By employing this result we can now formulate a more general kernel version. Consider a training sample $S = \left(\left(x_1, y_1 \right), \ldots, \left(x_\ell, y_\ell \right) \right)$ using the feature space implicitly defined by the kernel $K\left(x, z \right)$ and suppose that the parameters α^* solve the following quadratic optimization problem:

$$\text{maximize } W\left(\alpha \right) = \sum_{i=1}^{\ell} \alpha_i - \frac{1}{2} \sum_{i,=1}^{\ell} y_i y_j \alpha_i \alpha_j \left\langle x_i, x_j \right\rangle,$$

$$\text{subject to } \sum_{i=1}^{\ell} y_i \alpha_i = 0, C \geq \alpha_i \geq 0, \ i = 1, \ldots, \ell.$$

Let $f\left(x \right) = \sum_{i=1}^{\ell} y_i \alpha_i^* K\left(x_i, x \right) + b^*$, where b^* is chosen so that $y_i f\left(x_i \right) = 1$ for any i with $C > \alpha_i^* > 0$. Then the decision rule given by $sgn\left(f\left(x \right) \right)$ is equivalent to the hyperplane in the feature space implicitly defined by the kernel $K\left(x, z \right)$ which solves the abovementioned optimization problem, where the slack variables are defined relative to the geometric margin[17]

$$\gamma = \left(\sum_{i,j \in sv} y_i y_j \alpha_i^* \alpha_j^* K\left(x_i, x_j \right) \right)^{-\frac{1}{2}}.$$

The value of b^* is chosen from the primal constraints which by the KKT complimentarity conditions

$$\alpha_i \left[y_i \left(\left\langle w, x_i \right\rangle + b \right) - 1 + \xi_i \right] = 0, \ i = 1, \ldots, \ell,$$

$$\xi_i r_i = \xi_i \left(C - \alpha_i\right), \, i = 1, \ldots, \ell$$

must be equalities for non-zero α_i. For support vectors, both $\alpha_i \neq 0$ ($0 < \alpha_i < C$) and $\xi_i = 0$. Therefore we have $y_i f\left(x_i\right) = 1$ from the first KKT complementarity condition, from where b^* can be straightforwardly found.

The optimization problem above is equivalent to the maximal margin hyperplane, with the additional constraint that all the α_i are upper bounded by C. This gives rise to the name *box constraint* that is frequently used to refer to this formulation. The box constraints limit the influence outliers, which would otherwise allow for large Lagrange multipliers.

A Brief Summary of Previous Sections

Let us summarize main points of the previous sections:

- The training of a Support Vector Machine can be reduced to maximizing a convex quadratic form subject to linear constraints.
- Such convex quadratic programs have no local maxima and their solution can be always found.
- The dual representation of the problem allows the training even in the face of high dimensional data, e.g., by employing standard approaches from optimization theory.

In other words, training performs the role of learning the values of α^* and b^* via the solution of QP programme. Once training is done and support vectors are found in the training data, the optimal values of Lagrange multipliers and the bias together with the kernel function values are utilized in the test phase for classification.

MATLAB Code

Now we descended to a very practical question: how to implement the SVM algorithms? In other words, it is all about how to realize the optimization process. One can see that the effective solution of the optimization problem means finding the optimal values of the Lagrange multipliers $\alpha_i, i = 1, \ldots, \ell$ and the bias b, leading to the global maximum of the objective function, attained on the training data. Once these optimal values are found, they are used in the test phase in order to classify test data based on a sign of the decision function f.

There exist several approaches to solve the SVM optimization task, which are described in (Cristianini & Shawe-Taylor, 2000) and many other SVM-related books. We will only briefly explain them here because knowing their fine details is only necessary if one has a data set of large/huge size (say, 1000 or 1000000)[18]. Microarray data do not belong to this category since the training sets rarely exceed 100 data vectors. Another reason for omitting these details is that MATLAB® Bioinformatics Toolbox™ already includes the implementations of algorithms necessary for SVM training and testing, which we will show how to utilize.

I will nevertheless provide MATLAB® code relying on the built-in optimization functions of MATLAB®. This code can be useful if one does not have Bioinformatics Toolbox™. Besides, code including MATLAB® optimization functions demonstrates how theory can be transformed into practice, which is certainly advantageous if one wants to deeply understand SVMs and other kernel methods.

To begin with, one can apply the gradient ascent algorithm, iteratively arriving at the global maximum of the objective function, given the KKT constraints. This algorithm starts with some initial guess for the solution in terms of Lagrange multipliers and then iteratively updates values of Lagrange multipliers along the steepest ascent path, i.e. moving in the direction of the gradient of the objective function. The length of the update also known as the learning rate can be either fixed or variable so as to achieve the maximal gain in the objective function value at each iteration. In many cases, the gradient is evaluated for just one x_i, thus resulting in the update of α_i, associated with it. The gradient ascent algorithm is thus the online algorithm and can be useful if the data come incrementally, i.e. one vector at a time. Its drawback is however slow or oscillating convergence on some data sets.

Another approach to solving the optimization problem is to utilize decomposition of the original problem into smaller sub-problems. The decomposition algorithm only updates a fixed size subset (working set) of Lagrange multipliers, while keeping the rest of multipliers constant. Fixed size assumes that every time a new point enters the working set, another point should leave it. A decision on which point should be added to/removed from the working set is often based on checking the KKT conditions. If they are met for a given point, it implies that this point lies inside the feasible region and not on its boundary (such a point can hardly become a support vector). However, if the KKT constraints are violated, then the point can be kept in the working set since the final solution is not yet found (such a point can still become a support vector). The more severely KKT conditions are violated, the more likely the point needs to be included into the working set. Following this KKT conditions-based heuristic, the optimization of small sub-problems, for which generic optimization routines can be readily used, leads to an improvement in the overall objective function, hence, to solving the large original optimization problem.

In the well known Sequential Minimal Optimization (SMO) algorithm, the idea of decomposition is taken to its extreme: the optimization problem just for two points is solved at each iteration. Such an approach has an important advantage: the optimization problem for two points can be analytically solved and there is no need for a quadratic programme optimizer. The book (Cristianini & Shawe-Taylor, 2000) gives pseudo code of the SMO algorithm as well as its detailed explanation.

Armed with such a brief introduction, I first provide code of function *SVM_m*, based on the MATLAB® functions from Bioinformatics Toolbox™. In addition to training and test data, the MATLAB® structure, containing the optimization method, kernel type and kernel parameter values, is required to set as an input to *SVM_m*.

There are three optimization methods: SMO, quadratic programming (QP), and Least Squares (LS)[19]. For small-size microarray data, the choice of the optimization method does not seem to be an important issue, so any of the three methods can be selected as training time is not expected to be long. Selecting QP requires MATLAB® Optimization Toolbox™. If this toolbox is installed on your computer, QP is the default option; otherwise, SMO is the default.

Regarding the kernel type, in high dimensional space of gene expressions, linear separation of two classes is likely. Thus, the linear kernel is advised first to try. Another reason for choosing this type of kernel is that it does not require any extra parameter to set up. If results are not satisfactory, then a nonlinear kernel with a slight degree of nonlinearity (e.g., quadratic or Gaussian (RBF)) may work. It is assumed that a user took care of setting desired kernel parameter value(s) once he/she chose the kernel type.

Once all necessary parameters have been properly set up, SVM training starts by invoking MATLAB® function *svmtrain* that returns a special structure *SVMStruct* containing the learned bias and Lagrange multipliers to be used in the testing phase that immediately follows the training. MATLAB® function *svmclassify* performs test data classification.

Two other MATLAB® functions (*svmsmosest* and *optimset*) are used in order to set up various options for SMO and QP optimization methods, respectively. Function *svmtrain* can accept a lot of input parameters, some of which I pre-set up. For the complete list of parameters and their possible values, consult help related to this function. In many cases, I preferred to explicitly specify the default parameter values, e,g., normalization of the training data to zero mean and unit variance or box constraint value (parameter C in the described soft margin SVMs). I believe that such a style would ensure that all important parameters are correctly specified and it would facilitate changes of parameter values if necessary.

The output of SVMs is crisp labels. Sometimes it is necessary to have it probabilistic, e.g., for generating a Receiver Operating Characteristics (ROC) curve (Fawcett, 2006), (Pepe, 2004). Platt found the way how to do so in (Platt, 2000).

After training the SVM, another cycle of training occurs that learns two parameters, A and B, of a sigmoid function $p_i = \dfrac{1}{1 + exp\left(Af_i + B\right)}$, where $f_i = f\left(x_i\right)$ is the output (class label) of the SVM for the ith instance. The same training set as that used in the SVM training is typically utilized. Fitting the sigmoid can be done as a two-parameter minimization of the negative log (logarithm) likelihood[20] of the training data, which is a cross-entropy error function $-\sum_i t_i \log p_i + \left(1 - t_i\right)\log\left(1 - p_i\right)$, where t_i $(t_i \in \left\{0,1\right\})$ are target probabilities. Once the sigmoid parameters are learned, one obtains a mapping from the SVM outputs into probabilities. Platt argues that at least for linear SVMs (likely the first choice for microarray gene expression data), the repeated usage of the same training set will not lead toward severe bias in estimating the SVM output out of sample (i.e. outside the training set). For non-linear SVMs, this is not the case and Platt did not advise to try fitting the sigmoid with the training data as described above. In any case, extra training does not look attractive, given scarcity of training instances. In my opinion, making the SVM outputs probabilistic is better to be avoided. Perhaps, as one of the solutions could be a Relevance Vector Machine (RVM) (Tipping, 2001), which yields probability estimates at the cost of computationally heavier learning of sparse relevance vectors (they tend to be even more sparse than the conventional support vectors). As an RVM is a special case of Gaussian Processes (Rasmussen & Williams, 2006), it is advisable to consider the application of the latter for getting probability estimates.

```
function test_targets = SVM_m(train_patterns, train_targets,
test_patterns, params)

% Support Vector Machine classifier built on the functionality
from
% MATLAB Bioinformatics Toolbox
%
% Inputs:
%    train_patterns  - Training patterns (DxN matrix, where D -
data
%                      dimensionality, N - number of patterns)
%    train_targets   - Training targets (1xN vector)
%    test_patterns   - Test patterns (DxNt matrix, where D -
data
%                      dimensionality, Nt - number of patterns)
%
```

```
% Outputs:
%   test_targets    - Test targets (1xNt vector)
%
% Note: it requires Bioinformatis Toolbox 3, released in 2008



% Get a version number for Bioinformatics Toolbox
v = ver('bioinfo');
% If MATLAB release is before 2008, generate an error message
if isempty(v)
    error('No Bioinformatics toolbox was found!');
else
    if str2num(v.Version) < 3.1
        error('Version 3.1 (or higher) of Bioinformatics Tool-
box is needed to run this function!');
    end
end

% Unpack parameter values from structure
method = params.method;
krn_fnc = params.krn_fnc;
sigma = params.sigma;
polyord = params.polyord;
mlp_params = params.mlp_params;

% Train an SVM classifier
if strcmpi(method, 'SMO')
    % Sequential Minimal Optimization
    % One-norm soft-margin SVM
    opts = svmsmoset('Display', 'iter');
    switch krn_fnc
        case 'linear',
            fprintf(1, 'Linear kernel is used\n');
            SVMStruct = svmtrain(train_patterns', train_tar-
gets', ...
            'Kernel_Function', krn_fnc, 'SMO_Opts', opts, ...
            'BoxConstraint', 1, 'Autoscale', true, 'Showplot',
false);
        case 'quadratic',
```

```
            fprintf(1, 'Quadratic kernel is used\n');
            SVMStruct = svmtrain(train_patterns', train_tar-
gets', ...
            'Kernel_Function', krn_fnc, 'SMO_Opts', opts, ...
            'BoxConstraint', 1, 'Autoscale', true, 'Showplot',
false);
        case 'rbf',
            fprintf(1, 'RBF is used\n');
            SVMStruct = svmtrain(train_patterns', train_tar-
gets', ...
            'Kernel_Function', krn_fnc, 'RBF_Sigma', sigma, ...
            'SMO_Opts', opts, ...
            'BoxConstraint', 1, 'Autoscale', true, 'Showplot',
false);
        case 'polynomial',
            fprintf(1, 'Polynomial kernel is used/n');
            SVMStruct = svmtrain(train_patterns', train_tar-
gets', ...
            'Kernel_Function', krn_fnc, 'Polyorder', polyord,
...
            'SMO_Opts', opts, ...
            'BoxConstraint', 1, 'Autoscale', true, 'Showplot',
false);
        case 'mlp',
            fprintf(1, 'MLP kernel is used/n');
            SVMStruct = svmtrain(train_patterns', train_tar-
gets', ...
            'Kernel_Function', krn_fnc, 'MLP_Params', mlp_
params, ...
            'SMO_Opts', opts, ...
            'BoxConstraint', 1, 'Autoscale', true, 'Showplot',
false);
        otherwise
            error('Unsupported kernel!');
    end
end
if strcmpi(method, 'QP')
    % Quadratic Programming optimization
    % Two-norm, soft-margin SVM
    opts = optimset('Display', 'iter');
```

```
    switch krn_fnc
        case 'linear',
            fprintf(1, 'Linear kernel is used\n');
            SVMStruct = svmtrain(train_patterns', train_tar-
gets', ...
            'Kernel_Function', krn_fnc, 'QuadProg_Opts', opts,
...
            'BoxConstraint', 1, 'Autoscale', true, 'Showplot',
false);
        case 'quadratic',
            fprintf(1, 'Quadratic kernel is used\n');
            SVMStruct = svmtrain(train_patterns', train_tar-
gets', ...
            'Kernel_Function', krn_fnc, 'QuadProg_Opts', opts,
...
            'BoxConstraint', 1, 'Autoscale', true, 'Showplot',
false);
        case 'rbf',
            fprintf(1, 'RBF kernel is used\n');
            SVMStruct = svmtrain(train_patterns', train_tar-
gets', ...
            'Kernel_Function', krn_fnc, 'RBF_Sigma', sigma, ...
            'QuadProg_Opts', opts, ...
            'BoxConstraint', 1, 'Autoscale', true, 'Showplot',
false);
        case 'polynomial',
            fprintf(1, 'Polynomial kernel is used/n');
            SVMStruct = svmtrain(train_patterns', train_tar-
gets', ...
            'Kernel_Function', krn_fnc, 'Polyorder', polyord,
...
            'QuadProg_Opts', opts, ...
            'BoxConstraint', 1, 'Autoscale', true, 'Showplot',
false);
        case 'mlp',
            fprintf(1, 'MLP kernel is used/n');
            SVMStruct = svmtrain(train_patterns', train_tar-
gets', ...
            'Kernel_Function', krn_fnc, 'MLP_Params', mlp_
params, ...
```

```
                'QuadProg_Opts', opts, ...
                'BoxConstraint', 1, 'Autoscale', true, 'Showplot',
false);
        otherwise
            error('Unsupported kernel!');
    end
end
if strcmpi(method, 'LS')
    % Least Squares optimization
    switch krn_fnc
        case 'linear',
            fprintf(1, 'Linear kernel is used\n');
            SVMStruct = svmtrain(train_patterns', train_tar-
gets', ...
            'Kernel_Function', krn_fnc, ...
            'BoxConstraint', 1, 'Autoscale', true, 'Showplot',
false);
        case 'quadratic',
            fprintf(1, 'Quadratic kernel is used\n');
            SVMStruct = svmtrain(train_patterns', train_tar-
gets', ...
            'Kernel_Function', krn_fnc, ...
            'BoxConstraint', 1, 'Autoscale', true, 'Showplot',
false);
        case 'rbf',
            fprintf(1, 'RBF kernel is used\n');
            SVMStruct = svmtrain(train_patterns', train_tar-
gets', ...
            'Kernel_Function', krn_fnc, 'RBF_Sigma', sigma, ...
            'BoxConstraint', 1, 'Autoscale', true, 'Showplot',
false);
        case 'polynomial',
            fprintf(1, 'Polynomial kernel is used/n');
            SVMStruct = svmtrain(train_patterns', train_tar-
gets', ...
            'Kernel_Function', krn_fnc, 'Polyorder', polyord,
...
            'BoxConstraint', 1, 'Autoscale', true, 'Showplot',
false);
        case 'mlp',
```

```
            fprintf(1, 'MLP kernel is used/n');
            SVMStruct = svmtrain(train_patterns', train_tar-
gets', ...
            'Kernel_Function', krn_fnc, 'MLP_Params', mlp_
params, ...
            'BoxConstraint', 1, 'Autoscale', true, 'Showplot',
false);
        otherwise
            error('Unsupported kernel!');
    end
end


% Classify test data
test_targets = svmclassify(SVMStruct, test_patterns', 'Show-
plot', false);


return
```

Now I provide my own implementation of SVMs based on MATLAB® function *quadprog* from the Optimization Toolbox™. This implementation serves educational purpose; readers can see and understand how SVMs can be programmed. In my code, I split training and test procedures between two separate functions called *train_SVC* and *test_SVC*, respectively. Besides, there are separate functions for each kernel type.

Let us first consider the kernel-computing functions. These are *Linear* (linear kernel), *Poly* (polynomial kernel), *Gauss* (Gaussian (RBF) kernel), *RBF* (RBF (Gaussian) kernel), and *Sigmoid* (MLP or sigmoid kernel). Interface of all these functions is pretty straightforward: one should provide training data, test data, and kernel parameter or parameters as input. If the training data are used for kernel computation, then *test_patterns* should be set to [], so that the call of a function, e.g., *Linear*, looks like

$$K = Linear(train\_patterns, [], kernel\_param).$$

Functions *Gauss* and *RBF* are identical copies of each other. They both are presented only for user convenience. Of course, readers can easily add other kernels of their choice/liking.

Kernel computations form the basis for SVM training (*train_SVM*) and testing (*test_SVM*). In both cases, the input *params* is a structure containing certain parameters that are retrieved from it once either function is invoked. The quadratic optimization problem is solved by the MATLAB® function *quadprog* , hence it is advisable to read the description of this function in MATLAB® help. After the QP program has been solved, support vectors, Lagrange multipliers corresponding to them and the bias are returned by *train _ SVM* in order to be used in the test phase. By comparing logic of our SVM implementation and that of MATLAB®, one can see that in both cases the flow of operations is identical.

```
function K = Linear(train_patterns, test_patterns, kernel_
param)

% Linear kernel
%
% Inputs:
%    train_patterns             - Training patterns (DxN matrix: D
-
%                                 dimensionality, N- #points)
%    test_patterns              - Test patterns (DxM matrix)
%    kernel_param               - []
%
% Output:
%    K                          - Linear kernel




if isempty(test_patterns)
    % Compute kernel function
    K = train_patterns'*train_patterns;
else
    % Compute kernel function
    K = train_patterns'*test_patterns;
end

return
function K = Poly(train_patterns, test_patterns, kernel_param)

% Polynomial kernel
%
% Inputs:
%    train_patterns             - Training patterns (DxN matrix: D
```

```
%                                      dimensionality, N- #points)
%    test_patterns            - Test patterns (DxM matrix)
%    kernel_param             - Polynomial degree
%
% Output:
%    K                        - Polynomial kernel



if isempty(test_patterns)
    % Compute kernel function
    K = (train_patterns'*train_patterns + 1).^kernel_param;
else
    % Compute kernel function
    K = (train_patterns'*test_patterns + 1).^kernel_param;
end

return
function K = Gauss(train_patterns, test_patterns, kernel_param)

% RBF (Gaussian) kernel
%
% Inputs:
%    train_patterns           - Training patterns (DxN matrix: D

%                               dimensionality, N- #points)
%    test_patterns            - Test patterns (DxM matrix)
%    kernel_param             - Gaussian width
%
% Output:
%    K                        - RBF (Gaussian) Kernel



% Compute kernel function
if isempty(test_patterns)
    % Get the number of training patterns
    N = size(train_patterns,2);
    d2 = sum(train_patterns.^2,1);
```

```
    distance = repmat(d2,N,1)+repmat(d2',1,N)-2*train_
patterns'*train_patterns;
    K = exp(-distance/(2*kernel_param^2));
else
    % Get the number of support vectors
    L = size(train_patterns,2);
    % Get the number of test patterns
    M = size(test_patterns,2);
    K = zeros(L,M);
    for i = 1:M
        p = test_patterns(:,i);
        distance = sum((train_patterns - repmat(p,[1
L])).^2,1);
        distance = distance';
        K(:,i) = exp(-distance/(2*kernel_param^2));
    end
end


return
function K = RBF(train_patterns, test_patterns, kernel_param)

% RBF (Gaussian) kernel
%
% Inputs:
%    train_patterns                 - Training patterns (DxN matrix: D
-
%                                     dimensionality, N- #points)
%    test_patterns                  - Test patterns (DxM matrix)
%    kernel_param                   - Gaussian width
%
% Output:
%    K                              - RBF (Gaussian) Kernel



% Compute kernel function
if isempty(test_patterns)
    % Get the number of training patterns
    N = size(train_patterns,2);
    d2 = sum(train_patterns.^2,1);
```

```
    distance = repmat(d2,N,1)+repmat(d2',1,N)-2*train_
patterns'*train_patterns;
    K = exp(-distance/(2*kernel_param^2));
else
    % Get the number of support vectors
    L = size(train_patterns,2);
    % Get the number of test patterns
    M = size(test_patterns,2);
    K = zeros(L,M);
    for i = 1:M
        p = test_patterns(:,i);
        distance = sum((train_patterns - repmat(p,[1
L])).^2,1);
        distance = distance';
        K(:,i) = exp(-distance/(2*kernel_param^2));
    end
end


return
function K = Sigmoid(train_patterns, test_patterns, kernel_
param)

% Sigmoid kernel
%
% Inputs:
%   train_patterns          - Training patterns (DxN matrix: D
_
%                             dimensionality, N- #points)
%   test_patterns           - Test patterns (DxM matrix)
%   kernel_param            - [slope constant]
%
% Output:
%   K                       - Sigmoid kernel



if length(kernel_param) ~= 2
    error('This kernel needs two parameters to operate!')
end
```

```
if isempty(test_patterns)
    % Compute kernel function
    K = tanh(train_patterns'*train_patterns*kernel_
param(1)+kernel_param(2));
else
    % Compute kernel function
    K = tanh(train_patterns'*test_patterns*kernel_
param(1)+kernel_param(2));
end

return
function [sv, ind_sv, alpha_star, b_star] = train_SVC(train_
patterns, train_targets, params)

% Training an SVM classifier for two classes
%
% Inputs:
%    train_patterns          - Training patterns (DxN matrix: D
—
%                              dimensionality, N- #points)
%    train_targets           - Training targets (1XN vector)
%    params                  - Structure with the following
fields:
%                              kernel - 'Gauss', 'RBF' (the same
as
%                              'Gauss'), 'Poly', 'Sigmoid',
'Linear'
%                              kernel parameter - Gaussian width
(RBF
%                              kernel), polynomial degree (Poly-
nomial
%                              kernel), the slope and constant
of the
%                              sigmoid in the format [1 2] with
no
%                              separating commas, linear - none
%                              solver type - 'Quadprog'
%                              C - trade-off between classifier
capacity %                                      and the
number of errors (C=Inf for the
```

```
%                                      hard margin SVM)
%
% Outputs:
%   sv                      - Support vectors (DxL matrix,
L<=N)
%   ind_sv                  - Indices of support vectors in the
%                             training set
%   alpha_star              - Lagrange multipliers correspond-
ing to
%                             support vectors
%   b_star                  - Bias (scalar)



% Retrieve parameters of an SVM
kernel_function = params.kernel_function;
kernel_param = params.kernel_param;
solver = params.solver;
C = params.C;

% Precompute kernel function
K = feval(kernel_function, train_patterns, [], kernel_param);

% Get the number of points
N = length(train_targets);

% Check if there are only two classes
u = unique(train_targets);
if length(u) ~=2
    error('This function is intended for two-classes problems
only!');
    return
end

% Check if length of train_targets is equal to N
if length(train_targets) ~= N
    error('Not every training point has a label!');
    return
end
```

```
% Check C value
if C < 0
    error('C cannot be negative!');
    return
end


% Rescale values of training targets
i1 = find(train_targets == u(1));
i2 = find(train_targets == u(2));
train_targets(i1) = -1;
train_targets(i2) = +1;


% Initialize parameters (see help to 'quadprog' for explanation
of
% parameters)
% Initialization according to pseudo code on p.323 in Ralf
Herbrich's
% book "Learning Kernel Classifiers: Theory and Algorithms",
MIT Press,
% Cambridge, MA, 2002
%
% Aeq cannot have more rows than columns. In this case, a
medium-scale
% algorithm is called. A medium-scale algorithm is also called
when
% there are both equality and inequality constraints.
%
% If x has no upper bounds, set ub to Inf
% If x has no lower bounds, set lb to -Inf

Y = diag(train_targets);
H = Y*K*Y; % Signs H and f are changed since 'quadprog' solves
a minimization problem
f = -ones(1,N);
if C == Inf % Hard margin SVM
    Aeq = []; beq = [];
else
    Aeq = train_targets; beq = 0;
end
lb = zeros(1,N);
```

```
ub = C*ones(1,N);

% Solve the quadratic programming problem
options = optimset('MaxIter',1000);
[x,fval,exitflag,output,lambda] = quadprog(H,f,[],[],Aeq,beq,lb
,ub,[],options);
alpha_star = x';

% Find support vectors
ind_sv = find(alpha_star > 0.00001 & alpha_star < C); % Compar-
ison with zero doesn't work since many alphas have very small
values
sv = train_patterns(:,ind_sv);

fprintf(1,'\nNo.training points = %d, No.support vectors =
%d\n',N,length(ind_sv));

% Compute b_star (scalar)
if isempty(ind_sv)
    b_star = 0;
elseif C == Inf % Hard margin SVM
    % Compute w_star
    w_star = (alpha_star.*train_targets)*train_patterns'; %
w_star is of 1xD elements

    % Notice that here all training data are used
    % MIN is taken since the support vectors for +1 class lie
close to
    % or on the hyperplane <w*x>+b=1 and all patterns of this
class
    % lying in interior have <w*x>+b>1
    % MAX is taken since the support vectors for -1 class lie
close to
    % or on the hyperplane <w*x>+b=-1 and all patterns of this
class
    % lying in interior have <w*x>+b<1
    b_star = -0.5*(min(w_star*train_patterns(:,i1)) + max(w_
star*train_patterns(:,i2)));

    % This is an alternative way to compute b_star involving
```

```
only the
    % support vectors
    %i1 = find(train_targets(ind_sv) == +1);
    %i2 = find(train_targets(ind_sv) == -1);
    %b_star = -0.5*(sum(w_star*sv(:,i1)) + sum(w_
star*sv(:,i2)));
else
    % Yet another way based on the formula (7.31) in book of
Bernhard
    % Scholkopf and Alexander Smola "Learning with Kernels:
Support
    % Vector Machines, Regularization, Optimization, and Be-
yond", MIT
    % Press, MA, 2002
    t = (alpha_star.*train_targets)*K;
    b_star = mean(train_targets(ind_sv) - t(ind_sv));
end

if exitflag <= 0
    fprintf(1,'exitflag = %d\n',exitflag);
end

return
function test_targets = test_SVC(train_patterns, train_targets,
test_patterns, ind_sv, alpha_star, b_star, params)

% Test the trained SVM classifier for two classes
%
% Inputs:
%    train_patterns        - Training patterns (DxN matrix: D
-
%                            dimensionality, N- #points)
%    train_targets         - Training targets (1XN vector)
%    test_patterns         - Test patterns (DxM matrix)
%    ind_sv                - Indices of support vectors in the
%                            training set
%    alpha_star            - Lagrange multipliers correspond-
ing to
%                            support vectors
%    b_star                - Bias (scalar)
```

```
%    params                    - Structure with the following
fields:
%                              kernel - 'Gauss', 'RBF' (the same
as
%                              'Gauss'), 'Poly', 'Sigmoid',
'Linear'
%                              kernel parameter - Gaussian width
(RBF
%                              kernel), polynomial degree (Poly-
nomial
%                              kernel), the slope and constant
of the
%                              sigmoid in the format [1 2] with
no
%                              separating commas, linear - none
%
% Outputs:
%    test_targets             - Test targets (1xM vector)



% Retrieve parameters of an SVM
kernel_function = params.kernel_function;
kernel_param = params.kernel_param;

% Precompute kernel function
K = feval(kernel_function, train_patterns(:,ind_sv), test_pat-
terns, kernel_param);

alpha_star = alpha_star(ind_sv);
train_targets = train_targets(ind_sv);

% When this function is used for multiclass problems, it re-
turns real
% test_targets rather than binary integers so that for 2-class
problems % the calling function must take care of proper
thresholding
test_targets = (alpha_star.*train_targets)*K + b_star;

return
```

REFERENCES

Abe, S. (2005). *Support vector machines for pattern recognition*. London: Springer-Verlag.

Aizerman, M., Braverman, E., & Rozonoer, L. (1964). Theoretical foundations of the potential function method in pattern recognition learning. *Automation and Remote Control, 25*, 821–837.

Bishop, C. M. (2006). *Pattern recognition and machine learning*. New York, NY: Springer Science-Business Media.

Cristianini, N., & Shawe-Taylor, J. (2000). *An introduction to support vector machines (and other kernel-based learning methods)*. Cambridge, UK: Cambridge University Press.

Duda, R. O., Hart, P. E., & Stork, D. G. (2000). *Pattern classification* (2nd ed.). New York, NY: John Wiley & Sons.

Everitt, B. (2006). *The Cambridge dictionary of statistics* (3rd ed.). Cambridge, UK: Cambridge University Press.

Fawcett, T. (2006). An introduction to ROC analysis. *Pattern Recognition Letters, 27*(8), 861–874. doi:10.1016/j.patrec.2005.10.010

Furey, T. S., Cristianini, N., Duffy, N., Bednarski, D. W., Schummer, M., & Haussler, D. (2000). Support vector machine classification and validation of cancer tissue samples using microarray expression data. *Bioinformatics (Oxford, England), 16*(10), 906–914. doi:10.1093/bioinformatics/16.10.906

Guyon, I., Weston, J., Barnhill, S., & Vapnik, V. (2002). Gene selection for cancer classification using support vector machines. *Machine Learning, 46*(1-3), 389–422. doi:10.1023/A:1012487302797

Hamel, L. (2009). *Knowledge discovery with support vector machines*. Hoboken, NJ: John Wiley & Sons. doi:10.1002/9780470503065

Hastie, T., Tibshirani, R., & Friedman, J. (2003). *The elements of statistical learning: data mining, inference, and prediction*. New York, NY: Springer-Verlag.

Herbrich, R. (2002). *Learning kernel classifiers: theory and algorithms*. Cambridge, MA: MIT Press.

Izenman, A. J. (2008). *Modern multivariate statistical techniques: regression, classification, and manifold learning*. New York, NY: Springer Science+Business Media.

Kecman, V. (2001). *Learning and soft computing: support vector machines, neural networks, and fuzzy logic models.* Cambridge, MA: MIT Press.

Neuhaus, M., & Bunke, H. (2007). *Bridging the gap between graph edit distance and kernel machines.* Singapore: World Scientific.

Pepe, M. S. (2004). *The statistical evaluation of medical tests for classification and prediction* (1st paperback ed.). Oxford, UK: Oxford University Press.

Platt, J. C. (2000). Probabilities for SV Machines. In Smola, A. J., Barlett, P. L., Schölkopf, B., & Schuurmarns, D. (Eds.), *Advances in large margin classifiers* (pp. 61–74). Cambridge, MA: MIT Press.

Rasmussen, C. E., & Williams, C. K. (2006). *Gaussian processes for machine learning.* Cambridge, MA: MIT Press.

Schölkopf, B., Burges, C. J., & Smola, A. J. (Eds.). (1999). *Advances in kernel methods: support vector machine learning.* Cambridge, MA: MIT Press.

Schölkopf, B., & Smola, A. J. (2002). *Learning with kernels: support vector machines, regularization, optimization, and beyond.* Cambridge, MA: MIT Press.

Schölkopf, B., Tsuda, K., & Vert, J.-P. (Eds.). (2004). *Kernel methods in computational biology.* Cambridge, MA: MIT Press.

Shawe-Taylor, J., & Cristianini, N. (2004). *Kernel methods for pattern analysis.* Cambridge, UK: Cambridge University Press.

Smola, A. J., Bartlett, P. L., Schölkopf, B., & Schuurmans, D. (Eds.). (2000). *Advances in large margin classifiers.* Cambridge, MA: MIT Press.

Steinwart, I., & Christmann, A. (2008). *Support vector machines.* New York, NY: Springer Science+Business Media.

Suykens, J. A., van Gestel, T., De Brabanter, J., De Moor, B., & Vandewalle, J. (2002). *Least squares support vector machines.* Singapore: World Scientific. doi:10.1142/9789812776655

Tan, J.-Y., & Yang, Z.-X. (2008). Incorporating gene similarity into support vector machine for microarray classification and gene selection. In X.-S. Zhang, L. Chen, L.-Y. Wu, & Y. Wang (Ed.), *Proceedings of the 2nd International Symposium on Optimization and Systems Biology, Lijiang, China. Lecture Notes in Operations Research* (Vol. 9, pp. 350-357). Beijing, China: World Publishing Corporation.

Tipping, M. E. (2001). Sparse Bayesian learning and the relevance vector machine. *Journal of Machine Learning Research, 1,* 211–244. doi:10.1162/15324430152748236

Vapnik, V. N. (1995). *The nature of statistical learning theory*. New York, NY: Springer-Verlag.

Vapnik, V. N. (1998). *Statistical learning theory*. New York, NY: John Wiley & Sons.

Wang, Y., Makedon, F. S., Ford, J. C., & Pearlman, J. (2005). HykGene: a hybrid approach for selecting marker genes for phenotype classification using microarray gene expression data. *Bioinformatics (Oxford, England)*, *21*(8), 1530–1537. doi:10.1093/bioinformatics/bti192

Wu, X., & Kumar, V. (Eds.). (2009). *The top ten algorithms in data mining*. Boca Raton, FL: Chapman & Hall/CRC Press.

Zhang, Z., & Ke, H. (2000). ALL/AML cancer classification by gene expression data using SVM and CSVM approach. *Genome Informatics*, *11*, 237–239.

ENDNOTES

[1]   Euclidean space is the Euclidean plane where the axioms of Euclidean geometry are applicable. The term "Euclidean" is due to the Greek mathematician Euclid who lived and flourished around 300 BC in Alexandria in Ptolemaic Egypt, then governed by one of the trusted and gifted generals of Alexander the Great of Macedon - Ptolemy. After sudden death of Alexander in the prime of his life and conquests, Ptolemy was very quick to challenge Alexander's vision of the united empire (he was not alone to do so). He claimed Egypt as his kingdom and declared himself an independent king. However, this small excurse on ancient history should not convince you, my readers, that your most capable subordinates are your most dangerous competitors and enemies. I also abstain from any judgment of Ptolemy and his actions and leave this to the professional historians. The famous Cleopatra of Egypt, associated with the name of Julius Caesar, belonged to the Ptolemy's dynasty.

[2]   Consider a simple 2D example: the hyperplane crosses the horizontal axis at the point $\left(x_1 = 5, x_2 = 0 \right)$ and the vertical axis at the point $\left(x_1 = 0, x_2 = 4 \right)$. Hence, the hyperplane can be defined as $0.8x_1 + x_2 - 4 = 0$, where $w_1 = 0.8, w_2 = 1, b = -4$. Let us substitute the point $\left(x_1 = 4, x_2 = 0 \right)$ lying below the hyperplane into the previous equation. We obtain -0.8. On the other hand, for the point $(x_1 = 5, x_2 = 1)$ located above the hyperplane we have 1. In other words, $\left\langle w, x_i^{below} \right\rangle < \left\langle w, x_i^0 \right\rangle < \left\langle w, x_i^{above} \right\rangle$, given b is fixed.

3   This is also known as the kernel trick, which was first described in (Aizerman, Braverman, & Rozonoer, 1964).

4   However, this should not be perceived as the statement that kernel methods when applied to high dimensional data do not require dimensionality reduction before data classification. I believe that kernel methods will benefit from removing irrelevant or redundant features before these methods are applied.

5   The Cauchy-Schwarz inequality is $\left|\langle x, z\rangle\right| \leq \|x\|\|z\|$.

6   A square symmetric matrix is positive semi-definite if its eigenvalues are all nonnegative.

7   Optimization is a general term for extremum (maximum or minimum) searching.

8   It also implies that both α_i^* and $g_i(w^*)$ cannot be simultaneously equal to zero.

9   Unless otherwise stated, we assume that y_i can be either -1 or +1. Hence, correct classification occurs when both y_i and $\langle w, x_i\rangle + b$ have the same sign.

10   The geometric margin is equal to the functional margin if $\|w\| = 1$.

11   In general, one cannot certainly say in advance whether data are linearly separable in the feature space or not.

12   A difference is equal to $\left\langle w, \left(x^+ - x^-\right)\right\rangle = 2$.

13   It is useful to recall that on the hyperplane itself $\langle w, x_i\rangle + b = 0$.

14   The minus sign in front of the second summand appears because the constraint is in the form $g_i(w) \geq 0$, but recall that it should be $g_i(w) \leq 0$ to have the plus sign in the Lagrangian. The multiplier 1/2 is introduced for convenience when calculating partial derivatives of the Lagrangian; its introduction has no influence on the minimum we seek.

15   Since $K\left(x_i, x_j\right)$ is a kernel, it can be represented as $\left\langle \phi\left(x_i\right), \phi\left(x_j\right)\right\rangle$ according to the definition of a kernel. Therefore $y_i y_j K\left(x_i, x_j\right) = \left\langle y_i \phi\left(x_i\right), y_j \phi\left(x_j\right)\right\rangle$. That is, $y_i y_j K\left(x_i, x_j\right)$ is the kernel, too, and therefore this matrix is positive semi-definite.

16   Support vectors are associated with $\alpha_i > 0$.

17   For derivation of this formula see p.108 of (Cristianini & Shawe-Taylor, 2000).

18   Storage demands for a kernel matrix grow quadratically with the training set size. For large kernel matrices, this creates an obstacle for many standard optimization methods and prompts alternative methods, specifically tuned to SVMs.

19   Least Squares SVM have been introduced and fully described in (Suykens, van Gestel, De Brabanter, De Moor, & Vandewalle, 2002).

[20]   Likelihood is the probability of a set of observations given the value of some parameter or parameters (Everitt, 2006). For example, the likelihood of a random sample of n observations $x_1, \ldots, x_n$ with probabilityl distribution $f(x, \theta)$ is given by $L = \prod_{i=1}^{n} f(x_i, \theta)$ (Everitt, 2006).

Chapter 8
Introduction to Feature and Gene Selection

PROBLEM OF FEATURE SELECTION

One can claim that times when features were scarce in many branches of science and technology are nowadays in the past. It could be said that feature abundance is a blessing but on the contrary and to surprise of many, it is not. It could be thought that having more features brings more discriminating power, thus facilitating classification. However, in practice, this causes problems: redundant and irrelevant features lead to the increased complexity of classification and degrade classification accuracy.

Hence, some features have to be removed from the original feature set in order to mitigate these negative effects *before* a classifier is utilized. The task of redundant/ irrelevant feature removal is termed feature selection in machine learning and data mining literature. It is a data dimensionality reduction[1] when the original set of features $\mathcal{F}$ is reduced to another set $\mathcal{G} \subseteq \mathcal{F}$, where the symbol $\subseteq$ means 'subset of or equal to', implying that it is not impossible to have an irreducible set of features in certain cases[2]. However, for microarray data this is surely not the case, since there are thousands and tens of thousands of features (gene expression levels) in each

DOI: 10.4018/978-1-60960-557-5.ch008

dataset. In analyzing high dimensional microarray data, one is interested in retaining just a few genes out of many thousands. Thus gene selection becomes synonymous to feature selection, which means that many existing feature selection methods can be readily applied to gene selection.

There is huge interest to gene selection among researches working in bioinformatics (see, e.g., links to collected journal articles at http://www.nslij-genetics.org/microarray/). This is because gene selection represents a challenging and important task for both biology and machine learning.

By selecting a small fraction of genes from a microarray, one aims at finding the genes that can be used as indicators of a certain disease or even early predictors of that disease. Since different types of cancer threaten humankind with great persistence, the overwhelming majority of articles about gene selection apply theoretical ideas and methods to a very practical problem related to cancer.

From the machine learning point of view, feature selection removes meaningless, i.e. not related to a studied disease, genes, thus mitigating overfitting of a classifier on high dimensional microarray data. Overfitting is plague when there are a lot of features and only few samples or instances characterized by these features. Overfitting leads to very good and often perfect classification performance (zero or close to zero error rate) on the training data, but this seemingly wonderful result does automatically translate to new, out-of-sample data. Put it differently, a researcher neglecting the harmful effect of overfitting in the case of microarray data would find a small set of genes which he claims to predict a certain type of cancer. However, when biologists and/or doctors try to pay attention to expression levels of these genes when observing test volunteers and/or real patients, they see no value of those genes, because during machine learning stage, healthy and diseased patients were separated purely based on the noise present in microarray measurements rather than on the disease presence or absence. This happened because without prior removal irrelevant genes, the classification problem is known as the small sample size problem (the number of features far exceeds the number of samples in a dataset) in statistics and machine learning. For such problems, the lack of classifier generalization to new data is a norm rather than an exception, unless a data dimensionality is dramatically reduced.

Thus, the goal in microarray data classification is to identify the differentially expressed genes that can be used to predict class membership of new, unseen samples. The classification of gene expression data involves feature selection and classifier design. Feature selection identifies the subset of differentially-expressed genes that are good (useful, relevant) for distinguishing different classes of samples.

As I mentioned above, gene selection can mitigate the effect of overfitting, which means that it should not be considered as a "silver bullet" and one must be aware of certain problems remaining even after redundant genes have been eliminated. One of such problems is the lack of robustness, implying that different feature selection methods select different subsets of genes. This led Ein-Dor et al. to conclude in (Ein-Dor, Zuk, & Domany, 2006) with some justified pessimism that thousands of samples are needed for robust gene selection. However, the current microarray technology allows producing thousands of gene expression measurements. Collecting many proper samples from humans is not so easy. Hence, we must live with the shortage of biological samples and look for other ways to overcome this problem. One of such ways is to complement findings of gene selection with those of other (than microarrays) types of biological data. Yet another salvation may come from works on sparse decoding as pointed out by Smola et al. in (Song, Bedo, Borgwardt, Gretton, & Smola, 2007). These works, such as (Candes & Tao, 2005) and (Wainwright, 2006), suggest that for a very well defined family of inverse problems[3], asymptotically (i.e. about) only $d\left(1+\log D\right)$ observations are needed to accurately recover d features from D dimensions. For instance, if $D = 1000$ and *log* means logarithm of base 10, then in order to recover $d = 10$ features, one only needs 40 samples – very reasonable number for microarray datasets.

To say that feature/gene selection is a hot topic in bioinformatics and artificial intelligence is to say nothing. GoogleScholar[4] (www.googlescholar.com) can provide a better answer if one searches for the exact phrases "feature selection" and "gene selection". For instance, the search for "feature selection" returned 87,800 links while the search for "gene selection" returned 10,900 links[5]. These are impressive numbers, even though some links may duplicate each other. And do not forget GooglePatents (http://www.google.com/patents) if you are seriously thinking of patenting your own method for feature or gene selection one day. How can one "sail" in this sea of information? Unfortunately, a straightforward answer does not exist, because there are multiple ways leading to the same destination.

Perhaps, it is worthy to start from Wikipedia's page on feature selection at http://en.wikipedia.org/wiki/Feature_selection, which provides the wealth of information during a few minutes of reading. However, the next step may be not so easy to recommend. I can only say to look for surveys and books. Here come two sources that are, in my opinion, valuable to begin with: the survey of Guyon and Elisseeff (Guyon & Elisseeff, 2003) and the book edited by Liu and Motoda (Liu & Motoda, 2007). Again, this is only my personal opinion that may be completely different from opinions of other people involved in feature/gene selection research. I hope readers understand the difficult task of advising on this subject as there are so many good researchers and so many links found by GoogleScholar.

I myself, however, followed a different path when choosing feature/gene selection algorithms for this book. Instead of presenting those algorithms that are well described in the renowned sources, I tried to find "exotic" algorithms. By "exotic", I do not mean rarely cited algorithms, but those algorithms that were built on non-conventional principles and therefore the number of representative algorithms fitting to each of these principles is not too big. I call such a state as "sparsely crowded research niche". I tried to select as diverse principles as possible but clearly understand impossibility to cover all or 99% of the field. Next chapters represent results of my efforts, open to criticism from researchers whose methods were (unintentionally) left behind this book. Sorry, folks.

Some punctual readers may ask at this point: where is your taxonomy of feature/gene selection methods? Here it is, simple and cold. I assume that all or almost all methods are iterative or recursive, i.e. they starts from the empty set of features and recursively add features, or they have the full set of features at their disposal and try to iteratively remove features. In either case, the iterations halt when either a pre-defined number of features are found or a stopping criterion is satisfied. After that, everything else comes down to two dimensions shown in Table 1.

From this table, one can see that either one or several features can be added/deleted at a time. All that remained is to explain what filters and wrappers are. A filter is a feature/gene selection method that bases its conclusions about which feature or features to select solely on characteristics of the (training) data. In other words, a classifier is not involved in decision making; hence, feature/gene selection is independent of a classifier model. In contrast, a wrapper does rely on a classifier in order to say which feature or feature subset to pick; hence, feature/gene selection is classifier-driven. Each of the two feature/gene selection models has own advantages and drawbacks so that the decision of which model to prefer is very much problem-dependent (see, e.g., the overview in (Inza, Larrañaga, Blanco, & Cerrolaza, 2004)). I hope that after getting yourselves acquainted with the methods described in this book, you, dear readers, will compose your own recipe. And do not forget about fads and fallacies in gene selection (Braga-Neto, 2007). Happy reading!

Table 1. Taxonomy of feature/gene selection methods

	Filter	Wrapper
Individual evaluation of each feature		
Subset evaluation of a group of features		

REFERENCES

Braga-Neto, U. M. (2007). Fads and fallacies in the name of small-sample micro-array classification - A highlight of misunderstanding and erroneous usage in the applications of genomic signal processing. *IEEE Signal Processing Magazine, 24*(1), 91–99. doi:10.1109/MSP.2007.273062

Candes, E., & Tao, T. (2005). Decoding by linear programming. *IEEE Transactions on Information Theory, 51*(12), 4203–4215. doi:10.1109/TIT.2005.858979

Ein-Dor, L., Zuk, O., & Domany, E. (2006). Thousands of samples are needed to generate a robust gene list for predicting outcome in cancer. *Proceedings of the National Academy of Sciences of the United States of America, 103*(15), 5923–5928. doi:10.1073/pnas.0601231103

Guyon, I., & Elisseeff, A. (2003). An introduction to variable and feature selection. *Journal of Machine Learning Research, 3*, 1157–1182. doi:10.1162/153244303322753616

Inza, I., Larrañaga, P., Blanco, R., & Cerrolaza, A. J. (2004). Filter versus wrapper gene selection approaches in DNA microarray domains. *Artificial Intelligence in Medicine, 31*(2), 91–103. doi:10.1016/j.artmed.2004.01.007

Liu, H., & Motoda, H. (Eds.). (2007). *Computational methods of feature selection*. Boca Raton, FL: Chapman & Hall/CRC Press.

Song, L., Bedo, J., Borgwardt, K. M., Gretton, A., & Smola, A. J. (2007). Gene selection via the BAHSIC family of algorithms. *Bioinformatics (Oxford, England), 23*(13), i490–i498. doi:10.1093/bioinformatics/btm216

Wainwright, M. J. (2006). *Sharp thresholds for noisy and high-dimensional recovery of sparsity using l_1-constrained quadratic programming*. Technical Report 709, University of California, Department of Statistics, Berkeley.

ENDNOTES

[1]   Feature selection should not be confused with feature extraction. Although both feature selection and feature extraction perform dimensionality reduction, feature selection finds a subset of the original features while feature extraction produces a new (either linearly or nonlinearly) transformed set of features from the original features. With feature extraction it is not possible to determine the contribution of the original features to each transformed feature since the importance of the original features is lost during feature transformation.

In contrast, feature selection preserves the original features intact; features deemed unimportant/irrelevant/redundant are simply eliminated from further consideration while selecting only those features that significantly contribute to classification.

[2]   Typically, this can happen if the number of original features is small and features are dependent of each other.

[3]   An inverse problem is the problem of finding model parameters fitting to observed data. The word "inverse" implies the order of inference, opposite to the conventional inference consisting in prediction of observations, given the values of model parameters. Applied to gene selection from microarray data, we are presented with high dimensional data (observations) and we need to discover a linear or non-linear transformation (model) that best describes these data. This transformation maps the original high dimensional data to a low dimensional feature space.

[4]   I prefer GoogleScholar over Google, because the former returns mostly research articles and Master/PhD theses that are often freely available for download. In contrast, search with simply Google can return the links to web sites either containing some brief information about the subject of search or only a title and/or abstract of the article where the subject of search is mentioned. However, I do not enforce our preferences by assuming that every reader found his or her best way to get useful information from the Web.

[5]   Search was done on October 25, 2009.

Chapter 9
Feature Selection Based on Elements of Game Theory

FEATURE SELECTION BASED ON THE SHAPLEY VALUE

This chapter describes an approach to feature selection originating from the principles of game theory, in particular, in the context of coalition games (Cohen, Ruppin, & Dror, 2005), (Cohen, Dror, & Ruppin, Feature selection via coalition game theory, 2007). Feature selection is iteratively done through the Contribution-Selection algorithm aiming at optimizing classifier performance on unseen data. This algorithm combines both filter and wrapper approaches to feature selection. In contrast to the conventional filters, features are iteratively re-ranked by a classifier. The ranking itself employs the Shapley value in order to estimate the importance of each feature for the given task.

The algorithm assumes that two sets are available: training set and validation set. The former is employed for training a classifier while the latter is used to

DOI: 10.4018/978-1-60960-557-5.ch009

evaluate the trained classifier performance, given a subset of features - candidates for selection. The classifier performance is measured by classification accuracy, though, in principle, any other suitable characteristic, such as the area under the Receiver Operating Characteristic (ROC) curve or False Acceptance Rate (FAR), can be utilized as well.

Main Idea

Coalition games belonging to cooperative game theory involve a set of players and a reward associated with different groups or coalitions of players. The reward of a certain coalition depends on individual contributions of players composing this coalition to the game. The larger the contribution of a player is, the higher the benefit of having this player in a coalition. Coalitions with high reward are naturally preferable over those with small reward.

The contribution is based on the Shapley value (Shapley, 1953). This value is defined as follows. Let the marginal importance of player i to a coalition S ($i \notin S$) be

$$\Delta_i(S) = v\big(S \cup \{i\}\big) - v(S),$$

where $v(S)$ is the reward associated with coalition S. The reward can be negative, zero, or positive. The negative or zero reward implies no benefits of inclusion of player i into the current coalition.

The Shapley value is then defined as

$$\Phi_i(v) = \frac{1}{n!} \sum_{\pi \in \Pi} \Delta_i\big(S_i(\pi)\big),$$

where n is the total number of players, Π is the set of permutations over n, and $S_i(\pi)$ is the set of players appearing before player i in permutation π. Thus, the Shapley value of a given player is the mean of its marginal importance averaged over all possible coalitions of players.

Casting these game theory definitions into feature selection terminology maps the set of players into the set of features and the reward given to a coalition of players into the value of the classifier performance characteristic (here, the accuracy), attained with a certain subset of features. From now on, we will utilize feature selection terminology instead of game theory one.

As can be seen from the definition of the Shapley value, summing is done over all possible subsets of features, which is impractical. One solution is to devise a heuristic coupled with the limit on feature subset size, which corresponds to the fact that the number of significant interactions d among features is, in general, much smaller than the total number of features n. As the result, $\Phi_i(v)$ is approximated by

$$\varphi_i(v) = \frac{1}{|\Pi_d|} \sum_{\pi \in \Pi_d} \Delta_i \left(S_i(\pi) \right),$$

where Π_d is the set of sampled permutations on feature subsets of size d, and $|\cdot|$ means the cardinality of a set.

Since the number of possible permutations is large even for moderate values of d, we will replace sampling of permutations of size d with random sampling d features with replacement from the original set of features.

Algorithm Description

The Contribution Selection algorithm iteratively selects features through either forward selection or backward elimination. The forward selection variant iteratively adds a predefined number of features with the highest contribution to the classification accuracy as long as there are features with sufficiently large contribution values exceeding a preset contribution threshold. The backward elimination variant iteratively deletes a predefined number of features with the lowest contribution to the classification accuracy as long as the contribution values of all candidate features fall below a preset contribution threshold, i.e. such features add insignificant contribution and therefore they can be safely eliminated.

Pseudo code of the forward selection variant is shown in Figure 1. Indices of selected and remaining features are stored in vectors *Selected* and *Remaining*, respectively. Initially, the former is empty since no features have been yet selected while the latter contains indices of all features. Features are iteratively selected until there are still remaining features to choose from. A single iteration involves all remaining features and computation of the contribution of each remaining feature to the classification accuracy, independently of the other remaining features. This contribution is computed in function *Contribution* that will be discussed below in detail.

In the beginning, contribution of each original feature to the classification accuracy is found, independently of other features. In other words, classification is done using one feature only and the accuracy is computed. After that, if the maxi-

Figure 1. Pseudo code of the Contribution Selection Algorithm (forward selection variant)

Contribution Selection Algorithm (forward selection)

$Selected := \varnothing$;
$Remaining := \{1,...,D\}$;
While *Remaining* is not empty
  For $i := 1 : |Remaining|$
    If *Selected* is empty, then $c_i := Contribution(Remaining_i, \varnothing)$;
    Else { $s := 0$;
      For $j := 1 : t$
        Randomly sample with replacement d features from *Selected*;
        Let indices of sampled features be stored in *ii*;
        $s := s + Contribution(Remaining_i, Selected_{ii})$;
      End For
    $c_i := s/t$; }
  End For
  If $\max_i \{c\} \leq \Delta$, then stop;
  Find *nsf* features corresponding to the *nsf* highest values of *c*;
  Let their indices be stored in *ind*;
  $Selected := Selected \cup ind$;
  $Remaining := Remaining \setminus ind$;
End While

---

mum among all contribution values is less than or equal to threshold Δ, which implies that all contributions are too small to be treated as significant, the algorithm halts and the result is delivered in *Selected*.

If the maximal contribution strictly exceeds threshold Δ, which indicates that there are certain features that are able to significantly improve classification accuracy, then the *nsf* best features according to the magnitude of the contribution are retained and added to *Selected*. These features are simultaneously deleted from *Remaining*.

After the first features have been selected, random sampling (with replacement) of d features from the already selected features is performed, and the contribution of each yet remaining feature is found, given these sampled features. Sampling is repeated t times and each time the contribution is computed and accumulated. After t rounds of sampling, the average contribution approximating the Shapley value is calculated. Again, the maximum of all contributions is compared to threshold Δ,

Figure 2. Pseudo code of the Contribution function

Contribution function

Training a given classifier with pre-specified parameters on a training set by using only *Selected* features;
Assessing classification accuracy of the trained classifier on a validation set by using only *Selected* features;
Let the accuracy achieved on the validation set be acc_{old};

Training a given classifier with pre-specified parameters on the training set by using the union of *Selected* features and the given *Remaining* feature;
Assessing classification accuracy of the trained classifier on the validation set by using the union of *Selected* features and the given *Remaining* feature;
Let the accuracy achieved on the validation set be acc_{new};

Contribution, c, is equal to $acc_{new} - acc_{old}$.

and if the former does not exceed the latter, the algorithm stops, signaling that no more features can be selected. Otherwise, *nsf* most contributing features are added to the already selected and removed from still remaining features.

Pseudo code of *Contribution* is given in Figure 2. In essence, this function assesses classification accuracy for two cases: selected features alone and the union of selected features and the given remaining feature. The difference in accuracy corresponds to the desired contribution value. The training phase is optional for some classifiers such as k-nearest neighbors.

Pseudo code of the backward elimination variant of the Contribution Selection algorithm is presented in Figure 3. Unlike the forward selection variant, instead of *Remaining* we will use *Discarded* as the latter better illustrates the direction of the entire process. In the beginning, *Selected* contains indices of all features while *Discarded* is empty. Features are to be removed from *Selected* and stored in *Discarded*.

When no feature has been yet deleted, the individual contribution[1] of each original feature is determined. If the maximal value among all contributions does not fall below threshold Δ, which implies that all contributions are so significant that no feature can be removed without degrading classification accuracy, the algorithm terminates and *Selected* contains only those features that are deemed to be significant for classification. If the maximal contribution is strictly less than Δ, then *nef* features associated with the *nef* smallest contributions are removed from *Selected* while they are added to *Discarded*.

Figure 3. Pseudo code of the Contribution Selection Algorithm (backward elimination variant)

```
Contribution Selection Algorithm (backward elimination)

Selected := {1,...,D};
Discarded := ∅;
While Selected is not empty
  For i := 1:|Selected|

    If Discarded is empty, then c_i := Contribution(Selected_i, ∅);
    Else { s := 0;
      For j := 1:t
        Randomly sample with replacement d features from Discarded;
        Let indices of sampled features be stored in ii;
        s := s + Contribution(Selected_i, Discarded_ii);
      End For
      c_i := s/t; }
  End For
  If max{c} ≥ Δ, then stop;
       i
  Find nef features corresponding to the nef lowest values of c;
  Let their indices be stored in ind;
  Selected := Selected \ ind;
  Discarded := Discarded ∪ ind;
End While
```

When length of *Selected* is not anymore equal to *D*, *d* features are randomly sampled with replacement from *Discarded*, and the contribution of each discarded feature to the remaining features in *Selected* is computed. If the maximal contribution is less than threshold Δ, *nef* features with the smallest contributions are deleted from *Selected* and returned to *Discarded*. Otherwise, the algorithm stops since no feature is considered to be redundant.

As one can see, the backward elimination variant exchanged the roles of *Selected* and *Remaining*, compared to its forward selection cousin.

MATLAB Code

The forward selection variant, the contribution to classification accuracy, and the backward elimination variant are encoded in functions *ForwardContributionSelection*, *Contribution*, and *BackwardContributionSelection*, respectively. For completeness, code of the function *Contribution* is provided for both variants of the algorithm.

In theory, classifiers with and without training can be utilized with the Contribution Selection Algorithm. However, I prefer simple algorithms such as Naïve Bayes and Nearest Neighbors as they do not require training and hence, they are less probably to introduce a bias in feature selection.

In both variants, selection of a proper threshold (parameter *delta*) is of importance. I found that the best way to set up it is by trial-and-error. That is, you choose the first guess for this threshold and run either function. If no feature was added (*ForwardContributionSelection*) or eliminated (*BackwardContributionSelection*), your initial guess was too high or too low, respectively, and you need to adjust threshold accordingly.

The forward selection is much faster than the backward elimination, but it is said that the latter may be better to apply in practice than the former as many more feature subsets are evaluated during backward elimination, thus increasing the chance that important features will not be missed.

It is interesting to note that MATLAB® Statistics Toolbox™ includes the function *sequentialfs* that performs sequential forward or backward feature selection in the manner reminiscent to the Contribution Selection Algorithm. The only but crucial difference is that *sequentialfs* selects one feature at a time while the Contribution Selection Algorithm selects a group of features at each round. Although *sequentialfs* assumes that k-fold cross-validation can be utilized in order to split the original data into training and validation sets, I do not recommend this option because of very small size of microarray gene expression datasets. However, the readers are encouraged to read help description for *sequentialfs* in MATLAB® documentation in order to better understand the main idea behind the whole class of algorithms based on iterative addition or removal of one or several features at a time. As a starting point to study this family of feature selection algorithms, the article by Pudil et al (Pudil, Novovičová, & Kittler, 1994) as well as other works of these researchers are highly recommended.

```
function selected_features = ForwardContributionSelection(tra
in_patterns, train_targets, valid_patterns, valid_targets,
delta, d, t, nsf, classifier, varargin)

% Forward Contribution Selection Algorithm for feature selec-
tion
% Original sources:
% Shay Cohen, Eytan Ruppin, Gideon Dror, "Feature selection via
% coalitional game theory", Neural Computation 19(7), 1939-
1961, 2007
% Shay Cohen, Eytan Ruppin, Gideon Dror, "Feature selection
```

```
based on
% the Shapley value", In Leslie Pack Kaelbling and Alessandro
Saffioti % (Eds.), Proceedings of the 9th International Joint
Conference on
% Artificial Intelligence, Edinburgh, Scotland, UK, July 30-Au-
gust 5,
% pp. 665-670, Professional Book Center, Denver, CO, 2005
%
% Inputs:
%    train_patterns    - Training patterns (DxN matrix, where D
- data
%                        dimensionality, N - number of patterns)
%    train_targets     - Training targets (1xN vector)
%    valid_patterns    - Validation patterns (DxNt matrix, where D
- data %                 dimensionality, Nt - number of patterns)
%    valid_targets     - Validation targets (1xNt vector)
%    delta             - Contribution value threshold (nonnegative
real
%                        number)
%    d                 - Sample size for computing contribution
values
%    t                 - Number of random samples
%    nsf               - Number of features to be selected during
each
%                        round
%    classifier        - Classifier (string with function name
where a
%                        classifier is implemented)
%    varargin          - Parameters of a classifier
%
% Outputs:
%    selected_features    - Selected features (vector of feature
indices)
%
% Examples:
% selected_features =
%  ForwardContributionSelection(train_patterns, train_targets,
%  valid_patterns, valid_targets, 10, 10, 5, 5, 'NearestNeigh-
bor', 3);
% selected_features =
```

```
%  ForwardContributionSelection(train_patterns, train_targets,
%  valid_patterns, valid_targets, 10, 10, 5, 5, 'NaiveBayes');



% Get the total number of features
D = size(train_patterns, 1);

% Initialize sets of selected and remaining features
selected_features = [];
remaining_features = 1:D;

while ~isempty(remaining_features) % iterate until there are
remaining features

c = [];

% For each remaining feature, compute its contribution to
improving
% classification performance, based on the Sharpley value
for i = 1:length(remaining_features)
    if isempty(selected_features) % only in the very beginning
when no feature has been yet selected
        % Compute the contribution of the ith feature to im-
proving
        % classification performance
        c(i) = Contribution(remaining_features(i), [], train_
patterns, train_targets, valid_patterns, valid_targets, classi-
fier, varargin{:});
    else
        s = 0;
        % Compute the contribution of the ith remaining feature
to
  % improving classification performance
        for j = 1:t % do t times
            % Randomly sample d features with replacement from
the set
            % of already selected features
            idx = randsample(selected_features, d, true);
            % Compute the contribution of the ith remaining
```

```
feature to
% the already selected d features randomly sampled from the %
whole set of the selected features
            s = s + Contribution(remaining_features(i), idx,
train_patterns, train_targets, valid_patterns, valid_targets,
classifier, varargin{:});
        end
        % Compute the Shapley value
        c(i) = s/t;
    end
end

% Display a hint to set up delta
if isempty(selected_features)
    fprintf(1, 'This is the maximal value to be used to adjust
''delta'': %d\n', max(c));
    fprintf(1, 'The value of ''delta'' should be smaller than
this maximum!\n');
end

% If the contribution of all candidate features is less than or
equal
% to the preset threshold, i.e., the improvement of classifica-
tion is
% small, then stop
if max(c) <= delta
    break;
end

% Otherwise sort all contribution values in descending order of
% magnitude
[c, idx] = sort(c, 'descend');

% Add at most nsf new features to the already selected ones
if length(idx) > nsf
    idx = idx(1:nsf);
end
% Add features to the list of selected ones
selected_features = [selected_features remaining_
features(idx)];
```

```
% Remove added features from the list of remaining ones
remaining_features(idx) = [];

% Display the number of selected features
fprintf(1, '%d features selected, %d features remained...\n',
length(selected_features), length(remaining_features));

end

return

% Helper function computing the contribution of a given feature
to
% classification performance
function c = Contribution(candidate_feature, selected_features,
train_patterns, train_targets, valid_patterns, valid_targets,
classifier, varargin)

% If no feature has been yet selected, assign zero to acc_old,
% otherwise run a classifier on the selected features only
if isempty(selected_features)
    acc_old = 0;
else
    % Classification performance just with selected features
without
    % adding any new feature
    i = selected_features;
    predicted_targets = feval(classifier, train_patterns(i,:),
train_targets, valid_patterns(i,:), varargin{:});
    % Classification accuracy in per cent
    acc_old = 100 - class_error(valid_targets, predicted_tar-
gets);
end

% Next, check the classification performance if a candidate
feature is % added to the set of previously selected features
i = [selected_features candidate_feature];
predicted_targets = feval(classifier, train_patterns(i,:),
train_targets, valid_patterns(i,:), varargin{:});
% Classification accuracy in per cent
```

```
acc_new = 100 - class_error(valid_targets, predicted_targets);

% Compute the difference in performance after and before adding
the
% candidate feature
c = acc_new - acc_old;
% c can be also negative, meaning that no useful gain from
feature
% addition was attained

return
function selected_features = BackwardContributionSelection(tra
in_patterns, train_targets, valid_patterns, valid_targets,
delta, d, t, nef, classifier, varargin)

% Backward Contribution Selection Algorithm for feature selec-
tion
% Original sources:
% Shay Cohen, Eytan Ruppin, Gideon Dror, "Feature selection via
% coalitional game theory", Neural Computation 19(7), 1939-
1961, 2007
% Shay Cohen, Eytan Ruppin, Gideon Dror, "Feature selection
based on
% the Shapley value", In Leslie Pack Kaelbling and Alessandro
Saffioti % (Eds.), Proceedings of the 9th International Joint
Conference on
% Artificial Intelligence, Edinburgh, Scotland, UK, July 30-Au-
gust 5,
% pp. 665-670, Professional Book Center, Denver, CO, 2005
%
% Inputs:
%   train_patterns  - Training patterns (DxN matrix, where D
- data
%                     dimensionality, N - number of patterns)
%   train_targets   - Training targets (1xN vector)
%   valid_patterns  - Validation patterns (DxNt matrix, where D
- data %              dimensionality, Nt - number of patterns)
%   valid_targets   - Validation targets (1xNt vector)
%   delta           - Contribution value threshold (nonnegative
real
```

```
%                       number)
%    d                - Sample size for computing contribution
values
%    t                - Number of random samples
%    nef              - Number of features to be eliminated
during each
%                       round
%    classifier       - Classifier (string with function name
where a
%                       classifier is implemented)
%    varargin         - Parameters of a classifier
%
% Outputs:
%    selected_features   - Selected features (vector of feature
indices)
%
% Examples:
% selected_features =
%  BackwardContributionSelection(train_patterns, train_targets,
%  valid_patterns, valid_targets, 110, 10, 5, 100, 'Nearest-
Neighbor', 3);
% selected_features =
%  BackwardContributionSelection(train_patterns, train_targets,
%  valid_patterns, valid_targets, 110, 10, 5, 100, 'Naive-
Bayes');



% Get the total number of features
D = size(train_patterns, 1);

% Initialize sets of selected and discarded features
selected_features = 1:D;
discarded_features = [];

while ~isempty(selected_features) % iterate until there are
features to eliminate

c = [];
```

```
% For each remaining feature, compute its contribution to
improving
% classification performance, based on the Sharpley value
for i = 1:length(selected_features)
    if isempty(discarded_features) % only in the very beginning
when no feature has been yet eliminated
        % Compute the contribution of the ith feature to im-
proving
        % classification performance
        c(i) = Contribution(selected_features(i), [], train_
patterns, train_targets, valid_patterns, valid_targets, classi-
fier, varargin{:});
    else
        s = 0;
        % Compute the contribution of the ith selected feature
to
    % improving classification performance
        for j = 1:t % do t times
            % Randomly sample d features with replacement from
the set
            % of already eliminated features
            idx = randsample(discarded_features, d, true);
            % Compute the contribution of the ith selected
feature to
            % the already eliminated d features randomly sam-
pled from
            % the whole set of the eliminated features
            s = s + Contribution(selected_features(i), idx,
train_patterns, train_targets, valid_patterns, valid_targets,
classifier, varargin{:});
        end
        % Compute the Shapley value
        c(i) = s/t;
    end
end

% Display a hint to set up delta
if isempty(discarded_features)
    fprintf(1, 'This is the maximal value to be used to adjust
''delta'': %d\n', max(c));
```

```
    fprintf(1, 'The value of ''delta'' should be larger than
this maximum!\n');
end

% If the contribution of all candidate features is larger than
or equal % to the preset threshold, i.e., no feature can be
eliminated without
% significant loss in classification performance, then stop
if max(c) >= delta
    break;
end

% Otherwise sort all contribution values in ascending order of
% magnitude
[c, idx] = sort(c, 'ascend');

% Eliminate at most nef features whose contribution to classi-
fication
% performance is smallest
if length(idx) > nef
    idx = idx(1:nef);
end
% Add eliminated features to the list of discarded ones
discarded_features = [discarded_features selected_
features(idx)];
% Eliminate features
selected_features(idx) = [];

% Display the number of features to select from
fprintf(1, '%d features to select from, %d features
discarded...\n', length(selected_features), length(discarded_
features));

end

return

% Helper function computing the contribution of a given feature
to
% classification performance
```

```
function c = Contribution(candidate_feature, selected_features,
train_patterns, train_targets, valid_patterns, valid_targets,
classifier, varargin)

% If no feature has been yet selected, assign zero to acc_old,
% otherwise run a classifier on the selected features only
if isempty(selected_features)
    acc_old = 0;
else
    % Classification performance just with selected features
without
    % adding any new feature
    i = selected_features;
    predicted_targets = feval(classifier, train_patterns(i,:),
train_targets, valid_patterns(i,:), varargin{:});
    % Classification accuracy in per cent
    acc_old = 100 - class_error(valid_targets, predicted_tar-
gets);
end

% Next, check the classification performance if a candidate
feature is % added to the set of previously selected features
i = [selected_features candidate_feature];
predicted_targets = feval(classifier, train_patterns(i,:),
train_targets, valid_patterns(i,:), varargin{:});

% Classification accuracy in per cent
acc_new = 100 - class_error(valid_targets, predicted_targets);

% Compute the difference in performance after and before adding
the
% candidate feature
c = acc_new - acc_old;
% c can be also negative, meaning that no useful gain from
feature
% addition was attained

return
function errprcnt = class_error(test_targets,targets)
```

```
% Calculate error percentage based on true and predicted test
labels
errprcnt = mean(test_targets ~= targets);
errprcnt = 100*errprcnt;

return
```

REFERENCES

Cohen, S., Dror, G., & Ruppin, E. (2007). Feature selection via coalition game theory. *Neural Computation, 19*(7), 1939–1961. doi:10.1162/neco.2007.19.7.1939

Cohen, S., Ruppin, E., & Dror, G. (2005). Feature selection based on the Shapley value. In L. P. Kaelbling, & A. Saffiotti (Ed.), *Proceedings of the 9th International Conference on Artificial Intelligence, Edinburgh, Scotland, UK* (pp. 665-670). Denver, CO: Professional Book Center.

Pudil, P., Novovičová, J., & Kittler, J. (1994). Floating search methods in feature selection. *Pattern Recognition Letters, 19*(11), 1119–1125. doi:10.1016/0167-8655(94)90127-9

Shapley, L. (1953). A value for n-person games. In H. Kuhn, & A. Tucker (Eds.), *Contributions to the Theory of Games* (Vol. II of Annals of Mathematics Studies 28, pp. 307-317). Princeton: Princeton University Press.

ENDNOTE

[1]   Contributions are computed using the function *Contribution* described earlier.

Chapter 10

Kernel–Based Feature Selection with the Hilbert–Schmidt Independence Criterion

KERNEL METHODS AND FEATURE SELECTION

You already met with kernels in chapter on support vector machines. However, because of my claim that each chapter in this book is self-sufficient and self-contained, I would like to briefly reproduce the definition of kernel methods once again. Here, Wikipedia comes to our help (http://en.wikipedia.org/wiki/Kernel_methods).

Kernel methods are a class of pattern analysis (regression, clustering, correlation, classification) algorithms that approach the problem by projecting the data into a high dimensional feature space[1] $\mathcal{P}$, where relations hidden in the data under study can be better revealed than when doing pattern analysis in the original space $\mathcal{O}$ of features. A good thing is that one does not need to know the explicit form of this

DOI: 10.4018/978-1-60960-557-5.ch010

projection, since it can be done via the inner products[2] between the images[3] of all pairs of data in the high dimensional feature space $\mathcal{P}$. The inner products are embedded into the definition of a kernel function (hence, the name 'kernel methods'). The complete set of all inner products forms a kernel matrix. Another important thing about kernel methods is that computing the kernel function is cheaper than the explicit computation of the projection $\mathcal{O} \to \mathcal{P}$.

Readers who by some reason missed the chapter on support vector machines may wonder why the mapping to the space which dimension can be much higher than that of the original space is beneficial in our microarray setting. To address this worries and healthy doubts, I can say (after all the pioneers of kernel methods) that kernel matrix size is $N \times N$, where N is the number of instances in the training set. Since for gene expression data, it is common that the number of instances is much smaller than the number of genes whose expression levels were measured, we obtain a compact structure needed for pattern analysis.

In this chapter, by pattern analysis, we mean looking for dependence between the features and the class labels in the kernel-induced space. The key pre-assumption is that good (relevant for classification) features will maximize such dependence. The dependence itself is formulated through a mutual-information (mutual dependence of two variables) like quantity, which is introduced in the next section.

Dependence and the Hilbert-Schmidt Norm

Let us begin with the example of linear dependence, which will be then extended to the detection of more general types of dependence as done in (Song, Bedo, Borgwardt, Gretton, & Smola, 2007). Given the joint sample observations (x, y) of data (feature values) x and class labels y, let us define a covariance matrix as follows

$$C_{xy} = E_{xy}\left(xy'\right) - E_x\left(x\right)E_y\left(y'\right),$$

where E_{xy} is the expectation with respect to the join distribution of the data and the class labels, E_x is the expectation with respect to the marginal distribution of the data, and E_y is the expectation with respect to the marginal distribution of labels. The expectation or the expected value is the mean of a random variable. For example, if y is a discrete variable[4] with probability distribution $Pr\left(Y = y\right)$, then the expectation $E_y = \sum_y yPr\left(Y = y\right)$ (Everitt, 2006).

The covariance matrix encodes all second-order dependence between the random variables. A statistic that efficiently summarizes the content of this matrix is its Hilbert-Schmidt norm.

So, it is the time to introduce the precise definition of this norm. We resorted to (Steeb, 2006) for this. Let A and B be $N \times N$ matrices over $\mathbb{C}$ (set of complex numbers). A scalar product can be defined as

$$\langle A, B \rangle := tr\left(AB^*\right),$$

where B^* is a conjugate transpose of B and $tr\left(\cdot\right)$ is a trace of a square matrix, i.e., the sum of its diagonal elements[5]. Since, in our case, expression levels are real numbers, $B^* = B'$, i.e. the conjugate transpose coincides with the ordinary transpose.

The scalar product thus defined is called Hilbert-Schmidt inner product and it implies the square norm

$$\langle A, A \rangle = \|A\|^2 = tr\left(AA'\right).$$

This (square) norm is called the Hilbert-Schmidt norm.

We denote the Hilbert-Schmidt norm of the covariance matrix C_{xy} as $\|C_{xy}\|^2_{HS}$. This quantity is zero if and only if there exists no second-order dependence between x and y.

However, the second-order dependence is not the only kind of dependence. To cover more general dependence, let us introduce a mapping of the original data and labels to high dimensional feature space induced by a kernel function. This space is called a reproducing kernel Hilbert space (RKHS). Let $\varphi\left(x\right)$ and $\psi\left(y\right)$ be feature map to RKHS for the data x and the class labels y, respectively. Then we can define a cross-covariance operator between these feature maps such that

$$C_{xy} = E_{xy}\left[\left(\varphi\left(x\right) - \mu_x\right) \otimes \left(\psi\left(y\right) - \mu_y\right)\right],$$

where $\mu_x = E_x\left(\varphi\left(x\right)\right)$ and $\mu_y = E_y\left(\psi\left(y\right)\right)$ and $\otimes$ is the tensor product.

The square of the Hilbert-Schmidt norm of the cross-covariance operator, $\|C_{xy}\|^2_{HS}$, is then used as the feature selection criterion. In (Gretton, Bousquet, Smola, &

Schölkopf, 2005) it was shown that this operator can be expressed in terms of kernels as

$$\left\| C_{xy} \right\|_{HS}^2 = \left(N - 1 \right)^{-2} tr \left(KHLH \right),$$

where K and L are the kernel matrices for the data and the class labels, respectively, and $H_{ij} = \delta_{ij} - N^{-1}$ centers the data and the label features ($\delta_{ij} = 1$ when $i = j$, and zero otherwise).

As with second-order dependence, $\left\| C_{xy} \right\|_{HS}^2 = 0$ if and only if there is no dependence between the data and the class labels.

Let us define the Hilbert-Schmidt Independence Criterion (HSIC) as $\left\| C_{xy} \right\|_{HS}^2$.

The trace in the formula for HSIC can be further expanded as

$$tr \left(KHLH \right) = tr \left(KL \right) - 2N^{-1} I' KLI + N^{-2} tr \left(H \right) tr \left(L \right),$$

where I is an $N \times 1$ vector of ones.

Theorem 1 in (Gretton, Bousquet, Smola, & Schölkopf, 2005) says that HSIC is a biased estimate. However, the bias is negligible in the overall process. Nevertheless, in the next section and in the algorithm implementation in MATLAB®, I will utilize the unbiased estimate of HSIC. Definitions of a biased and unbiased estimator are given in the next section, too.

Hilbert-Schmidt Independence Criterion

In a series of work, see, e.g., (Song, Smola, Gretton, Borgwardt, & Bedo, 2007), (Song, Bedo, Borgwardt, Gretton, & Smola, 2007), the groups of German and Australian researchers headed by Alexander Smola (this name is well familiar to those who work with kernel methods) proposed feature/gene filtering based on the Hilbert-Schmidt Independence Criterion (HSIC) that measures dependence between features and class labels. They called their approach supervised because labels are utilized. Nevertheless, in our taxonomy of feature/gene selection methods, their method belongs to filters rather than wrappers, because it does not use a classifier to judge on feature relevance/importance.

Smola and his colleagues and coauthors argued that in order for any measure to be considered eligible for feature relevance, it should meet two conditions: (I) it should be capable of detecting functional dependence between features and class labels, and (II) it should lead to relevance concentration that would guarantee that the detected functional dependence is preserved in the test (i.e. previously unseen) data. Smola et al. further argue that few feature selection criteria explicitly took both conditions into account. As a result, they propose own remedy – a mutual information (mutual dependence of two variables) like quantity which they and we after them abbreviated as HSIC. Two variables are, of course, features and labels. History of HSIC originates in not very distant past (Gretton, Bousquet, Smola, & Schölkopf, 2005).

HSIC (exactly its unbiased estimate) is defined as follows. Let K and L be $N \times N$ kernel matrix whose ijth elements defined as

$$K_{ij} = \left(1 - \delta_{ij}\right) k\left(x_i, x_j\right),$$

$$L_{ij} = \left(1 - \delta_{ij}\right) l\left(y_i, y_j\right),$$

$$\delta_{ij} = \begin{cases} 1, if\ i = j \\ 0, if\ i \neq j \end{cases}$$

$$k\left(x_i, x_j\right) = \left\langle x_i, x_j \right\rangle,$$

$$l\left(y_i, y_j\right) = \left\langle y_i, y_j \right\rangle.$$

In all these formulas, $x_i \left(x_j\right)$ is a vector while $y_i \left(y_j\right)$ is a scalar! As a result, the inner products like $\left\langle x_i, x_j \right\rangle$ can be computed as

$$x_i' x_j = \sum_{k=1}^{n} x_i^{(k)} x_j^{(k)},$$

where $x_i^{(k)} \left(x_j^{(k)}\right)$ denotes the kth element of vector $x_i \left(x_j\right)$, respectively.

This formula defines the so called *linear* kernel, but there are several nonlinear kernels, one of which will be considered below. Different kernels incorporate different prior knowledge into the dependence estimation. For instance, a linear kernel concentrates only on second-order dependence. I will talk more about different kernel types in the next sections.

You can see that the diagonal of both kernel matrices is set to zero. Then, the unbiased estimate[6] of HSIC is computed as in (Song, Smola, Gretton, Borgwardt, & Bedo, 2007):

$$HSIC = \frac{1}{N(N-3)} \left[tr(KL) + \frac{I'KII'LI}{(N-1)(N-2)} - \frac{2}{N-2} I'KLI \right],$$

where I is $N \times 1$ vector of ones, i.e. it consists of all ones arranged in a column fashion whereas I' is a row vector (transposed version of I), and $tr(\cdot)$ is a trace of a square matrix. Despite of crowded matrix products in the formula for HSIC, it is easy to verify with paper and pen that all summands are numbers, not matrices. Thus, HSIC is a number, too. For those who might find it not very straightforward, I provide a real-world example:

$$\underbrace{I'KII'LI}_{1 \times 1} = \underset{1 \times N \times N \times N \times N \times 1}{\overset{I' \quad K \quad I \quad I' \quad L \quad I}{1 \times 1 \times N \times N \times N \times 1}}$$

Although it looks extremely simple and may be even naïve to some of you for whom mathematics is everyday business, it is good to do such paper-and-pen exercises in order to check if (1) you understand what will be a result, and (2) a formula you are about to implement in code is correct. I believe that following these simple principles would save a lot of your valuable time and nerves, which, as we all know too well from ordinary life, cannot be completely repaired, as well as it would train you in not-very-sophisticated math time after time.

Returning back to the properties of HSIC, Smola et al. proved (see, e.g., (Song, Smola, Gretton, Borgwardt, & Bedo, 2007)) that HSIC is an unbiased estimator. According to (Everitt, 2006), bias is deviation of results from the truth. In estimation of a certain quantity we would like to know how much an obtained estimate $\hat{\theta}$ of this quantity differs from its expected value $E(\hat{\theta})$. An estimator for which $E(\hat{\theta}) = \theta$ is called unbiased. The variance of such an estimator is, however, larger than that of a biased estimator, for which $E(\hat{\theta}) \neq \theta$, due to the so called bias-variance dilemma or trade-off. It should be remembered that bias is the property of the estimator, not of the estimate.

HSIC is also concentrated within a certain range as shown in (Gretton, Bousquet, Smola, & Schölkopf, 2005), (Song, Smola, Gretton, Borgwardt, & Bedo, 2007). So, unseen data will likely have HSIC values falling within the same range, which makes conclusions drawn from HSIC well grounded.

The facts that HSIC is unbiased and concentrated imply that the empirical HSIC closely reflects its population counterpart. Put it differently, a model faithfully reflects reality.

Note that HSIC is simple to compute once two kernel matrices K and L are computed. In fact, L can be pre-computed, which makes the entire algorithm even simpler.

Main Idea

Smola et al. chose greedy feature selection approaches (forward selection and backward elimination) as a basis for demonstrating the HSIC usefulness. To remind you, forward selection iteratively adds features according to a certain criterion until the desired number of features is collected or a termination condition is satisfied, while backward elimination iteratively removes features according to a certain criterion until the desired number of features is collected or a termination condition is satisfied (Guyon & Elisseeff, 2003). Smola et al. argued that although forward selection is computationally more efficient (in order to select 10 out of 100 features, it is much faster to select 10 out of 100 features than to delete 90 out of 100 features), backward elimination provides better features, in general, since the candidate-features are assessed within the context of the others (forward selection may leave contributions of many features un-assessed). To remedy mutual drawbacks of both feature selection approaches while maximizing their mutual benefits, a mix of forward selection and backward elimination can be tried. It should be noted that since HSIC computation does not involve a classifier, feature selection using HSIC belongs to the filter family of algorithms. Decoupling the feature selection process and classification speeds up the computation over wrapper methods of feature selection where a classifier is used as a judge of candidate-features.

In terms of feature selection, HSIC is zero if and only if features and labels are independent. Therefore, nonzero values of HSIC will indicate the strength of dependence between features and class labels. Nothing prevents HSIC from being negative. However, negative values are out of interest for feature selection, where strong positive dependence is sought. Hence, the minimum of HSIC will point to a feature or a feature subset that needs to be eliminated while the maximum of HSIC will be associated with a feature or a feature subset that needs to be preserved. Given that the entire interval of HSIC has lower and upper bounds, say, $-B$ and $+B$, the subinterval $[-B, 0)$ (0 is excluded) corresponds to irrelevant features, zero means independence of features and labels, and the subinterval $(0, +B]$ (0 is excluded) is associated with relevant features.

Hence, features are added or eliminated based on HSIC corresponding to them. If elimination is sought, then the minimum value of HSIC will point to a feature or a feature subset than should be deleted. If addition is chosen, then a feature or a feature subset with the maximal HSIC needs to be selected.

Algorithm Description

Given reasons provided above, Smola et al. preferred backward elimination strategy for feature selection. Their algorithm is called backward elimination using HSIC (BAHSIC). The forward selection using HSIC (FOHSIC) is straightforward and therefore omitted here and left as an exercise for readers. BAHSIC is directly applicable to binary, multi-class, and regression problems.

To speed up backward elimination, Smola et al. decided to remove a subset of features at each round of the algorithm, rather than a single feature. Hence, starting from the list S containing indices of all available features, subset after subset is iteratively removed from the list. At each round, a subset with the minimal overall HSIC is removed, which corresponds to the least relevant features.

Pseudo code of BAHSIC is shown in Figure 1. In the beginning, each feature is normalized (independently of other features) to zero mean and unit variance. This operation makes all features to have the same domain and neutralizes scale difference between features. As the part of initialization, the kernel matrix of labels is computed, too, since it does not change in later stages.

The algorithm appends the features deleted from S to another list $S^{\dagger}$. In the beginning, S includes the full set of features (expression levels of genes). Because the most relevant features will join $S^{\dagger}$ last, the feature selection problem can be simply solved by taking the last M elements from $S^{\dagger}$. BAHSIC fills $S^{\dagger}$ recursively, eliminating the least relevant features from the current S. Since removing one feature at a time would be inefficient when there are a lot of irrelevant features, Smola et al. advocate to delete a feature subset at a time (its size is 10% of the current features, i.e. $0.1|S|$, which is a good trade-off between speed and the risk to lose relevant features).

The important question not explicitly discussed by Smola et al. is how to choose subsets for elimination. Pseudo code in (Song, Smola, Gretton, Borgwardt, & Bedo, 2007) nevertheless implicitly assumes that there are several subsets – candidates for elimination. However, only one of them is eventually removed from S. I complemented BAHSIC with such an option by providing an extra parameter ℓ to set up. This parameter defines the number of subsets, each of size $0.1|S|$, to be randomly sampled from the current S.

The contribution of the ith subset $\mathcal{T}_i, i = 1, \ldots, \ell$ is composed of contributions of the individual features, belonging to it. In particular, the jth feature ($j = 1, \ldots, 0.1|S|$) is temporarily removed from the current S and HSIC is computed for the remaining features. This operation is repeated for each feature in $\mathcal{T}_i$, and all HSIC values thus obtained are summed up together to produce a single composite value of HSIC

Figure 1. Pseudo code of BAHSIC

BAHSIC

Add indices of all features to the list S.
Set to empty the list of selected features: $S^\dagger \leftarrow \emptyset$.
Set the number of feature subsets ℓ to check for elimination.
Set the number of features t to be selected.

Normalize each feature to zero mean and unit standard deviation.

Compute the kernel matrix L for the class labels.

While $(S \neq \emptyset)$

    Adjust σ if a Gaussian (RBF) kernel is chosen: $\sigma := 1/(2d), d = |S| - 1$.
    For $i := 1 : \ell$
        Randomly sample the ith feature subset $\mathcal{T}_i$ of size $0.1|S|$ from S.
        $s_i := 0$; /* Initialize accumulator */
        For every j belonging to $\mathcal{T}_i$ **do**
            Temporarily remove the jth feature from S, and
            compute $HSIC(\sigma, L, S \setminus \{j\})$ without this feature.
            Add the obtained HSIC to accumulator s_i.
    End
    End
    Find a subset $\mathcal{T}$ such that

$$\mathcal{T} = arg \min_{i, 1 \leq i \leq \ell} s_i$$

    Delete $\mathcal{T}$ from S: $S \leftarrow S \setminus \mathcal{T}$.
    Add $\mathcal{T}$ to $S^\dagger$: $S^\dagger \leftarrow S^\dagger \cup \mathcal{T}$.

End While

Pick t last elements of $S^\dagger$ as indices of selected features.

---

for the whole subset $\mathcal{T}_i$. The minimum of the composite HSIC values is then found, pointing to the subset to be removed from S and be added to $S^\dagger$.

The algorithm iterates until S is empty. After that, last t features appended to $S^\dagger$ are considered as the most relevant features and they are selected.

Regarding the kernel choice when computing HSIC, Smola et al. indicated that when using a linear kernel on both the data and labels, FOHSIC and BAHSIC are equivalent: the objective function decomposes into individual components, and feature selection can be therefore done without recursion. Microarray data are likely linearly separable in the original feature space due to their high dimensionality. Thus, a linear kernel for the data would likely work well. Besides linear kernel, a Gaussian or Radial Basis Function (RBF) kernel could be used as Smola et al. demonstrated in (Song, Smola, Gretton, Borgwardt, & Bedo, 2007). The parameter of the RBF kernel can be adjusted as shown in Figure 1, where it increases as the size of S decreases. This adjustment is adaptation to the potential scale of the nonlinearity present in the (feature-reduced) data. In order to emphasize this adaptation of the RBF kernel parameter σ, it was explicitly included into the definition of HSIC in pseudo code. At each round of BAHSIC, only the data kernel matrix K is re-computed while the labels kernel matrix L remains the same all the time.

Following extensive experiments, Smola et al. (Song, Bedo, Borgwardt, Gretton, & Smola, 2007) concluded that the linear kernel outperforms many comparative methods (see their article for details) both for binary and multiclass problems. Only if nonlinearities are known to exist in data (for example, when one class contains two or more overlapping subclasses), then the RBF kernel can provide superior classification results to those achieved by employing the linear kernel. It could be also said that if gene interactions are important for cancer prediction, then the RBF kernel may be a better choice than its linear counterpart. However, RBF success, of course, depends on the correct choice of the kernel parameter. Determining such an optimal value can be a non-trivial task.

Given pros and cons of both types of kernels, Smola et al. (Song, Bedo, Borgwardt, Gretton, & Smola, 2007) suggested two rules of thumb:

- always apply the linear kernel for general purpose gene selection;
- apply the RBF kernel if nonlinear effects are present, such as overlapping subclasses within a more general class or complementary effects of different genes.

In principle, other than linear and RBF kernels can be utilized as well, but I am not interested to investigate this question here since the answer will be problem-dependent, and it is difficult to judge if the projection into the kernel-induced space would linearly separate two classes.

Similar Approaches and Methods That are Instances of BAHSIC

Smola et al. in (Song, Smola, Gretton, Borgwardt, & Bedo, 2007) noted that for binary classification, an alternative criterion for feature selection is to check if the conditional distributions $P\left(x|y = +1\right)$ and $P\left(x|y = -1\right)$ differ or not[7]. For this purpose, Maximum Mean Discrepancy (MMD) (Borgwardt, Gretton, Rasch, Kriegel, Scholkopf, & Smola, 2006) and Kernel Target Alignment (KTA) (Cristianini,

Kandola, Elisseeff, & Shawe-Taylor, 2003) can be used to test if there is correlation between data and labels.

In (Song, Bedo, Borgwardt, Gretton, & Smola, 2007), it was demonstrated that several well-known feature selectors commonly used in bioinformatics research are, in fact, the instances of BAHSIC. Among them are Pearson's correlation (Ein-Dor, Zuk, & Domany, 2006), (van 't Veer, et al., 2002), centroid (Hastie, Tibshirani, & Friedman, 2001), (Bedo, Sanderson, & Kowalczyk, 2006), shrunken centroid (Tibshirani, Hastie, Narasimhan, & Chu, 2002), (Tibshirani, Hastie, Harasimhan, & Chu, 2003), t-score (Hastie, Tibshirani, & Friedman, 2001).

MATLAB Code

Python implementation of BAHSIC is available at http://elefant.developer.nicta. com.au as a part of the Elefant package. I, however, coded BAHSIC myself in MATLAB®.

My MATLAB® implementation of BAHSIC consists of three functions: *BAHSIC* (main function), *compute _ kernel*, and *compute _ HSIC*.

The main function contains code of feature selection based on HSIC. Among input parameters is *nsets* defining the number feature subsets to be randomly sampled at each round of elimination. Since the size of each subset constitutes 10% of the whole set of features, the value of *nsets* can be between 3 and 5 in order to keep the number of HSIC computations moderately low. In this way, the gain in speed, compared to single-feature elimination, can be achieved. Subsets sampled at each round may overlap, but there are no two features with the same index, belonging to one subset. In other words, each feature within any given subset is unique. Indices of sampled features are stored in a cell array, each element of which is a real-valued row-vector storing features belonging to a certain subset.

Indices of selected genes are put into the row-vector *selected _ features*. These genes are recursively deleted from another row-vector S, initially containing indices of all genes. It should be noted that if the length of S is less than or equal to the number of genes, M, to select, the remaining genes in S automatically added to the tail of *selected _ features*.

HSIC computation for genes in the current S when one gene is temporarily taken out is done in the function *compute _ HSIC*. Both kernels are computed by means of function *compute _ kernel*. Kernel size is $N \times N$ elements (N is the number of samples taken from patients), which is rather small for typical microarray datasets, compared to the number of gene expressions. Currently, only two kernels are supported: linear and RBF. However, users can easily add their kernels as well. Pay attention to the efficient way of the distance matrix computation borrowed from MATLAB® code of manifold learning method, called Locally Linear Embedding (http://www.cs.toronto.edu/~roweis/lle/code/lle.m).

```
function selected_features = BAHSIC(train_patterns, train_tar-
gets, nsets, M, kernel_x, krn_x_params, kernel_y, krn_y_params)

% Backward elimination with the Hilbert-Schmidt Independence
Criterion % (HSIC)
% Original sources:
% Le Song, Alex Smola, Arthur Gretton, Karsten Borgwardt, Jus-
tin Bedo,
% "Supervised feature selection via dependence estimation", in
Zoubin
% Ghahramani (ed.), Proceedings of the 24th International Con-
ference on
% Machine Learning, Corvallis, OR, pp.823-830, 2007
% Le Song, Justin Bedo, Kirsten M. Borgwardt, Arthur Gretton,
Alexander % J. Smola, "Gene selection via the BAHSIC family of
algorithms",
% Proceedings of the 15th International Conference on Intelli-
gent
% Systems for Molecular Biology and 6th European Conference on
% Computational Biology, Vienna, Austria, pp.490-498, 2007
%
% Inputs:
%    train_patterns  - Training patterns (DxN matrix, where D -
data
%                                   dimensionality, N - number
of patterns)
%    train_targets   - Training targets (1xN vector)
%    nsets           - The number of feature subsets to check at
each
%                         iteration
%    M               - The number of features to select
%    kernel_x        - Kernel type for data
%    kernel_y        - Kernel type for class labels
%    krn_x_params    - Parameters of a data kernel
%    krn_y_params    - Parameters of a class label kernel
%
% Outputs:
```

```
%    selected_features    - Selected features (vector of feature
indices)



% Get data dimensionality and size
[D,N] = size(train_patterns);

% Compute the mean
m = mean(train_patterns, 2); % m is a column vector
% Compute the standard deviation
s = std(train_patterns, 0, 2); % s is a column vector
% Normalize each feature to zero mean and unit standard devia-
tion
train_patterns = (train_patterns - repmat(m, 1, N))./repmat(s,
1, N);
% Output the mean and standard deviation for visual check
mean(train_patterns, 2), std(train_patterns, 0, 2)

% Precompute L
L = compute_kernel(train_targets, kernel_y, krn_y_params);
L = L - diag(diag(L));

% Initialize a set S with indices of all features
S = 1:D;
% Initialize the list of selected features
selected_features = [];

% Pre-allocate memory
ind = cell(1,nsets);

while ~isempty(S)
    % Display the number of features remaining to choose from
    fprintf(1, 'Feature pool size: %d\n', length(S));

    % If size of S is less than or equal to the number of fea-
tures
    % to be selected, add remaining features in S to the list
of
    % selected features and halt
```

```
    if length(S) <= M
        selected_features = [selected_features S];
        break;
    end

    s = zeros(1,nsets);
    for i = 1:nsets
        % Sample without replacement the set S containing indi-
ces of
        % not yet selected features
        % Each thus formed subset contains about 10% of fea-
tures in S
        % Subsets may overlap, i.e., several subsets may in-
clude the
        % same feature or features. However, no subset includes
        % multiple instances of a feature
        n = floor(0.1*length(S));
        if n == 0, n = 1; end
        ind{i} = randsample(S, n, false);
        for j = 1:length(ind{i})
            % In case of the RBF kernel, adapt the kernel pa-
rameter,
            % depending of size of S
            if strcmpi(kernel_x, 'rbf')
                d = length(S) - 1;
                krn_x_params(1) = 1/(2*d);
            end
            % Temporarily exclude one feature...
            indices = S; indices(S == ind{i}(j)) = [];
            % and then estimate HSIC for the remaining features
            % Estimate is added to those of other features in
the ith
            % subset
            s(i) = s(i) + compute_HSIC(train_
patterns(indices,:), kernel_x, krn_x_params, L);
        end % end for j
    end % end for i

    % Find a subset with the minimal HSIC
    [smin,ii] = min(s);
```

```
    % Check if there are several subsets with the same minimal
value of
    % HSIC
    k = find(s == smin);
    if length(k) > 1
        % Randomly choose one of them
        ii = randsample(k,1);
    end

    % Exclude the iith subset from S
    for j = 1:length(ind{ii})
        S(S == ind{ii}(j)) = -1;
    end
    S(S < 0) = [];

    % Add the iith subset to the list of selected features
    selected_features = [selected_features ind{ii}];
end % end while

% Form the list of M selected features
selected_features = selected_features(end-M+1:end);

return




function HSIC = compute_HSIC(X, kernel_x, krn_x_params, L)

% Compute the HSIC value
% See p.825 (Property II) of "Supervised feature selection via
% dependence estimation"
m = size(X,2);
I = ones(m,1);
K = compute_kernel(X, kernel_x, krn_x_params);
K = K - diag(diag(K));
HSIC = trace(K*L) + (I'*K*I*I'*L*I)/((m-1)*(m-2)) -
2*(I'*K*L*I)/(m-2);
HSIC = HSIC/(m*(m-3));

return
```

```
function K = compute_kernel(X, kernel, krn_params)

% Compute kernel matrix for data X, given kernel parameter krn_
params.
% Kernel matrix K is of NxN elements
% For high-dimensional microarray data, K is rather small

switch lower(kernel)
    case 'linear', % linear kernel
        K = X'*X;
    case 'rbf', % Gaussian or RBF (Radial Basis Function) ker-
nel
        % First, compute a matrix of square distances
        % See http://www.cs.toronto.edu/~roweis/lle/code/lle.m
        X2 = sum(X.^2,1);
        N = size(X,2);
        distance = repmat(X2,N,1) - repmat(X2',1,N) + 2*X'*X;
        % Now compute kernel matrix itself
        K = exp(-krn_params(1)*distance);
    otherwise,
        error('Unsupported kernel!');
end

return
```

REFERENCES

Bedo, J., Sanderson, K., & Kowalczyk, A. (2006). An efficient alternative to SVM based recursive feature elimination with applications in natural language processing and bioinformatics. In A. Sattar, & B. H. Kang (Ed.), *Proceedings of the 19th Australian Joint Conference on Artificial Intelligence, Hobart, TAS. Lecture Notes in Computer Science 4304* (pp. 170-180). Berlin/Heidelberg: Springer-Verlag.

Borgwardt, K., Gretton, A., Rasch, M., Kriegel, H.-P., Scholkopf, B., & Smola, A. (2006). Integrating structured biological data by kernel maximum mean discrepancy. *Bioinformatics (Oxford, England)*, *22*(14), e49–e57. doi:10.1093/bioinformatics/btl242

Cristianini, N., Kandola, J., Elisseeff, A., & Shawe-Taylor, J. (2003). *On optimizing kernel alignment*. UC Davis, Department of Statistics.

Ein-Dor, L., Zuk, O., & Domany, E. (2006). Thousands of samples are needed to generate a robust gene list for predicting outcome in cancer. *Proceedings of the National Academy of Sciences of the United States of America*, *103*(15), 5923–5928. doi:10.1073/pnas.0601231103

Everitt, B. (2006). *The Cambridge dictionary of statistics* (3rd ed.). Cambridge, UK: Cambridge University Press.

Gretton, A., Bousquet, O., Smola, A., & Schölkopf, B. (2005). Measuring statistical dependence with Hilbert-Schmidt norms. In S. Jain, H.-U. Simon, & E. Tomita (Ed.), *Proceedings of the 16th International Conference on Algorithmic Learning Theory, Singapore* (LNCS 3734, pp. 63-77). Berlin/Heidelberg: Springer-Verlag.

Guyon, I., & Elisseeff, A. (2003). An introduction to variable and feature selection. *Journal of Machine Learning Research*, *3*, 1157–1182. doi:10.1162/153244303322753616

Hastie, T., Tibshirani, R., & Friedman, J. (2001). *The elements of statistical learning*. New York, NY: Springer-Verlag.

Song, L., Bedo, J., Borgwardt, K. M., Gretton, A., & Smola, A. J. (2007). Gene selection via the BAHSIC family of algorithms. *Bioinformatics (Oxford, England)*, *23*(13), i490–i498. doi:10.1093/bioinformatics/btm216

Song, L., Smola, A., Gretton, A., Borgwardt, K., & Bedo, J. (2007). Supervised feature selection via dependence estimation. In Z. Ghahramani (Ed.), *Proceedings of the 24th International Conference on Machine Learning, Corvallis, OR* (pp. 823-830).

Steeb, W.-H. (2006). *Problems and solutions in introductory and advanced matrix calculus*. Singapore: World Scientific.

Tibshirani, R., Hastie, T., Harasimhan, B., & Chu, G. (2003). Class prediction by nearest shrunken centroids, with applications to DNA microarrays. *Statistical Science*, *18*(1), 104–117. doi:10.1214/ss/1056397488

Tibshirani, R., Hastie, T., Narasimhan, B., & Chu, G. (2002). Diagnosis of multiple cancer types by shrunken centroids of gene expression. *Proceedings of the National Academy of Sciences of the United States of America, 99*(10), 6567–6572. doi:10.1073/pnas.082099299

van 't Veer, L. J., Dai, H., van de Vijver, M. J., He, Y. D., Hart, A. A., & Mao, M. (2002). Gene expression profiling predicts clinical outcome of breast cancer. *Nature, 415*, 530–536. doi:10.1038/415530a

ENDNOTES

[1]   In pattern recognition, a feature space is an abstract space where pattern is represented as a point (http://en.wikipedia.org/wiki/Feature_space). Each pattern is described by an n dimensional vector of features, with features playing the role of coordinates of the space; hence the name 'feature space'.

[2]   An inner product is a mathematical structure associating each pair of vectors x and y with a scalar computed as $\|x\|\|y\|\cos\alpha$, where $\|.\|$ is the norm of a vector, α is the angle between x and y, and $\cos$ is the cosine function. The norm is the Euclidean length of a vector and should not be confused with another length counting the number of elements in a vector. The norm of an n dimensional vector x is defined as $\sqrt{x_1^2 + x_2^2 + \cdots + x_n^2}$. The inner product of vectors x and y is denoted as $\langle x, y \rangle$. In MATLAB®, the norm of a vector is computed by the built-in function *norm* while vector length can be found with the help of the function *length*. So simple!

[3]   Here, an image means the image of an element of some domain, which, in its turn, means the output of a certain function evaluated at that element. If x is an element and f is a function, then $f(x)$ is the image of x under f.

[4]   For binary classification considered in this book, y takes two values: 0 and 1. Hence, computing the expectation amounts to the summation of two products: 0 times $Pr(Y = 0)$ and 1 times $Pr(Y = 1)$. As a result, the expectation is reduced to $Pr(Y = 1)$, i.e. one needs to count the fraction of times when y took the value of 1.

[5]   For an $N \times N$ matrix X, its trace is computed as $\sum_{i=1}^{N} X_{ii}$; X can be the product of two or more matrices. If both A and B are $N \times N$ matrices, then their product is also an $N \times N$ matrix.

[6]   Pay attention that in the previous section, the formula for a biased estimator is given. The present formula is therefore a slight modification of that formula for the unbiased estimator.

[7] $P\left(x|y=c\right)$ is the probability of observing feature x, given label $y=cy=c$

.

Chapter 11
Extreme Value Distribution Based Gene Selection

BLEND OF ELEMENTS OF EXTREME VALUE THEORY AND LOGISTIC REGRESSION

(Li, Sun, & Grosse, 2004) introduced the idea of using extreme value distribution theory for gene selection based on logistic regression. Each gene is modeled by means of logistic regression separately from other genes. This discriminant method is preferred over such a simple method as the *t*-test because logistic regression does not assume that gene expression levels are normally distributed. As a result, logistic regression is more robust to outliers than the *t*-test.

In statistical modeling, one of the central notions is the likelihood which is the probability of a set of observations given some parameter or parameters θ (Everitt, 2006). For a random sample consisting of n observations, $x_1, x_2, \ldots, x_n$ with probability distribution $f(x, \theta)$, the likelihood is defined as (Everitt, 2006)

DOI: 10.4018/978-1-60960-557-5.ch011

$$L = \prod_{i=1}^{n} f\left(x_i, \theta\right).$$

The maximum likelihood principle says that out of all possible values of θ one should choose the one maximizing the likelihood L. This means that one needs first to write down the mathematical expression for L, then to take the derivative $\dfrac{dL}{d\theta}$ and set the result to zero. Good examples of how to utilize the maximum likelihood principle in practice are provided in Chapter 2 in (Roff, 2006).

In statistical models such as logistic regression, the typical way to perform gene selection is to compare the maximum likelihood of the model given the real data and the expected maximum likelihood of the model given an ensemble of surrogate data with randomly permuted labels[1]. The computational bottleneck is the second likelihood due to the very large number of possible permutations. Li et al. proposed to replace this step with another one involving extreme value statistics based on which two gene selection criteria are introduced. In other words, in the approach of Li et al. the maximum likelihood of each gene in the real data is compared with the maximum likelihood of the *top-ranking* (hence, the extreme value theory emerges here) gene in the label-permuted data. Therefore, numerous calculations of single-gene likelihoods in the surrogate data are replaced with the calculation of the top-ranking gene likelihood carried out only once.

As follows from its name, this gene selection method rests on three components requiring definition: extreme values, extreme value distribution, and logistic regression.

Elements of Extreme Value Theory

Extreme values are the smallest and largest values in a given sample of observations. Analysis of extreme values is important in such areas as financial risk estimation and prediction of natural disasters. Extreme values are often associated with rare events (Falk, Hüsler, & Reiss, 2004), (Albeverio, Jentsch, & Kantz, 2006), (de Haan & Ferreira, 2006). For instance, the world financial crisis that I together with the rest of humankind eyewitness in time of writing this book fortunately does not happen every second year. Normally it is a quite rare event as can be seen from human history. On the other hand, exceptional prosperity (another extreme which is opposite to crisis) is also not everlasting ("nothing good lasts forever"). What lies in between these two extremes is ordinary, routine life that makes us all either happy or unhappy.

Extreme value distribution (EVD) is therefore the probability distribution of extreme values. Emil Julius Gumbel, a German mathematician, laid foundations of EVD theory in the 1950s (Gumbel, 1954), (Gumbel, Statistics of extremes, 1958). Extreme value distributions are often used to model the smallest or largest value among a large set of independent, identically distributed random variables. These extreme values are located in left and right "tails" of a distribution. Since the density in such regions is usually low, parametric distribution models might poorly fit to the data in the tails. This is where EVD apparatus comes to rescue. The Gumbel distribution, the Generalized Extreme Value distribution and the Generalized Pareto Distribution are just tips of this quickly growing branch of statistics. For more information about the EVD and its applications, see MATLAB® Statistics Toolbox™ 7.0 and *Extremes*, Springer-Verlag published journal (http://www.springer.com/statistics/journal/10687) as well as the books cited in this chapter.

Three examples of the probability density function (PDF) for the EVD of type I also known as the Gumbel distribution are shown in Figure 1 for several values of location (μ) and scale (σ) parameters[2]. One can notice that the PDF remains skewed to the left, regardless parameter values. Another observation that can come into mind is a certain similarity to the skewed normal distribution. The MATLAB® function *evpdf* was used for computing the PDF. MATLAB® code for producing Figure 1 is given in section "MATLAB Code" (*plotEVD*). The mean and variance of each EVD are also calculated there, given location and scale parameters.

Another MATLAB® function *gevpdf* (Generalized EVD) was used to generate the PDFs for the EVD of type III (Weibull-like distribution) in Figure 2 for different values of location and scale parameters and shape parameter set to -1 (type II EVD can be obtained with positive values of the shape parameter). One can observe the abrupt fall of each PDF to zero: this fall always happens if a random variable exceeds a certain threshold determined by location, scale, and shape parameters (see p.4 in (Kotz & Nadarajah, 2000) for details). MATLAB® code for producing Figure 2 can be found in section "MATLAB Code" (*plotGEVD*). The mean and variance of each EVD are also calculated there, given location, scale and shape parameters.

Finally, three Generalized Pareto PDFs are generated using the MATLAB® function *gppdf* and plotted in Figure 3 for different values of location, scale, and shape (tail index) parameters. The Generalized Pareto is a right-skewed distribution. MATLAB® code for producing Figure 3 can be found in section "MATLAB Code" (*plotGP*). The mean and variance of each EVD are also calculated there, given location, scale and shape parameters. If the shape parameter is larger than or equal to 1, the mean of the Generalized Pareto distribution is infinite. If the shape parameter is larger than or equal to 0.5, the variance of the Generalized Pareto distribution

Figure 1. Examples of the PDF for the EVD of type I (Gumbel distribution)

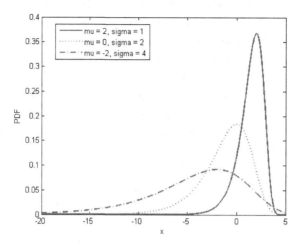

Figure 2. Examples of the PDF for the EVD of the type III (Weibull-like distribution)

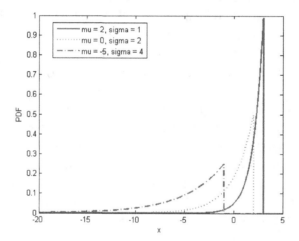

is infinite. Infinite values of the mean and variance are produced in the third example in *plotGP*, where the tail index is set to 5.

Logistic Regression

Let us introduce some useful notation. Let us index the samples by i and genes by j. Let N be the number of samples in a dataset, p be the total number of genes, x

Figure 3. Examples of the Generalized Pareto PDF

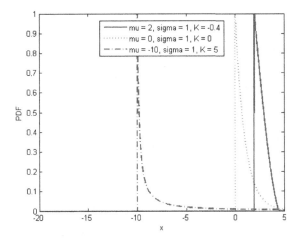

be the expression level, and y be the sample label value, which can be either 0 or 1.

Logistic regression is a form of regression analysis used when the response variable is binary (Everitt, 2006). It can be written as

$$\log \frac{\mu}{1-\mu} = b_0 + b_1 x_1 + \cdots + b_p x_p,$$

where $\mu = Pr\left(response\, variable = 1\right)$ and $x_1, x_2, \ldots, x_p$ are predictor variables. Parameters $b_0, b_1, \ldots, b_p$ of the regression model can be estimated by means of maximum likelihood.

The single-gene logistic regression model M_j of gene j is defined by the conditional probabilities of the sample label y_i, given the expression levels x_{ij}:

$$Pr\left(y_i = 1 | x_{ij}\right) = \frac{1}{1 + e^{-a_j - b_j x_{ij}}},$$

for $i = 1, 2, \ldots, N$ and $j = 1, 2, \ldots, p$. Here a_j and b_j are parameters of the logistic regression to be estimated from all samples $i = 1, 2, \ldots, N$. The maximum likelihood below measures how well the model M_j fits to the data

$$\hat{L}\left(D|M_j\right) = \max_{a_j,b_j} \prod_{i=1}^{N}\left[Pr\left(y_i = 1|x_{ij}\right)\right]^{y_i}\left[1 - Pr\left(y_i = 1|x_{ij}\right)\right]^{1-y_i},$$

where D denotes the data.

Since each gene is assigned a logistic regression model, the gene selection problem is transformed into the task of selection of single-gene logistic regression models associated with large maximum likelihoods.

Maximum Likelihood for the Surrogate Data

Let $\hat{L}\left(D_0|M_j\right)$ be the maximum likelihood under the single-gene logistic regression model M_j and D_0 is the surrogate data. The Wilks theorem[3] (Wilks, 1938) says given that the data are generated by the null model M_0, the asymptotic distribution of the 2-log-likelihood ratio is the χ^2 distribution with df degrees of freedom, where df is equal to the difference of the number of parameters in models M_j and M_0. Applied to the task considered, theorem states that when $N \to \infty$ (one should keep in mind that here Li et al. made an extrapolation to the large sample size since for microarray data the number of samples is rather small; whether such an extrapolation is valid in practice or not depends on the data at hand),

$$2\log\hat{L}\left(D_0|M_j\right) = 2\log\hat{L}\left(D_0|M_0\right) + t,$$

where t is a random variable sampled from a χ^2 distribution with df degrees of freedom.

We choose the model M_0 to be the same for all genes, i.e. $Pr\left(y_i = 1 \mid x_{ij}\right) = c$ for all $j = 1,..., p$. The maximum likelihood estimate of c is simply the percentage of samples labeled as 1, i.e. $\hat{c} = N_1 / N$. The maximum likelihood under M_0 is therefore

$$\hat{L}\left(D_0 \mid M_0\right) = \hat{c}^{N_1}\left(1 - \hat{c}\right)^{N-N_1}$$

according to the definition of the likelihood.

Taking the logarithm of this expression leads to

$$\log\hat{L}\left(D_0 \mid M_0\right) = -NH,$$

where

$$H = -\frac{N_1}{N} \log \frac{N_1}{N} - \frac{N - N_1}{N} \log \frac{N - N_1}{N}.$$

Because the ratio N_1 / N is the same in D and D_0 (label permutation does not change the number of samples with label 1), $\hat{L}(D \mid M_0) = \hat{L}(D_0 \mid M_0)$.

Applying the logistic regression model to the surrogate data (see the first equation in this section) and assuming $N \to \infty$, we obtain for the best single-gene maximum log-likelihood that

$$\max_{j} \left[\log \hat{L}(D_0 \mid M_j) \right] = \max_{j} \left[\log \hat{L}(D_0 \mid M_0) + \frac{t_j}{2} \right] = -NH + \frac{1}{2} \max_{j} \left[t_1, t_2, \ldots, t_p \right],$$

where t_1, t_2,..., t_p are p random variables sampled from a χ^2 distribution with df degrees of freedom. In our case, $df = 2\text{-}1 = 1$ since M_j has 2 parameters (a_j and b_j), and M_0 has only one parameter c.

Main Idea

In this section, we present theoretical arguments reproduced from (Li, Sun, & Grosse, 2004) that link the material in the previous section with EVD.

Let us define $T_p \equiv \max\left[t_1, t_2, \ldots, t_p \right]$ (here is the extreme value! gotcha!!!). Based on inequality in (Casella & Berger, 2002):

$$\sqrt{\frac{2}{\pi}} \frac{\sqrt{t}}{1+t} e^{-\frac{t}{2}} \leq Pr\left(t_i \geq t \right) \leq \sqrt{\frac{2}{\pi}} \frac{1}{\sqrt{t}} e^{-\frac{t}{2}}$$

and by defining

$$c_p \equiv \log \frac{p^2}{\pi \log p},$$

one can find that for asymptotically large p, the cumulative distribution of $v_p = \left(T_p - c_p \right) / 2$ converges to the double exponential function[4]:

$$F_v\left(x\right) = \lim_{p\to\infty} F_{v_p}\left(x\right) \equiv \lim_{p\to\infty} Pr\left(\frac{T_p - c_p}{2} \le x\right) = exp\left(-e^{-x}\right).$$

The double exponential function is nothing but the standard (location parameter = 0, scale parameter = 1) Gumbel distribution of the χ^2 distributed variables.

From the asymptotic distribution $F_v\left(x\right)$, one can compute the mean, $E\left[v\right]$, which is equal to γ (the Euler or Euler-Mascheroni constant[5] ≈ 0.5772)[6]. If one runs the MATLAB® function *evstat* with the location parameter set to 0 and the scale parameter set to 1, then the mean will be

-0.5772 (sign '-' results from a slightly different than in other sources (see, for example, http://en.wikipedia.org/wiki/Gumbel_distribution) way of computing the cumulative density function in MATLAB® *evcdf*, and it should not bother us).

For $T_p = c_p + 2v_p$, we have that

$$E\left[T_p\right] \approx 2\log p - \log\log p - \log\pi + 2\gamma.$$

Based on the EVD of T_p, two gene selection criteria are proposed in the next section.

Algorithm Description

Gene Selection Based on the E-Value of the EVD

The first criterion is called the E-criterion, because as you will see, it relies on the expected value. It compares the maximum likelihood of each gene obtained from the real data with the expected value of the maximum likelihood of the top-ranking gene from the surrogate data. When transformed into a criterion for the log-likelihood ratio, it becomes for each gene $j = 1, 2, \ldots, p$:

$$t_j = 2\log\frac{\hat{L}\left(D|M_j\right)}{\hat{L}\left(D|M_0\right)} = 2\log\frac{\hat{L}\left(D|M_j\right)}{\hat{L}\left(D_0|M_0\right)} = 2\log\hat{L}\left(D|M_j\right) + 2NH.$$

Here we used the rule saying that the logarithm of the ratio is equal to the difference of two logarithms:

$$2 \log \frac{\hat{L}\left(D|M_j\right)}{\hat{L}\left(D_0|M_0\right)} = 2 \log \hat{L}\left(D|M_j\right) - 2 \log \hat{L}\left(D_0|M_0\right),$$

and the following results obtained earlier in this chapter:

$$\hat{L}\left(D|M_0\right) = \hat{L}\left(D_0|M_0\right),$$

$$\log \hat{L}\left(D_0 \mid M_0\right) = -NH.$$

After computing all t_j's, we order them such that $t_{(1)} \geq t_{(2)} \geq t_{(3)} \cdots \geq t_{(p)}$ and declare genes $j = 1, 2, \ldots, J$ as differentially expressed if

$$t_{(J)} \geq E\left[T_p\right] = 2 \log p - \log \log p - \log \pi + 2\gamma > t_{(J+1)}.$$

In other words, the expected value of T_p acts as a threshold separating two types of genes represented by the sorted t_j values. Pay attention to the fact that though the logistic regression model has been fitted for each individually, a subset of genes is selected once the E-criterion is satisfied.

Gene Selection Based on the P-Value of the EVD

The second criterion for gene selection is called the P-criterion since it compares the P-value[7] of the maximum likelihood of each gene obtained from the real data based on the distribution of the maximum likelihood of the top-ranking gene from the surrogate data. Usually, one fixes the P-value to a small constant P_0, which can be, for example, 0.01 or 0.001.

That is, for each gene $j = 1, 2, \ldots, p$, we first calculate t_j as in the E-criterion, i.e.,

$$t_j = 2 \log \hat{L}\left(D|M_j\right) + 2NH.$$

and order them to $t_{(j)}$, then convert them to $v_{(j)} = \left(t_{(j)} - c_p\right) / 2$. Genes $j = 1, 2, \ldots, J$ are declared as differentially expressed if

$$1 - exp\left(-e^{-v_{(J)}}\right) \leq P_0 \leq 1 - exp\left(-e^{-v_{(J+1)}}\right).$$

Again, as in case of the E-criterion, a subset of genes is selected despite the individual gene modeling. Choosing P_0 very small focuses attention on the tail of the EVD. This is in contrast to the E-criterion, which considers the middle-range of the EVD due to the mean value. As a result, using the P-criterion will lead to fewer selected genes, compared to the E-criterion. Nevertheless, I would like readers to exercise caution when relying on the P-value, since this criterion might be less useful than it appears. Hsing et al. carried out research on the relation between the P-value and classifier error estimates for microarray data (Hsing, Attoor, & Dougherty, 2003). They studied gene expression-based classification when the label permutation was done as in (Li, Sun, & Grosse, 2004). I give the exact citation from (Hsing, Attoor, & Dougherty, 2003) on page 12 to state their point:

Is the p value useful in ranking the genes in a common study using a common classifier?

In variable-selection problems in regression, it is common to use a p value as a basis for identifying the predictors that have superior predictive powers in predicting the response. A small p value corresponds to high predictive power. ... In our setting, genes in a common study can be ranked according to p values based on a common classifier; however, the issue is whether this ranking is significantly different from that based on error rates. Conceivably these rankings will be somewhat different in some problems, in which case one has to provide a careful judgment as to whether it makes sense theoretically or practically to include p values as part of the selection process.

Thus, in my opinion, a reader must carefully think if the P-criterion would lead to informative results or not before applying it in practice. Of course, in case of any doubts, it is always better to abstain from using this criterion and restrict attention only to the E-criterion.

Based on experimental results in (Li, Sun, & Grosse, 2004), it might be good to apply either criterion considered in this chapter as a pre-filter prior to a more sophisticated algorithm, since both criteria return quite many genes among selected.

MATLAB Code

The first three MATLAB® functions (*plotEVD*, *plotGEVD* and *plotGP*) serve only explanatory purpose as one could already see. Thus, I do not comment them anymore.

Our interest is ***EVDGeneSelection*** implementing the ideas described. In this function, the key point to pay attention to is how the logistic regression parameters and conditional probabilities are computed. The parameters a_j and b_j of the logistic regression model for gene j are returned by the MATLAB® function ***glmfit*** with parameters 'logit' (logistic regression) and 'binomial' (default distribution type for y_i; recall that y_i is a binary variable). This function implements generalized linear model regression. Another MATLAB® function, ***glmval***, is then invoked in order to calculate the conditional probabilities $Pr\left(y_i = 1 \mid x_{ij}\right)$ of the sample label y_i, given the expression levels x_{ij}, based on the result of ***glmfit***. Finally, in order to quickly calculate the maximum likelihood, the call to the MATLAB® function ***prod*** is used, which multiplies all elements of its input parameter, which is a vector in this case. In other words, if a vector Z consists of n elements $z_1, z_2, \ldots, z_n$, then the result of the call $prod(Z)$ is $z_1 \times z_2 \times \ldots \times z_n$, where symbol $\times$ stands for ordinary multiplication (its more common meaning is vector product but here it was utilized for another purpose in order to simplify the exact meaning of what ***prod*** does).

```
function plotEVD
% Plot the three PDFs for the extreme value distribution with
different
% location and scale parameter values

x = [-20:.01:5];
plot(x, evpdf(x,2,1),'-', x, evpdf(x,0,2),':', x,
evpdf(x,-2,4),'-.', 'LineWidth', 2);
legend({'mu = 2, sigma = 1', 'mu = 0, sigma = 2', 'mu = -2,
sigma = 4'}, 'Location','Best')
xlabel('x');
ylabel('PDF');

% Get the mean and variance of the extreme value distribution
[M,V] = evstat(2,1);
fprintf(1, 'Mean = %f, Variance = %f\n', M,V);
[M,V] = evstat(0,2);
fprintf(1, 'Mean = %f, Variance = %f\n', M,V);
[M,V] = evstat(-2,4);
fprintf(1, 'Mean = %f, Variance = %f\n', M,V);

return
function plotGEVD
```

```
% Plot the three PDFs for the generalized extreme value distri-
bution
% with different location and scale parameter values

% Pay attention that in contrast to evpdf, the third parameter
is sigma % and the fourth parameter is mu!
x = [-20:.01:5];
plot(x, gevpdf(x,-1,1,2),'-', x, gevpdf(x,-1,2,0),':', x,
gevpdf(x,-1,4,-5),'-.', 'LineWidth', 2);
legend({'mu = 2, sigma = 1', 'mu = 0, sigma = 2', 'mu = -5,
sigma = 4'}, 'Location','Best')
xlabel('x');
ylabel('PDF');

% Get the mean and variance of the generalized extreme value
% distribution
[M,V] = gevstat(-1,1,2);
fprintf(1, 'Mean = %f, Variance = %f\n', M,V);
[M,V] = gevstat(-1,2,0);
fprintf(1, 'Mean = %f, Variance = %f\n', M,V);
[M,V] = gevstat(-1,4,-5);
fprintf(1, 'Mean = %f, Variance = %f\n', M,V);

return
function plotGP

% Plot the three PDFs for the generalized Pareto distribution
with
% different location, scale, and shape parameter values

x = [-20:.01:5];
plot(x, gppdf(x,-.4,1,2),'-', x, gppdf(x,0,1,0),':', x,
gppdf(x,5,1,-10),'-.', 'LineWidth', 2);
legend({'mu = 2, sigma = 1, K = -0.4', 'mu = 0, sigma = 1, K =
0', 'mu = -10, sigma = 1, K = 5'}, 'Location','Best')
xlabel('x');
ylabel('PDF');

% Get the mean and variance of the generalized extreme value
```

```
% distribution
[M,V] = gpstat(-.4,1,2);
fprintf(1, 'Mean = %f, Variance = %f\n', M,V);
[M,V] = gpstat(0,1,0);
fprintf(1, 'Mean = %f, Variance = %f\n', M,V);
% If the first parameter is greater than or equal to 1, the
mean is not
% finite, i.e., M = Inf
% If the first parameter is greater than or equal to 0.5, the
variance % is not finite, i.e., V = Inf
[M,V] = gpstat(5,1,-10);
fprintf(1, 'Mean = %f, Variance = %f\n', M,V);


return
function selected_genes = EVDGeneSelection(train_patterns,
train_targets, criterion, p0)


% Extreme Value Distribution based gene selection using logis-
tic
% regression
% Original source:
% Wentian Li, Fengzhu Sun, Ivo Grosse, "Extreme value distribu-
tion
% based gene selection criteria for discriminant microarray
data
% analysis using logistic regression", Journal of Computational
Biology % 11(2-3), pp.215--226, 2004
%
% Inputs:
%   train_patterns      - Training patterns (DxN matrix, where
D - data
%                                    dimensionality, N -
number of patterns)
%   train_targets       - Training targets (1xN vector)
%   criterion           - Criterion for gene selection (charac-
ter):
%                                    either 'E' or 'P'
%   p0                  - P-value (small real number, e.g. 0.01
or
%                                    0.001)
```

```
%
% Outputs:
%    selected_features   - Selected genes (vector of gene indi-
ces)



% Default values
if nargin < 4
    p0 = 0.01;
end
if nargin < 3
    criterion = 'E';
end

% Check if the criterion for gene selection is correct
switch criterion
    case 'E'
        fprintf(1, 'E-value criterion is chosen\n');
    case 'P'
        fprintf(1, 'P-value criterion is chosen\n');
    otherwise
        error('Unknown criterion!');
end

% Transform to the form used in the original article
train_patterns = train_patterns';
train_targets = train_targets';

% Check if train_targets is a binary vector
% This check is needed for glmfit function below
u = unique(train_targets);
if ~isequal(u, [0 1]')
    error('Elements of train_targets must be either 0 or 1!');
end

% Get the number of samples, n, and the number of genes, p
[n,p] = size(train_patterns);

% Compute entropy
```

```
n1 = length(find(train_targets == 1));
h = -(n1/n)*log(n1/n) - ((n-n1)/n)*log((n-n1)/n);

% Preallocate memory for regression coefficients and condition-
al
% probabilities
b = zeros(2,p);
pr = zeros(n,p);
% Preallocate memory for likelihood and log-likelihood ratio
values
L = zeros(1,p);
t = zeros(1,p);

% Build the single-gene logistic regression models
for j = 1:p
    % Compute regression coefficients
    b(:,j) = glmfit(train_patterns(:,j), train_targets, 'bino-
mial', 'logit');
    % Compute the conditional probability according to Eq.1 of
Li et
    % al. article
    pr(:,j) = glmval(b(:,j), train_patterns(:,j), 'logit');
    % Compute the maximum likelihood estimate
    L(j) = prod(pr(:,j).^train_targets)*prod((1-pr(:,j)).^(1-
train_targets));
    t(j) = 2*log(L(j)) + 2*n*h;
end

% Sort log-likelihood values in descending order of magnitude
[t,idx] = sort(t, 'descend');

% Compute the mean of the extreme value distribution of
% chi-square-distributed random variables, where 0.5772 is the
Euler
% constant (see Eq.9 in Li et al. article)
m = 2*log(p) - log(log(p)) - log(pi) + 2*0.5772;

% Select genes, depending on the criterion
switch criterion
    case 'E'
```

```
        for j = 1:p-1
            if t(j) >= m && t(j+1) < m
                % Select genes with indices idx(1) to idx(j)
                selected_genes = idx(1:j); break;
            end
        end
    case 'P'
        cp = log(p^2/(pi*log(p)));
        v = 0.5*(t - cp);
        for j = 1:p-1
            if 1-exp(-exp(-v(j))) <= p0 && p0 < 1-exp(-exp(-
v(j+1)))
                % Select genes with indices idx(1) to idx(j)
                selected_genes = idx(1:j); break;
            end
        end
end

return
```

REFERENCES

Albeverio, S., Jentsch, V., & Kantz, H. (Eds.). (2006). *Extreme events in nature and society*. Berlin, Heidelberg: Springer-Verlag. doi:10.1007/3-540-28611-X

Casella, G., & Berger, R. L. (2002). *Statistical inference* (2nd ed.). Belmont, CA: Duxbury Press.

de Haan, L., & Ferreira, A. (2006). *Extreme value theory: an introduction.* New York, NY: Springer Science+Business Media.

Everitt, B. (2006). *The Cambridge dictionary of statistics* (3rd ed.). Cambridge, UK: Cambridge University Press.

Falk, M., Hüsler, J., & Reiss, R.-D. (2004). *Laws of small numbers: extremes and rare events* (2nd ed.). Basel: Birkhäuser Verlag.

Gumbel, E. J. (1954). *Statistical theory of extreme values and some practical applications* (Vol. 33). Washington, DC: National Bureau of Standards, US Government Printing Office.

Gumbel, E. J. (1958). *Statistics of extremes*. New York: Columbia University Press.

Hsing, T., Attoor, S., & Dougherty, E. (2003). Relation between permutation-test P values and classifier error estimates. *Machine Learning*, *52*(1-2), 11–30. doi:10.1023/A:1023985022691

Kotz, S., & Nadarajah, S. (2000). *Extreme value distributions: theory and applications*. Singapore: World Scientific. doi:10.1142/9781860944024

Li, W., Sun, F., & Grosse, I. (2004). Extreme value distribution based gene selection criteria for discriminant microarray data analysis using logistic regression. *Journal of Computational Biology*, *11*(2-3), 215–226. doi:10.1089/1066527041410445

Roff, D. A. (2006). *Introduction to computer-intensive methods of data analysis in biology*. Cambridge, UK: Cambridge University Press. doi:10.1017/CBO9780511616785

Wilks, S. S. (1938). The large sample distribution of the likelihood ratio for testing composite hypothesis. *Annals of Mathematical Statistics*, *9*(1), 60–62. doi:10.1214/aoms/1177732360

ENDNOTES

[1] The motivation behind label permutation in gene selection and classification problems is that if a gene is indeed important for discrimination between classes of data, the permutation would lead to increased error rate. On the other hand, if a gene is irrelevant, classification error would stay relatively unchanged. However, as will be discussed later in this chapter, caution must be exercised regarding the validity of this assumption in practice.

[2] In case of EVD, location and scale parameters should not generally be confused (despite of similar notation) with the mean and standard deviation commonly used to describe other distributions such as the normal or binomial distribution.

[3] For those who are curious to read the Wilks' paper written long ago, i.e. in the last century, there is the following link: http://projecteuclid.org/DPubS/Repository/1.0/Disseminate?view=body&id=pdf_1&handle=euclid.aoms/1177732360

[4] I decided to omit the detailed proof that can be anyway found in (Li, Sun, & Grosse, 2004).

[5] See http://mathworld.wolfram.com/Euler-MascheroniConstant.html for detail information.

[6] In general, the mean of the EVD is equal to the location parameter plus the product of the Euler constant and the scale parameter (Everitt, 2006).

[7] The P-value is the probability of the observed data (or data showing a more extreme departure from the null hypothesis) when the null hypothesis is true (Everitt, 2006).

Chapter 12

Evolutionary Algorithm for Identifying Predictive Genes

EVOLUTIONARY SEARCH FOR OPTIMAL OR NEAR-OPTIMAL SET OF GENES

The overview of different feature selection methods would not be complete without mentioning evolutionary methods built on the principles of biological evolution involving natural selection and survival of the fittest. Evolutionary techniques perform stochastic search and optimization in the solution space. In contrast to many traditional optimization techniques, they involve a parallel search through the population of potential solutions.

In this chapter, we consider the evolutionary algorithm proposed in a number of articles (Deutsch J. M., 2001), (Deutsch J. M., 2003), (Jirapech-Umpai & Aitken, 2005). The algorithm in (Jirapech-Umpai & Aitken, 2005) is a modification of the

DOI: 10.4018/978-1-60960-557-5.ch012

statistical replication algorithm in (Deutsch J. M., 2001), (Deutsch J. M., 2003). Hence, we base our discussion on the former since it was the latest to appear.

The algorithm from (Jirapech-Umpai & Aitken, 2005) addresses the multi-class classification of microarray data, i.e. when the data are split into more than two classes. It should be noted, however, that it can be safely applied to binary classification problems as well.

Main Idea

The main idea of (Jirapech-Umpai & Aitken, 2005) and related methods is based on the genetic principles of evolution in the nature: selection, mutation, and survival of the fittest.

The evolutionary algorithm maintains a population of predictors (gene subsets) whose goodness for classification of microarray data is attested by a classifier. In the beginning, each predictor is created by means of the random selection of genes from the initial pool of genes. The initial pool of genes may include all genes in the study or just a subset of all genes, pre-selected by another technique, as was done in (Jirapech-Umpai & Aitken, 2005). As a classifier, any suitable algorithm can be utilized; in (Jirapech-Umpai & Aitken, 2005), k-nearest-neighbor (KNN) with the Euclidean distance function was chosen because of its simplicity and competitive performance, compared to more sophisticated algorithms (Dudoit & Fridlyand, 2003). The classifier returns the strength of each predictor, related to classification error attained when classifying the validation data with this predictor only.

At each iteration (generation) of the evolutionary algorithm, the replication of a particular predictor depends on how the mutation affects predictor's strength (scoring function). Mutation includes three alternatives: 1) adding one gene, randomly chosen from the initial pool of genes, 2) deleting one gene from a predictor, and 3) doing nothing, i.e. keeping the predictor unchanged. The selection process uses a statistical replication algorithm. The predictors which have a higher score after mutation survive in the next generation, while the worst predictor in the current generation is replaced with the best predictor from the previous generation. Iterations terminate when either the maximum number of generations is reached or all predictors deliver similar classification performance over a specific number of generations. After termination, the best scoring (most accurate) predictor is the one to be used for classification of test data.

Algorithm Description

The pseudo code of the evolutionary algorithm for gene selection is presented in Figure 1.

Figure 1. Pseudo code of the evolutionary algorithm for gene selection

---

Evolutionary Algorithm

Create the initial gene pool;
Let E_0 be the initial population of *NP* predictors $\{G_1, G_2, ..., G_{NP}\}$, where each predictor G_i is a gene subset of size *m*, and each subset is randomly sampled from the initial gene pool;
For $k := 1 : GEN$ /* for each generation */
  For $i := 1 : NP$ /* for each predictor */
    Compute the scoring (fitness) function of the current predictor $Fit(G_i)$;

    Create a new predictor G_i' by randomly mutating with probability p_m the genes in G_i according to one of the following possibilities:
Add (with probability p_c) an extra gene randomly chosen from the initial gene pool to G_i;
Delete (with probability $1 - p_c$) a randomly selected gene from G_i;
Keep the feature set the same, i.e. $G_i' := G_i$;

    Compute the scoring (fitness) function of the new predictor $Fit(G_i')$;

    Compute the difference, Δ_i, of the old and new scoring functions:
$\Delta_i := Fit(G_i') - Fit(G_i)$;

    Compute the weight for G_i': $w_i := e^{\beta \Delta_i}$, where β is the inverse temperature;
  End For

If the score of all predictors gives a standard deviation of less than 0.01 for n_g consecutive generations, then halt computations;

    Let *Z* be the sum of the weights, i.e. $Z := \sum_{i=1}^{NP} w_i$;

    Create a new population E_k by replicating all new predictors G_i' according to their normalized weight $w_i NP / Z$. With this normalized weight *w*, a predictor is replicated $[w]$ times and additional time with probability $w - [w]$, where $[w]$ is the largest integer less than *w*;

    Using the elitism technique, the worst predictor of the new population E_k is replaced with the best predictor from the previous generation;
End For

The first step, which can be optional, in my opinion, is to form the initial pool of genes[1]. In (Jirapech-Umpai & Aitken, 2005) it is done by using the RankGene software (http://genomics10.bu.edu/yangsu/rankgene/) written in C++. This software computes a number of feature ranking measures, including information gain, two-ing rule, Gini index, sum minority, max minority, and sum of variances. They quantify the usefulness of a gene by evaluating the strength of class predictability when the prediction is made by splitting the full range of expression values for a given gene into two non-overlapping regions (Su, Murali, Pavlovic, Schaffer, & Kasif, 2003). The split point is chosen so as to optimize the corresponding measure. Intuitively, a gene is likely to be diagnostic if its expression value in a disease state is different from its expression value in a normal state. In other words, the gene's predictive power to distinguish between different classes (e.g., normal versus cancer) is high. For each gene, RankGene computes a value that measures the ability of the gene to distinguish between the classes. It outputs the best k genes according to this measure, where k is specified by the user. Note that RankGene can miss dependencies between genes, and gene selection into the initial pool involves only training data.

Out of all measures, Jirapech-Umpai and Aitken chose information gain since it provided best ranking. From the experiments, they determined that the best size of the initial gene pool is 100.

Population size was fixed to NP predictors, and each predictor in any population consisted of m genes, randomly selected from the initial gene pool GP. In any single experimental set-up feature size (the number of genes per predictor) was, however, the same for all predictors.

In each generation, a scoring (fitness) function value is computed for each predictor G_i. This value is related to accuracy[2] attained by a trained classifier (k-nearest neighbors with the Euclidean distance to measure similarity) on a validation set. After that, mutation follows with probability p_m, which results in either one gene added to or removed from a predictor. The decision of whether to add or to delete a gene is made with probability p_c: if a random number is less than or equal to this probability, then an extra gene is added to the current predictor; otherwise, one gene is removed from the current predictor. A scoring (fitness) function is evaluated for a new (mutated) predictor G_i', and the difference of scores after and before mutation is calculated. This difference is then used to weigh the predictor G_i': the higher weight w_i is, the more times the predictor will be replicated, i.e. the higher its chances to survive in the next generation. Once all predictors were replicated, the worst (i.e. least accurate) predictor in the current generation is replaced with the best (i.e. most accurate) predictor in the previous generation, according to the elitism principle (only the fittest survives).

When the algorithm terminates (either because the maximum number of generations, *GEN*, has been achieved or because all predictors give a similar score over a number of consecutive generations), the best predictor from the last generation contains genes selected. These genes are to be used in order to classify the independent test set.

Reproducibility of Gene Selection Results

When an algorithm relies on random initialization like the one in this chapter, it is important to know how stable its outcome is. In other words, after a single run of the Evolutionary Algorithm one is still left with uncertainty if selected genes are indeed relevant or they were chosen purely by chance. In such cases, it is advisable to repeatedly run the algorithm and observe the frequency with which each selected gene appears. If a certain gene is consistently chosen as a member of a predictive set, it would suggest that the gene selection is reproducible despite the random initialization.

That is, one needs to find the best predictor, say 100 times (each time with a different initialization) in order to judge on the reproducibility of the results. Then the *z*-test for a population proportion (Sheskin, 2004), (see Test 9) can be used to statistically decide which gene was consistently chosen and which gene was selected by chance.

This test can be also considered as the binomial sign test for a single sample, which implies that an underlying test population comprises of two categories. The word 'binomial' means that this test is based on the binomial distribution. The assumption of this distribution is that each of independent observations randomly selected from a population can be classified in one of two mutually exclusive categories. The following simple example illustrates when the *z*-test can be used:

An experiment is conducted to determine if a coin is biased. The coin is flipped 100 times with 48 heads and 52 tails. Does this result indicate that the coin is biased?

Rephrasing this example for our case, we can formulate it as follows:

An experiment is conducted to determine if a certain gene is selected because of its relevance for classification rather than by chance. Running a gene selection algorithm 100 times led to 48 times when it was included into the best predictor and 52 times when it was not. Does this result indicate that this gene was selected due to its relevance rather than by chance?

Selecting a gene by chance means its observed proportion p among all selected genes is equal to 0.5 (in other words, the gene has fifty-fifty chances to be either relevant or irrelevant). Then the null hypothesis H_0 of the z-test can be stated as

$$H_0 : p < 0.5.$$

The alternative hypothesis H_1 of this test is then

$$H_1 : p > 0.5.$$

The test statistic z is given as

$$z = \frac{p - \pi_1}{\sqrt{\dfrac{\pi_1 \pi_2}{n}}},$$

where π_1 and π_2 are likelihoods of that a selected gene is relevant and irrelevant, respectively and n is the number of selection trials.

If we multiply both parts of a fraction by n, we obtain

$$z = \frac{np - n\pi_1}{\sqrt{n\pi_1 \pi_2}},$$

where $n\pi_1$ is the mean of the binomial distribution and $\sqrt{n\pi_1 \pi_2}$ is its standard deviation.

Since we assume that chances that a given gene is relevant are equal to chances that it is irrelevant, i.e. $\pi_1 = \pi_2 = 0.5$, then the last equation is turned to

$$z = \frac{pn - 0.5n}{0.5\sqrt{n}}.$$

For convenience of the test result interpretation, we will utilize the following expression for z:

$$z = \frac{p - 0.5}{\frac{0.5}{\sqrt{n}}}.$$

Then, it is said that one rejects[3] the null hypothesis if both following conditions are fulfilled:

the sign of z is positive and
the value of z is equal to or greater than the tabled critical one-tailed value z_α^{crit} at
    the pre-specified significance level α.

Usually, the significance level $\alpha = 0.05$. Therefore $z_{0.05}^{crit} = 1.65$ (Sheskin, 2004) for the one-tailed test. Hence, in order to be able to reject the null hypothesis, given that $n = 100$, we need to solve the following simple inequality:

$$\frac{p - 0.5}{\frac{0.5}{\sqrt{100}}} \geq 1.65.$$

After simplification, we obtain that p must be equal to or greater than 0.5825 in order to rule out a chance in selection of a certain gene. For the chosen value of n, it means that a gene must be included into the best predictor at least 59 times out of 100 in order to be treated as relevant for classification.

The z-test for a population proportion is implemented in the function *ztestpp*. The function *check_reproducibility* invokes a user-specified gene selection algorithm multiple times and returns the list of genes that passed the z-test, i.e. those genes that cannot be considered as selected by chance; in other words, these genes can be called consistently reproducible in the best predictor. Both functions are described in Section 'MATLAB Code'. It is worthy to mention that not only the Evolutionary Algorithm discussed in this chapter, but any suitable feature or gene selection algorithm can be employed to check gene reproducibility.

Wrap-Up Discussion on Feature Ranking and Evolutionary Algorithms for Feature Selection

Gene ranking is naturally not restricted to RankGene software: MATLAB® Bioinformatics Toolbox™ includes the function *rankfeatures* that ranks features by a given class separability criterion. This function assumes two-class problems. Possible class separability measures include the absolute value of a statistic of a two-sample *t*-test with pooled variance, Kullback-Leibler distance, minimum attainable classification error or Chernoff bound, area between the empirical Receiver Operating Characteristic (ROC) curve and the random classifier[4] slope[5], the absolute value of the statistic of a two-sample unpaired Wilcoxon test. Out of them, the first three assume normal distributed classes while the other two are nonparametric tests. However, all measures are based on distributional characteristics of classes (e.g., mean, variance) for a given feature. The detailed description of these measures is out of scope of this chapter. A brief yet good introduction to Kullback-Leibler divergence and Chernoff bound in two-class classification tasks is given in Section 10.1 of (Ripley, 2007). As a pointer to many statistical tests, the "Bible of statistical tests"[6] by David Sheskin (Sheskin, 2004) is warmly suggested as a very comprehensive source. As for ROC curves and various characteristics based on them, I advise to consult (Pepe, 2004). Readers are also encouraged to read help description of the following MATLAB® functions: *ttest2* (two-sample *t*-test), *ranksum* (Wilcoxon nonparametric test), *roc* (ROC), and *plotroc* (plotting ROC).

The function *rankfeatures* also allows ranking features, depending on correlation information and/or weight assigned to each feature. In these cases, features that are highly correlated with the features already selected and/or features that are close (by distance) to the already picked features will be less likely included into the output list of selected features. Three types of feature normalization (to ensure compatibility among different features) are also provided as an option in *rankfeatures*. For more details, see the description of *rankfeatures* in MATLAB® Bioinformatics Toolbox™. For the use of this function for cancer detection, readers are recommended to search for 'Cancer detection' in MATLAB® help and follow the link.

The papers of Josh Deutsch (Deutsch J. M., 2001), (Deutsch J. M., 2003) describe an alternative way to rank genes in order to reduce the number of genes from many thousands down to few hundreds. Deutsch's approach assumes two-class problems. For each gene in the initial gene pool, expression levels of this gene across training samples are considered. Training samples are then ranked (e.g,. by ordered from the largest expression level to the smallest one) and given the known class label associated with each sample, one looks for the pattern of intra-class separation of a gene. For example, one can find the following pattern:

*class*1, *class*2, *class*2, *class*1, *class*1, *class*2.

It can be seen that two classes alternate in this pattern and class separation is rather poor. If, however, one encounters a different pattern like

*class*1, *class*1, *class*1, *class*2, *class*2, *class*2,

then two classes are perfectly separated by expression levels of a given gene. We are highly interested in picking such genes to form the initial population for the Evolutionary Algorithm.

The concrete solution formulation is left to readers with a hint of checking statistical tests from (Sheskin, 2004).

If there are more than two classes of data, Deutsch advises to consider all distinct combinations of two classes and pick the best *m* genes from each combination, where *m* is a user-defined parameter. After that, all lists are combined together and common genes in several lists are replaced with one gene in order to exclude repetitions.

The Evolutionary Algorithm considered in this chapter is not the only representative of its family. The standard Genetic Algorithm (GA) has been successfully applied too (Li, Weinberg, Darden, & Pedersen, 2001), (Ooi & Tan, 2003), (Paul & Iba, 2004), (Huerta, Duval, & Hao, 2006), (Agrawal & Bala, 2007), (de Souza, Carvalho, & Ticona, 2007), (Hernandez, Duval, & Hao, 2007), (Alba, García-Nieto, Jourdan, & Talbi, 2007), (Dessì & Pes, 2009). The GA encodes a weight of each gene in a chromosome. This weight is frequently either 0 (gene is not selected) or 1 (gene is selected). Such binary chromosomes are then subjected to crossover and mutation. A classifier is then used in the same manner as explained above in order to judge if a new chromosome should survive or not.

In addition, a survey of evolutionary algorithms for classifying gene expression data (Wahde & Szallasi, 2006) can be useful to read.

MATLAB Code

This section contains comments on five functions: *create_initialpopulation*, *EvolutionaryGeneSelection*, *fitness_fcn*, *ztestpp*, and *check_reproducibility*.

The main function is *EvolutionaryGeneSelection*, which expects a lot of parameters to be defined. In addition to training data, validation data (*valid_patterns* and *valid_targets*) are required too, since a classifier decides on gene relevance. A classifier defined by a character string *classifier* is generic, e.g., it can be the name of the MATLAB® function such as *classify* or *knnclassify* or the name of a user-defined function. Classifier's parameters are given in *varargin*, which implies that there can be a variable number of parameters (see two examples in *EvolutionaryGeneSelection* after description of inputs/outputs). The values of other parameters as recommended in (Jirapech-Umpai & Aitken, 2005) are summarized in Table 1.

The initial gene pool *GP* is first used in order to create the initial population of *NP* predictors, each predictor containing *M* genes randomly sampled from *GP*. *GP* itself is a one-dimensional cell array, each element of which stores a vector containing indices of genes. If it is empty in the call of *EvolutionaryGeneSelection*, then all available genes form *GP*. Gene sampling is done in the function *create_initial-*

Table 1. Default values of input parameters of EvolutionaryGeneSelection

Parameter	Values
GEN	200
NP	10-50
M	10-50
prob_mut	0.7
prob_add	0.5
sigma_thr	0.01
ng_thr	10

population without replacement so that no gene is included in a predictor twice. However, different predictors may include the same gene or genes. The function *create_initialpopulation* thus returns a cell array, each element of which is a predictor (vector of gene indices).

Given the initial population of predictors, the fitness function (***fitness_fcn***) is called for each of them in order to assign a score to each predictor based on its classification performance on a validation set. The best predictor in the entire population is found and stored for the future use.

After this initialization step, the main iterations begin, with outer loop over generations and inner loop over predictors in each generation. First, each predictor is mutated with the probability ***prob_mut***, which means that a predictor might get or might not get mutated. That is, we are presented with two possible outcomes: mutation or its absence (in this case, a predictor is left unchanged). In its turn, this implies that each time when we decide whether to mutate or not, we carry out the Bernoulli trial with ***prob_mut*** being the success probability (Rubinstein & Kroese, 2008), p.63. Since the Bernoulli distribution is a special case of the binomial distribution when the number of trials is equal to 1, the MATLAB® function ***binornd*** can be utilized for random number generation. If the obtained random number is less than or equal to ***prob_mut***, then we achieved "success", which results in predictor mutation; otherwise, no mutation occurs.

The alternative way to trigger mutation is described in Algorithm 2.4.7 in (Rubinstein & Kroese, 2008). This algorithm given below generates a Bernoulli random variable:

Generate $U \sim U(0,1)$, where U is a uniform random number between 0 and 1.
If $U \le prob\_mut$, return $X = 1$ ('mutation'); otherwise, return $X = 0$ ('no mutation').

In (Jirapech-Umpai & Aitken, 2005), the probability of mutation is quite high (0.7), which implies that mutations are expected to be frequent in order to explore as much feature space as possible in the evolutionary search.

Mutation consists in either adding or deleting one gene from a predictor. It was advised in (Jirapech-Umpai & Aitken, 2005) to choose equal probabilities of addition or removal in order to make mutation in either direction equally likely. As a result, one of the two probabilities needs to be specified. We opted to specify the probability *prob_add* of addition of an extra gene, randomly selected from the initial gene pool *GP*. As with mutation, a uniform random number is generated and compared with this probability. If it is less than or equal to *prob_add*, then we add a gene to the current predictor; otherwise, we remove randomly chosen gene from this predictor. When adding a new gene, we exclude those genes that are already included into a predictor.

After mutation, the fitness function is invoked again and its value is returned in order to be compared with the value corresponding to the predictor state before mutation.

As the system evolves, the fitness function gives similar answers for all predictors. In order to improve convergence, it is useful to borrow the concept of "temperature" from the annealing techniques as done in (Deutsch J. M., 2001), (Deutsch J. M., 2003), where "temperature" is a function of the spread in scores. Deutsch found after testing a variety of schedules for the "temperature" that good results are obtained if this "temperature" is taken to be $\beta = 2 / (\triangle s)$, where $\triangle s$ is the maximum spread in scores between different predictors in the current population. In other words, $\triangle s = \max_i s_i - \min_i s_i, i = 1, \dots M$. Since it is not uncommon that $\triangle s$ may be equal to zero, in order to avoid program crushes due to 'Division by zero' error/warning, I added a very small positive constant ε to $\triangle s$.

The difference δ in scores after and before mutation and "temperature" are combined in definition of a weight w for each predictor: $e^{\beta \delta}$. Weights are used at later step for replicating predictors.

Once all predictors in the current population have been processed, the termination condition relying on two thresholds, *sigma_thr* and *ng_thr*, is checked. If the termination is reached, the algorithm halts. If either condition is not met, then the predictors are replicated to form the population for the next generation.

The statistical replication mechanism is borrowed from Deutsch's works. The replication is dependent on the normalized (between 0 and 1) weight w of a predictor. The higher the weight, the more times the predictor is replicated. Each predictor is replicated $[w]$ times and additional time with probability $w - [w]$, where $[w]$ is the largest integer less than w. It is worthy to mention that after replication, the number of predictors in the new population may be slightly less or slightly larger than *NP*. Hence, the *NP* correction is needed and done.

Finally, the worst predictor of a new population is replaced with the best predictor of the current population.

After algorithm termination, the best predictor in the last population is meant to contain relevant genes for classification of unseen test data. If several candidate predictors provide the same classification accuracy, choose the one with the fewest number of genes. If there are several such predictors, randomly select one of them.

Due to random initialization of the algorithm, it is not unnatural to expect that the composition of the best predictor would change from run to run, thus making results unstable and somewhat difficult to trust. To cope with this fact, one can run the algorithm multiple times, say 100, each time storing the best predictor as done in the function *check_reproducibility*. This function, as follows from its name, check if each gene got into the pool of the best predictors by chance due to random initialization of the gene selection algorithm or it is indeed important and relevant. Accumulating genes from all runs into one vector and removing duplicates provides the input for another function *ztestpp*, invoked from *check_reproducibility*. The function *ztestpp* implements the *z*-test for a population proportion that outputs either 0 (gene was selected by chance) or 1 (gene was not selected by chance). Its inputs are a proportion of times a given gene has been selected to the best predictor, population size, and statistical significance level (optional; default value is 0.05). Excluding genes that got zero leaves only reproducible genes for further analysis.

As an addition source of knowledge about MATLAB® implementation of evolutionary strategies for gene selection, a good example of using the GA for gene selection is given in MATLAB® Digest – November 2005 (Roberts, 2005). This example shows MATLAB® code relying on MATLAB® Statistics Toolbox™, MATLAB® Bioinformatics Toolbox™, and MATLAB® Genetic Algorithm and Direct Search Toolbox™. I highly recommend to look into this code to get yourself familiarized with main steps of the GA-based gene selection without the immediate need to procure this knowledge from books or/and long articles.

```
function G = create_initialpopulation(GP, NP, M)

% Create the initial population of predictors
%
% Inputs:
%   GP      - Initial gene pool (vector of indices)
%   NP      - The number of predictors
%   M       - The number of genes in each predictor
%
```

```
% Output:
%   G        - Initial population (cell array of predictors,
with each
%                predictor containing indices of genes belonging
to it)



% Allocate memory for the output
G = cell(1,NP);

% Form the initial population as a set of NP predictors
for i = 1:NP
    % Randomly select M genes from the initial gene pool
    % Sampling done uniformly but without replacement so that
no gene
    % is included into any predictor more than once
    % Each predictor G(i) is thus a subset of M genes
    G{i} = randsample(GP, M);
end

return
function selected_features = EvolutionaryGeneSelection(train_
patterns, train_targets, valid_patterns, valid_targets, GP,
GEN, NP, M, prob_mut, prob_add, sigma_thr, ng_thr, classifier,
varargin)

% Evolutionary feature selection
% Original sources:
% Thanyaluk Jirapech-Umpai and Stuart Aitken, "Feature selec-
tion and
% classification for microarray data analysis: evolutionary
methods for % identifying predictive genes", BMC Bioinformatics
6:148, doi:10.1186/1471-2105-6-148, 2005
% J. M. Deutsch, "Algorithm for finding optimal gene sets in
microarray % prediction", Retrieved February 1, 2009, from
arXiv.org:
% http://arxiv.org/abs/physics/0108011, 2001
% J. M. Deutsch, "Evolutionary algorithms for finding optimal
gene sets % in microarray prediction", Bioinformatics 19:1,
```

```
45-52, 2003.
%
% Inputs:
%   train_patterns   - Training patterns (DxN matrix, where D
- data
%                      dimensionality, N - number of patterns)
%   train_targets    - Training targets (1xN vector)
%   valid_patterns   - Validation patterns (DxNt matrix, where D
- data %                 dimensionality, Nt - number of patterns)
%   valid_targets    - Validation targets (1xNt vector)
%   GP               - Initial pool of genes (list of gene
indices)
%   GEN              - The number of generations
%   NP               - The number of predictors per generation
%   M                - The number of genes per predictor
%   prob_mut         - Probability of mutation of a predictor
%   prob_add         - Probability of gene addition to a predic-
tor
%   sigma_thr        - Sigma-threshold for algorithm termination
%   ng_thr           - Generation-number-threshold for algorithm
%                      termination
%   classifier       - Classifier (string with function name
where a
%                      classifier is implemented)
%   varargin         - Parameters of a classifier
%
% Output:
%   selected_features  - Selected features (vector of feature
indices)
%
% Examples:
% selected_features = EvolutionaryGeneSelection(train_patterns,
% train_targets, valid_patterns, valid_targets, '',
% 10000, 20, 15, 0.05, 0.1, 1, 100, 'NearestNeighbor', 3);
% selected_features = EvolutionaryGeneSelection(train_patterns,
% train_targets, valid_patterns, valid_targets, '',
% 10000, 20, 15, 0.05, 0.1, 1, 100, 'NaiveBayes');


% Get the number of genes
```

```
D = size(train_patterns,1);

% If the pool is empty, then fill it with indices of all genes
in a
% dataset
if isempty(GP)
    GP = 1:D;
end

% Create the initial population from the initial gene pool
G = create_initialpopulation(GP, NP, M);

% Set up flag used to check algorithm termination
flag = 0;

% Compute initial scores
for j = 1:NP
    s_prev(j) = fitness_fcn(train_patterns(G{j},:), train_tar-
gets, valid_patterns(G{j},:), valid_targets, classifier, vara-
rgin{:});
end
% Find the predictor with the best performance in the current
% generation
[s_best, j_best] = max(s_prev);
G_best = G{j_best};

% Iterate until termination
for i = 1:GEN % over generations
    fprintf(1, 'Generation %d...\n', i);
    for j = 1:NP % over predictors
        if binornd(1, prob_mut) % alternative is if rand <=
prob_mut
            % Mutate by either adding one gene from the initial
gene
            % pool or by deleting one gene from the current
predictor
            if rand <= prob_add
                % Add a gene to the predictor by randomly
sampling GP
                jj = randsample(GP,1);
```

```
                % Search if the same gene has been already
included
                % into the predictor. If yes, do nothing
                if isempty(find(G{j} == jj))
                    G{j} = [G{j} jj];
                end
            else
                % Delete a randomly selected gene from the
predictor
                jj = randsample(length(G{j}),1);
                G{j}(jj) = [];
            end
            % Compute score after mutation
            if ~isempty(G{j})
                s_cur(j) = fitness_fcn(train_patterns(G{j},:),
train_targets, valid_patterns(G{j},:), valid_targets, classi-
fier, varargin{:});
            else
                s_cur(j) = intmax;
            end
        else % no changes
            s_cur(j) = s_prev(j);
        end
        % Compute the difference of scores after and before
mutation
        delta(j) = s_cur(j) - s_prev(j);
        % Compute the annealing factor (so called inverse
        % "temperature")
        % See Deutsch's papers for explanation
        beta = 2/(max(s_prev) - min(s_prev) + eps);
        % Compute predictor's weight
        w(j) = exp(beta*delta(j));
    end % end for j

    % Check for algorithm termination
    if std(s_cur) < sigma_thr
        if flag == 0
            flag = 1;
            ng = 1;
        else
```

```
            ng = ng + 1;
            if ng == ng_thr
                break;
            end
        end
    else
        flag = 0;
    end


    % Normalize weights
    w = w*NP/sum(w);
    % Create a new population by replicating the current pre-
dictors
    % according to normalized weights
    jj = 0; % set the counter of predictors to zero
    for j = 1:NP
        % Replicate the predictor [w] times where [w] is the
largest
        % integer less than w
        for k = 1:floor(w)
            % Increase the counter of predictors
            jj = jj + 1;
            % Replicate the jth predictor
            Gt{jj} = G{j}; s_prev(jj) = s_cur(j);
        end
        % Replicate the predictor once with probability w-[w]
        if rand <= w - floor(w)
            % Increase the counter of predictors
            jj = jj + 1;
            % Replicate the jth predictor
            Gt{jj} = G{j}; s_prev(jj) = s_cur(j);
        end
    end
    % Update the number of predictors
    % This number can slightly change from generation to gen-
eration,
    % i.e. it can be a bit less than the original NP, equal to
the
    % original NP, or be a bit larger than the original NP
    NP = jj;
```

```
    % Copy predictors to a new population
    for j = 1:NP
        G{j} = Gt{j};
    end


    % Elitism
    % Find the worst performing predictor in the current gen-
eration
    [dummy, j] = min(s_prev);
    % Replace it with the best performing predictor from the
previous
    % generation
    G{j} = G_best; s_prev(j) = s_best;


    % Find the best performing predictor in the current genera-
tion
    [s_best, j_best] = max(s_prev);
    G_best = G{j_best};


end % end for i


% Find the predictor with the best performance
[s_opt, i_opt] = max(s_cur);
% Check if there are several predictors with the same perfor-
mance
ind = find(s_cur == s_opt);
if length(ind) > 1
% If it is the case, choose the predictor with the smallest
number    % of genes
    for jj = 1:length(ind)
        len(jj) = length(G{ind(jj)});
    end
    [len_min, i] = min(len);
    i_opt = ind(i);
    % Check if there are several predictors with the same
number of
    % genes
    ind2 = find(len == len_min);
    if length(ind2) > 1
        % If it is the case, randomly choose one predictor
```

```
        i = randsample(length(ind2),1);
        i_opt = ind(ind2(i));
    end
end

% Selected features
selected_features = G{i_opt};

return


% Fitness function
function fit = fitness_fcn(train_patterns, train_targets,
valid_patterns, valid_targets, classifier, varargin)

% Inputs:
%    train_patterns  - Training patterns (DxN matrix, where D
- data
%                      dimensionality, N - number of patterns)
%    train_targets   - Training targets (1xN vector)
%    valid_patterns  - Validation patterns (DxNt matrix, where D
- data %                dimensionality, Nt - number of patterns)
%    valid_targets   - Validation targets (1xNt vector)
%    classifier      - Classifier (string with function name
where a
%                      classifier is implemented)
%    varargin        - Parameters of a classifier
%
% Output:
%    fit             - Fitness value (classification accuracy in
per
%                      cent)

% Get predictions of a classifier for validation data
predicted_targets = feval(classifier, train_patterns, train_
targets, valid_patterns, varargin{:});
% Compute classification accuracy as a fitness value
fit = 100 - class_error(valid_targets, predicted_targets);

return
```

```
function errprcnt = class_error(test_targets,targets)

% Calculate error percentage based on true and predicted test
labels
errprcnt = mean(test_targets ~= targets);
errprcnt = 100*errprcnt;

return
function h = ztestpp(p, n, alpha)

% Z-test for a population proportion to find out whether a
certain gene % is selected by chance or not
% The null hypothesis: a population proportion is smaller than
0.5
% The alternative hypothesis: a population proportion is great-
er than
% 0.5
% Original source: David J. Sheskin, "Handbook of parametric
and non-
% parametric statistical procedures", 3rd edition, Chapman &
Hall/CRC, % Boca Raton, FL, 2004
%
% Inputs:
%    p                  - Observed population proportion (real
number
%                         between 0 and 1)
%    n                  - The number of observations
%    alpha              - Significance level
%
% Output:
%    h                  - Binary outcome of the z-test:
%                         h = 0: null hypothesis that a gene is
selected
%                         by chance cannot be rejected,
%                         h = 1: null hypothesis that a gene is
selected
%                         by chance is rejected
```

```
% Default value
if nargin < 3
    alpha = 0.05;
end

% Check that p is within limits
if p < 0 || p > 1
    error('p should be between 0 and 1!');
end

% Find the critical one-tailed value of z at the pre-specified
% significance level
z_crit = norminv(1-alpha, 0, 1);
% norminv - normal inverse cumulative distribution function

% The null hypothesis is rejected if
% 1) z > 0 and
% 2) z > z_crit at given alpha
if p <= 0.5
    h = 0;
else
    if p >= 0.5 + 0.5*z_crit/sqrt(n);
        h = 1;
    else
        h = 0;
    end
end

return
function reproducible_features = check_reproducibility(nruns,
algorithm, alg_params)

% Check for reproducibility of feature selection results of a
given
% feature selection algorithm
%
% Inputs:
%    nruns                       - The number of runs of feature
%                                          selection
%    algorithm                   - Name of a feature selection
```

```
algorithm
%    alg_params                    - Parameters of a feature
selection
%                                       algorithm
%
% Output:
%    reproducible_features         - List of indices of reproduced
%                                       features



% Initialization
list_of_features = [];
reproducible_features = [];

% Repeat nruns times
for i = 1:nruns
    % Select features with a given algorithm
    selected_features = feval(algorithm, alg_params);
    % Transpose a vector of feature indices if necessary
    if size(selected_features,1) > 1
        selected_features = selected_features';
    end
    % Add indices of selected features to the list
    list_of_features = [list_of_features selected_features];
end

% Find unique indices of features
u = unique(list_of_features);
% Count how many times each feature was included into the best
% predictor
h = histc(list_of_features, u);
n = sum(h);
% Get a proportion each feature appears in the population
p = h/n;

% Perform a z-test for a population proportion
o = ones(1,length(p));
for i = 1:length(p)
    o(i) = ztestpp(p(i), n);
```

```
end

% Remove features that were filtered out by the z-test for a
population
% proportion
u(o == 0) = [];
% Remaining features can be reliably reproduced
reproducible_features = u;

return
```

REFERENCES

Agrawal, R., & Bala, R. (2007). A hybrid approach for selection of relevant features for microarray datasets. *Proceedings of World Academy of Sciences. Engineering and Technology*, *23*, 281–287.

Alba, E., García-Nieto, J., Jourdan, L., & Talbi, E.-G. (2007). Gene selection in cancer classification using PSO/SVM and GA/SVM hybrid algorithms. *Proceedings of the IEEE Congress on Evolutionary Computation, Singapore* (pp. 284-290). Los Alamitos, CA: IEEE Computer Society.

de Souza, B. F., Carvalho, A. C., & Ticona, W. C. (2007). Applying genetic algorithms and support vector machines to the gene selection problem. *Journal of Intelligent and Fuzzy Systems*, *18*(5), 435–444.

Dessì, N., & Pes, B. (2009). An evolutionary method for combining different feature selection criteria in microarray data classification. *Journal of Artificial Evolution and Applications*, *2009*, 803973..doi:10.1155/2009/803973

Deutsch, J. M. (2001, August 8). *Algorithm for finding optimal gene sets in microarray prediction.* Retrieved February 1, 2009, from arXiv.org: http://arxiv.org/abs/physics/0108011

Deutsch, J. M. (2003). Evolutionary algorithms for finding optimal gene sets in microarray prediction. *Bioinformatics (Oxford, England)*, *19*(1), 45–52. doi:10.1093/bioinformatics/19.1.45

Dudoit, S., & Fridlyand, J. (2003). Classification in microarray experiments. In Speed, T. (Ed.), *Statistical analysis of gene expression microarray data* (pp. 93–158). Boca Raton, FL: Chapman & Hall/CRC Press. doi:10.1201/9780203011232.ch3

Hernandez, J. C., Duval, B., & Hao, J.-K. (2007). A genetic embedded approach for gene selection and classification of microarray data. In E. Marchiori, J. H. Moore, & J. C. Rajapakse (Ed.), *Proceedings of the 5th European Conference on Evolutionary Computation, Machine Learning and Data Mining in Bioinformatics, Valencia, Spain.* (LNCS 4447, pp. 90-101). Berlin/Heidelberg, Germany: Springer-Verlag.

Huerta, E. B., Duval, B., & Hao, J.-K. (2006). A hybrid GA/SVM approach for gene selection and classification of microarray data. In F. Rothlauf, J. Branke, S. Cagnoni, E. Costa, C. Cotta, R. Drechsler, et al. (Ed.), *Proceedings of the 4th European Workshop on Evolutionary Computation and Machine Learning in Bioinformatics, Budapest, Hungary* (LNCS 3907, pp. 34-44). Berlin/Heidelberg, Germany: Springer-Verlag.

Jirapech-Umpai, T., & Aitken, S. (2005). Feature selection and classification for microarray data analysis: evolutionary methods for identifying predictive genes. *BMC Bioinformatics, 6*(148). doi:.doi:10.1186/1471-2105-6-148

Li, L., Weinberg, C. R., Darden, T. R., & Pedersen, L. G. (2001). Gene selection for sample classification based on gene expression data: study of sensitivity to choice of parameters of the GA/KNN method. *Bioinformatics (Oxford, England), 17*(12), 1131–1142. doi:10.1093/bioinformatics/17.12.1131

Ooi, C., & Tan, P. (2003). Genetic algorithms applied to multi-class prediction for the analysis of gene expression data. *Bioinformatics (Oxford, England), 19*(1), 37–44. doi:10.1093/bioinformatics/19.1.37

Paul, T. K., & Iba, H. (2004). Selection of the most useful subset of genes for gene expression-based classification. *Proceedings of the 2004 Congress on Evolutionary Computation, Portland, OR* (pp. 2076-2083). Los Alamitos, CA: IEEE Computer Society.

Pepe, M. S. (2004). *The statistical evaluation of medical tests for classification and prediction* (1st paperback ed.). Oxford, UK: Oxford University Press.

Ripley, B. (2007). *Pattern recognition and neural networks* (1st paperback ed.). Cambridge, UK: Cambridge University Press.

Roberts, S. (2005, November). *Using genetic algorithms to select a subset of predictive variables from a high-dimensional microarray dataset.* Retrieved November 18, 2008, from MATLAB Digest - November 2005: http://www.mathworks.com/company/newsletters/digest/2005/nov/genalgo.html

Rubinstein, R. Y., & Kroese, D. P. (2008). *Simulation and the Monte Carlo method* (2nd ed.). Hoboken, NJ: John Wiley & Sons.

Sheskin, D. J. (2004). *Handbook of parametric and nonparametric statistical procedures* (3rd ed.). Boca Raton, FL: Chapman & Hall/CRC.

Su, Y., Murali, T., Pavlovic, V., Schaffer, M., & Kasif, S. (2003). RankGene: identification of diagnostic genes based on expression data. *Bioinformatics (Oxford, England)*, *19*(12), 1578–1579. doi:10.1093/bioinformatics/btg179

Wahde, M., & Szallasi, Z. (2006). A survey of methods for classification of gene expression data using evolutionary algorithms. *Expert Review of Molecular Diagnostics*, *6*(1), 101–110. doi:10.1586/14737159.6.1.101

ENDNOTES

[1]   Still, this step can significantly reduce the number of candidate features to be used by the evolutionary algorithm, i.e. it can dramatically speed up this algorithm. However, since this step needs to be very fast compared to the main feature selection procedure, it usually evaluates each feature separately from other features. As a result, certain features that are not relevant for classification on their own but become relevant when considered in combination with other features may not be selected. Therefore one needs to consider a trade-off between the number of features to choose and time required by the evolutionary algorithm to terminate.

[2]   If a fitness function returns classification error, then this quantity should be transformed into accuracy in order to avoid any changes in MATLAB® code. The transformation is 1-error if error is between 0 and 1 or 100-error if error is expressed in per cent.

[3]   It is the way of statistical logic that truth is established via impossibility to reject a certain statement. The word "truth" should not be interpreted in absolute sense: a failure to reject something does not necessarily imply its unconditional acceptance. Perhaps, such a way of thinking is deeply rooted in antiquity when the famous Greek philosopher Socrates questioned everything he met (his famous phrase is "I know that I know nothing"). It is probably because of him, we first state hypotheses and then try to eliminate them by looking for contradictions.

[4]   A random classifier as follows from its name is the one which assigns class labels randomly. For instance, flipping a coin would perfectly work as a randomly guessing classifier.

[5]   In ROC plots, the random classifier slope is the line originating at (0,0) and bisecting the ROC plane. The area under this line is equal to 0.5 while the

area under the ROC curve of the absolutely perfect classifier is equal to 1. As a result, by looking at a ROC plot containing the random classifier slope line, one can easily spot a better than random classifier: all useful classifiers will have their ROC curves strictly above that line.

[6] I call it the Bible since it presents more than 130 statistical procedures and tests on about 1200 pages. And it keeps growing from edition to edition.

Chapter 13
Redundancy–Based Feature Selection

REDUNDANCY OF FEATURES

A typical microarray dataset suffers from the shortage of samples while having features in abundance. It is obvious that not all measured features are indeed good predictors of a disease. Since there are a plenty of redundant features, there should be the way to detect them. (Yu, 2008) argues that wrapper models for gene selection, utilizing a classifier to judge on gene subset goodness, are computationally very expensive, and there is always a selection bias due to the learning algorithm used. Yu suggested applying the filter models for gene selection instead since they tend to be much faster than the wrapper ones.

However, many filter models treat each gene separately from other genes. The outcome of a typical filter is the ranked list of genes ordered according to a certain measure of relevance to the target class. From this list, only a few top genes are then selected. Such an approach has linear time complexity with respect to the total number of genes, but it cannot remove all redundant genes, since, for example, two highly ranked genes may duplicate each other because they are highly correlated (Ding

DOI: 10.4018/978-1-60960-557-5.ch013

& Peng, Minimum redundancy feature selection from microarray gene expression data, 2003), (Ding & Peng, Minimum redundancy feature selection from microarray gene expression data, 2005), (Peng, Long, & Ding, 2005)[1]. As a result, no gain in performance can be achieved, which sounds pretty much frustrating to you, right?

The keen reader might also notice a hidden problem with filters: how to determine threshold separating relevant and irrelevant genes in the ranked list? Sadly to say, but it is often heuristically chosen or its exact value can be only found with the help of domain-specific information, which might not be always available (no biologist, no chemist, no medical doctor nearby, and your machine learning/data mining/computer science/electrical engineering/mathematical/physical/any relevant to you knowledge unfortunately stops here). But nothing to do: this is the price to pay if we are in hurry and refused the help of a learning algorithm in this tough work. Thus, perhaps faced with similar questions and problems, Yu decided that automatic gene selection is absolutely necessary that would satisfy the following requirements: it belongs to the filter model for feature selection; it does not need to specify any threshold for gene selection; it removes both irrelevant and redundant genes, thus delivering a small set of biologically relevant genes. Nice intentions! Let us see how Yu succeeded to accomplish them.

Feature Relevance and Redundancy

Before describing Yu's main idea, let us introduce a number of useful definitions.

Let $P(C|F)$ be the probability distribution of the class C given the features F. In other words, P is the conditional distribution, because the probability of the class C depends or conditions on F. Let F be a full set of features, F_i a feature i, and $S_i = F \setminus F_i$, which means the set F with the feature F_i removed from F.

Kohavi and John (Kohavi & John, 1997) classified features into three non-overlapping groups: strongly relevant, weakly relevant, and irrelevant. These notions can be defined as follows.

Definition 1 (**Strong relevance**) A feature F_i is strongly relevant iff

$$P(C|F_i, S_i) \neq P(C|S_i).$$

Definition 2 (**Weak relevance**) A feature F_i is weakly relevant iff

$$P(C|F_i, S_i) = P(C|S_i)$$

and $\exists S_i' \subset S_i$, such that $P\left(C|F_i, S_i'\right) \neq P\left(C|S_i'\right)$.

Definition 3 (**Irrelevance**) A feature F_i is irrelevant iff

$$\forall S_i' \subseteq S_i, P\left(C|F_i, S_i'\right) = P\left(C|S_i'\right).$$

If a feature is strongly relevant, this feature is always necessary for the optimal subset of features, because its deletion from the optimal subset results in loss of discriminative power. This is reflected by the symbol ' $\neq$ ' between the probability of the class C when F_i is given in addition to S_i and the probability of the class C when just S_i is given with no knowledge about F_i. That is, the sign that a feature is strongly relevant is the change (typically, decrease) in probability if a strongly relevant feature is excluded from a subset of features.

Weak relevance implies that the feature is not always necessary to have among selected features in the optimal subset. However, under certain circumstances such a feature may become relevant, e.g., in combination with another feature. Compared to strong relevance, the weak relevance definition looks totally different. In other words, the fact that a weakly relevant feature is included or not into a given subset of already selected features (it is useful to think of S_i as the optimal subset) does not affect the probability value. However, among features in S_i there are probably certain features that change the status of weak relevance to strong relevance when combined together with a weakly relevant feature. This state transition can be seen in the second line of the weak relevance definition.

Finally, an irrelevant feature is the feature that can be safely omitted from the optimal subset, since it does not influence on classification performance in any way. Hence, adding an irrelevant feature to other previously selected features does not change the probability value.

The important implication of all three definitions above is that features are not considered in isolation from each other anymore! This allows taking into account interactions among features; hence, more complex relations between features can be discovered.

The optimal subset of features must include all strongly relevant and some weakly relevant features. However, the definition of weak relevant is rather vague; it does not indicate how one can distinguish between important weakly relevant and unimportant weakly relevant features. Hence, the concept of feature redundancy is necessary.

Redundancy can be defined in terms of correlation. Indeed, two features are redundant if their values are completely correlated. However, such an ideal case is quite rare in practice. In order to efficiently utilize feature redundancy, the following definition of a feature's Markov blanket is used (Pearl, 1988).

Definition 4 (**Markov blanket**) Given a feature F_i, let $M_i \subset F\left(F_i \notin M_i\right)$. M_i is a Markov blanket for F_i iff

$$P\left(F \setminus \left\{M_i, F_i\right\}, C \mid F_i, M_i\right) = P\left(F \setminus \left\{M_i, F_i\right\}, C \mid M_i\right).$$

The first thing to understand about a Markov blanket is that it has nothing to do with that linen or cotton thing which we all use in kitchen or bathroom. A Markov blanket is a feature associated with F_i. However, it is still useful to think of it as a kind of "blanket" that "covers" or "does not cover" a certain feature. Notice that a Markov blanket does not exist per se; it is only useful if linked to the feature of interest. Intuitively, you might already determine when a given feature is redundant or not: if a Markov blanket exists for it, it is redundant; otherwise, it is not. However, we must proceed toward a more formal explanation of the above mentioned definition.

It is easy to see that if M_i is a Markov blanket of F_i (pay attention that both M_i and F_i are excluded from the original feature set F when calculating probabilities), the class C is conditionally independent of F_i given M_i, i.e. knowing F_i does not bring extra information about C.

Moreover, the Markov blanket condition is stronger than conditional independence, because it requires that M_i subsumes not only the information F_i has about C, but also about all of the other features. In (Koller & Sahami, 1996) it was pointed out that an optimal subset can be obtained by using Markov blankets as the basis for feature elimination. Let G be the current set of features ($G = F$ in the beginning). At any step, if there exists a Markov blanket for F_i within G, F_i is deemed to be unnecessary for the optimal subset and thus it can be removed from G. It is proved that this process guarantees that a feature eliminated at an earlier stage will still have a Markov blanket at any later stages (Koller & Sahami, 1996). As a result, this means that if a feature has been removed at an earlier stage, one does not have to worry that this feature might be needed to check at a later stage.

Based on the feature relevance definitions, it is clear that a strongly relevant feature does not have the corresponding Markov blanket. On the other hand, a Markov blanket always exists for an irrelevant feature; therefore, such a feature will be always detected and removed from further analysis. This implies that we are left with ambiguous weakly relevant features, which may include both useful and useless features.

Definition 5 (**Redundant feature**) Let G be the current set of features. Then a feature is redundant and should be removed from G iff it is weakly relevant and has a Markov blanket within G.

After removal irrelevant and weakly relevant but redundant genes, only strongly relevant and weakly relevant but non-redundant features form an optimal subset of features we are interested in. However, the number of genes in microarray data is so huge that the task of finding the exact Markov blanket seems to be impractical due to its combinatorial complexity. Hence, Yu suggested searching for a suboptimal subset of genes by resorting to an approximate solution for a Markov blanket.

Main Idea

There are two types of correlation between genes and the class which Yu's algorithm is based on: individual C-correlation and combined C-correlation.

Definition 6 (**Individual C-correlation**) The correlation between gene F_i and the class C is called individual C-correlation.

Definition 7 (**Combined C-correlation**) The correlation between the pair of genes F_i and F_j $(i \neq j)$ and the class C is called combined C-correlation.

In combined C-correlation, two genes are treated as a single gene. For simplicity it is assumed that gene expression levels are represented by nominal values instead of continuous values. It would, of course, require discretization of the whole range $]-\infty, +\infty[$ into the small number of discrete values, each associated with its own subinterval. In order to avoid any ambiguity, there is no overlap between adjacent subintervals. Under such conditions, if both genes take nominal values -1, 0, and +1, there will be nine pairs of values covering all possible states: (-1,-1), (-1,0), (-1,+1), (0,-1), (0,0), (0,+1), (+1,-1), (+1,0), (+1,+1). In this case, the combined correlation measure the correlation between each of the nine states and the class C. Of course, individual C-correlations are also calculated for nominal data.

The method proposed by Yu determines if a single gene F_i can be an approximate Markov blanket for another gene F_j based on both individual C-correlations and the combined C-correlation. It assumes that a gene with a larger individual C-correlation possesses more information about the class than a gene with a smaller individual C-correlation. For two genes F_i and F_j, if the individual C-correlation for F_i is larger than or equal to the individual C-correlation for F_j, then the next check is to evaluate whether gene F_j is approximately redundant to gene F_i. In addition, if combining two genes F_i and F_j does not give more information about the class than F_i alone, it is heuristically decided that F_i forms an approximate Markov blanket for F_j. Therefore, an approximate Markov blanket is defined as follows.

Definition 8 (**Approximate Markov blanket**) For two genes F_i and F_j, F_i forms an approximate Markov blanket for F_j iff (1) the individual C-correlation for F_i is larger than or equal to the individual C-correlation for F_j and (2) the individual C-correlation for F_i is larger than or equal to the combined C-correlation for F_i, F_j.

Consequences of gene removal based on an approximate Markov blanket are not the same as those with the exact Markov blanket. In particular, if F_j is the only gene forming an approximate Markov blanket for F_k, and F_i forms an approximate Markov blanket for F_j, then after removing F_k based on F_j, further removal of F_j based on F_i will result in no approximate Markov blanket for F_k (this implies that one is not certain that there is no need to check F_k for relevance again; hence, the whole backward elimination scheme may collapse). Fortunately, it is possible to circumvent this problem by removing a gene only if an approximate Markov blanket exists for it that formed by a predominant gene defined below.

Definition 9 (**Predominant gene**) A gene is predominant iff it does not have any approximate Markov blanket in the current set.

Predominant genes (there can be several of them) will not be removed at any stage. This is what distinguishes them from other genes. If a certain gene is removed based on a predominant gene, it is guaranteed that a Markov blanket for it will still exist at any later stage when another gene is removed. To find predominant genes, all genes are sorted in descending order of their individual C-correlation. The gene with the highest value of C-correlation does not have an approximate Markov blanket, and hence, it is one of the predominant genes that will be used to filter out unimportant genes. In other words, the goal of Yu's method is to discover all predominant genes while eliminating all other genes.

Compared to the approaches treating each gene separately from the others, this method offers the following advantages:

• Redundancy among relevant genes is efficiently handled.

- Gene-to-gene interactions (though only pairwise, but not higher-order) are taken into account.
- Removal of both irrelevant and weakly relevant but redundant genes.

The last characteristic makes it unnecessary to determine a threshold value for partitioning all genes into important and unimportant for classification, since all predominant genes will be among the selected and iterations will stop when the list of gene ends. It is possible, of course, to extend the method so that it can handle higher-order gene interactions, but there are two reasons why it might not be a good idea: (1) dramatic increase in computational time and (2) over-searching problem (Jensen & Cohen, 2000) due to the combination of small sample size and high-dimensional data. Now almost everything is ready to start describing Yu's algorithm itself.

Algorithm Description

I first show how to compute certain measures employed in the algorithm as well as how discretization of continuous values has been done, followed by the algorithm itself in detail.

Discretization and Calculation of C-Correlations

Discretization of continuous values into a number of nominal states is usually intended to reduce noise. Various techniques for doing so can be employed (see, for instance, (Liu, Hussain, Tan, & Dash, 2002), but in (Yu, 2008) the following approach was exercised.

The number of different nominal values was set to 3: -1, 0, and +1, representing under-expression, baseline, and over-expression, respectively. These nominal values correspond to $]-\infty, \mu - \sigma / 2[$, $[\mu - \sigma / 2, \mu + \sigma / 2]$, and $]\mu + \sigma / 2, +\infty[$, respectively, where μ and σ are the mean and standard deviation of expression levels for a given gene across all samples. The algorithm of discretization is thus straightforward: if a continuous expression level comes into $]-\infty, \mu - \sigma / 2[$, it is replaced with -1; if it lies within $[\mu - \sigma / 2, \mu + \sigma / 2]$, it is assigned to 0; if it falls into $]\mu + \sigma / 2, +\infty[$, it is set to +1.

Once deciding upon discretization, the individual and combined C-correlations are defined in terms of information-theoretic measures based on entropy[2].

For nominal variables, entropy and conditional entropy are defined as

$$H(X) = -\sum_{i=1}^{m} P(x_i) \log_2 P(x_i),$$

$$H(X|C) = -\sum_{k=1}^{m} P(c_k) \sum_{i=1}^{m} P(x_i|c_k) \log_2 P(x_i|c_k),$$

where $m = 3$ because of three possible nominal values (-1, 0, +1), $P(x_i)$ is the probability that the nominal variable X takes the value x_i, $P(c_k)$ is the class c_k probability, and $P(x_i|c_k)$ is the probability that the nominal variable X takes the value x_i, given that the class is c_k.

For combined C-correlation, the formulae above are changed into

$$H(X) = -\sum_{i,j=1}^{m} P(x_i, x_j) \log_2 P(x_i, x_j),$$

$$H(X|C) = -\sum_{k=1}^{m} P(c_k) \sum_{i,j=1}^{m} P(x_i, x_j|c_k) \log_2 P(x_i, x_j|c_k).$$

Interpretation of conditional probabilities in these formulas is essentially the same as in those for the individual C-correlation; the only difference is that x_i is now replaced with the pair (x_i, x_j). Since the nominal variable in our case can take only three distinct values, there are nine pairs for which probabilities and hence, entropies need to be calculated. These pairs are (-1,-1),

(-1,0), (-1,+1), (0,-1), (0,0), (0,+1), (+1,-1), (+1,0), (+1,+1).

Having defined entropies, both types of correlation are finally expressed through symmetrical uncertainty *SU* as

$$SU(X,C) = 2\left[\frac{IG(X|C)}{H(X) + H(C)}\right],$$

where X stands for the nominal values of a single gene (individual C-correlation) or the nominal values of a pair of genes (combined C-correlation). $IG(X|C) = H(X) - H(X|C)$ is information gain from knowing the class information, meaning that the more information we know about the relation between X and C, the larger the entropy decrease is, i.e. the less uncertain we are about X. SU is a normalized characteristic whose values lie between 0 and 1, where 0 indicates that X and C are independent, i.e. knowing C does not affect our knowledge about X. It should be noted that instead of symmetrical uncertainty, other definitions for correlation can also be applied.

Algorithm

The algorithm abbreviated as RBF is shown in Figure 1, where the list S includes genes that have been selected by the algorithm; the end of this list is marked with NULL, which is borrowed from programming languages, where it usually means the end of a file or other structure storing data.

The algorithm starts from computing individual C-correlation for each gene and sorting all correlations in descending order of magnitude. The gene with the largest correlation is considered as predominant (no approximate Markov blanket exists for it) and hence it is put to the list S of the selected genes.

After that, the iteration begins with picking the first gene F_i from S in order to filter out other genes which are not in S. For all remaining genes, if F_i forms an approximate Markov blanket for any gene F_j, the latter is removed from further analysis. The following conditions must be satisfied for this to happen:

Individual C-correlation for F_i must be larger than or equal to individual C-correlation for F_j, which implies that a gene with a larger individual correlation provides more information about the class than a gene with a smaller individual correlation. This condition is automatically fulfilled after gene sorting based on individual C-correlation.

Individual C-correlation for F_i must be larger than or equal to combined C-correlation for genes F_i and F_j, which means that if combining genes F_i and F_j does not provide more discriminating power than F_i alone, F_j is decided to be useless.

After one iteration of filtering based on F_i, the algorithm takes the next (according to the magnitude of individual C-correlation) still unfiltered gene and the filtering process is repeated again, etc. The algorithm halts if there are no more predominant genes to be selected.

Figure 1. Pseudo code of Redundancy-Based Filter

Redundancy-Based Filter

Compute the individual C-correlation for each gene;
Sort correlations in descending order;
Put genes into the list S in the sorting order;
Get the first element F_i from S;
While F_i is not NULL
  Get the next element F_j from S;
  While F_j is not NULL
    If the individual correlation for F_i is larger than or equal to the combined
      correlation for F_i and F_j, then $S := S \backslash F_j$; /* remove F_j from S */
    Get the next element F_j from S;
  End While
  Get the next element F_i from S;
End While

Since a lot of genes are typically removed in each round (recall that gene expression data contain a lot of redundancy) and removed genes do not appear in the next iterations, RBF is much faster than the typical hill climbing (greedy forward selection or backward elimination where one gene at a time is either added to or removed from the current subset of selected genes). Only in the worst case when no gene can be redundant or irrelevant, which is extremely rare or even impossible in case of microarray data, the computational complexity of RBF is the same of the greedy search algorithms. Hence, one does not need to be scared of double loop in pseudo code in Figure 1.

Wrap-Up Discussion on Markov Blanket-Based Feature Selection

The Markov blanket based gene selection considered in (Yu, 2008) is not the only method relying on the Markov blanket concept. For instance, (Xing, Jordan, & Karp, 2001) employed a two-component Gaussian unconditional mixture model for feature discretization, followed by the information gain based ranking to select a small subset of genes (about 5% of the total number of genes). After that, the following algorithm in Figure 2 was applied in (Koller & Sahami, 1996), (Knijnenburg, Reinders, & Wessels, 2006), (Xing, Jordan, & Karp, 2001) to further reduce the number of candidate genes.

Figure 2. Another Markov blanket based feature selection algorithm

---

Another Markov blanket Filter (Koller & Sahami, 1996)

Let G be an initial subset of features to select from;
Let k^* be the number of features to preserve;
While $|G| \neq k^*$
  For each feature $f_i \in G$, let M_i be the set of k features $f_j \in G - \{f_i\}$ for which the correlations between non-quantized f_i and f_j are the highest;

  Compute the coverage score, $\Delta(f_i|M_i)$, for each feature f_i:

$$\Delta(f_i|M_i) = \sum_{n=1}^{N} P_n(M_{Di}, f_{Di}) \sum_{l=1}^{C} \left[P_n(c_l|M_{Di}, f_{Di}) \log \left(\frac{P_n(c_l|M_{Di}, f_{Di})}{P_n(c_l|M_{Di})} \right) \right],$$

  where N is the number of samples (examples), C is the number of classes, c_l is the lth class of the data, P stands for probability, M_{Di} and f_{Di} are the binary discretized versions of M_i and f_i, respectively, i.e. $M_{Di}, f_{Di} \in \{0,1\}$;

  Find i^* that minimizes $\Delta(f_i|M_i)$ and put $G := G - \{f_{i^*}\}$; /* remove f_{i^*} from G */
End While

The algorithm begins with finding k (k is a small integer) features that have the highest linear (Pearson) correlations with a given feature. Correlations are defined between original, non-quantized feature vectors. For each feature $f_i \in G$ (G is the initial set of features), these k features form the Markov blanket M_i for this feature. For each feature, the coverage of its blanket is then computed. The feature that has the lowest coverage score (absolute minimum is zero, but it may not be observed in practice) is considered to be the most redundant and is hence removed from G. This process iterates until the desired number of features, k^*, is left in G. In (Knijnenburg, Reinders, & Wessels, 2006), $k^* = k$.

This heuristic sequential method is far more efficient than methods that conduct an extensive combinatorial search over subsets of the feature set. The heuristic method only requires computation of quantities of the form $P_n\left(M_{Di}, f_{Di}\right)$, $P_n\left(c_l|M_{Di}, f_{Di}\right)$ and $P_n\left(c_l|M_{Di}\right)$, which can be easily computed using the discretization discussed in (Xing, Jordan, & Karp, 2001). If M_i is indeed the Markov blanket for f_i, then $P_n\left(c_l|M_{Di}, f_{Di}\right) = P_n\left(c_l|M_{Di}\right), \forall l, n$. That is, in this case, a feature is perfectly covered by its blanket and such a feature is conditionally independent of the class given the Markov blanket for that feature.

The crucial question is about blanket size k. In (Knijnenburg, Reinders, & Wessels, 2006), authors experimented with two values for k: 2 and 5. They found the results in terms of removed features are contradictory for these two values. However, one should not be discouraged by these findings.

Firstly, for small sample sizes, it is highly likely that many conditional and joint probabilities are equal to or very close to zero not because they are indeed zero, but because too many combinations of feature values are not represented or severely underrepresented in a small sample. For instance, for $k = 5$, six-dimensional probabilities must be estimated, which is not easy to accurately do if there are only few dozens of examples. Hence, I advocate relying on very small k values, such as 1 and 2 if dataset size is small.

Secondly, 10-fold cross-validation was utilized for gene selection and microarray data classification, which as shown in several recent works (Braga-Neto & Dougherty, 2004), (Isaksson, Wallman, Göransson, & Gustafsson, 2008), cannot provide reliable classification error estimation.

I do not argue, however, with Knijnenburg et al. that the coverage score and feature removal strategy are not optimal, since local optimization is only done at each round of the algorithm in Figure 2. Incorporating Markov blanket strategy into the global search algorithm could probably achieve better classification results. And, of course, on some microarray datasets, a simple heuristic, treating genes independently of each other and scoring them, e.g., based on t-test statistic, could occasionally deliver better results, which should not nevertheless be generalized as the superiority of a simple technique over a more sophisticated one.

As a potential solution, a beam search (Furcy & Koening, 1995), (Xu & Fern, 2007) suggested in (Knijnenburg, Reinders, & Wessels, 2006) or the combination of Markov blanket and genetic algorithm (Zhu, Ong, & Dash, 2007) can be tried.

MATLAB Code

MATLAB® code comprises not only RBF code but also auxiliary code for computing entropies. In order to convert original continuous expression levels to nominal scale, temporal array X is introduced. Please notice that memory for this array is released once the conversion is done.

Indices of genes to be selected are stored in vector *ind*. Initially, all genes are assumed to be valid for selection. However, as the algorithm is executed, many of them become redundant. If a certain gene is considered as redundant, its value in *ind* is set to -1 in order to distinguish it from non-redundant genes. When RBF halts, negative entries are simply removed from *ind*.

The main function implementing the RBF algorithm relies on calls to auxiliary functions *entropy_nominal* (to compute such quantities as $P\left(x_i\right)$ and $P\left(c_k\right)$), *condentropy_nominal* (to compute $P\left(x_i|c_k\right)$), *entropy2_nominal* (to compute $P\left(x_i, x_j\right)$), and *condentropy2_nominal* (to compute $P\left(x_i, x_j|c_k\right)$). In all these functions, the variable *p* denoting the probability is set to a very small positive value ε if its original value is less than ε; this is done in order to correctly process the case $0\log_2 0$ that can happen in practice[3]. When computing entropies in all auxiliary functions, the MATLAB® function *unique* delivers unique elements of a given vector or matrix, arranged in ascending order of magnitude. For instance, if $A = \left[1100 - 101 - 1 - 1\right]$, the outcome of *unique(A)* is $\left[-1, 0, 1\right]$. If the input of *unique* is a matrix, then this function returns the unique rows of this matrix, which is useful to exploit when computing entropies for pairs of genes.

Since entropy computation requires computation of the probability of a certain event, the latter is defined in the traditional frequentist sense as the ratio of the number of cases when the event occurred to the total number of cases. The MATLAB® function *find* that finds all required occurrences is useful in computing probabilities. When coupled with another MATLAB® function *length* returning length of a vector, computing entropy in *entropy_nominal* becomes straightforward. In *entropy2_nominal*, calculations are similar to those in *entropy_nominal*, except the fact that instead of counting the occurrence of a single number, the occurrence of a pair of numbers is counted; for example, we are interested to know how many times the combination $(0, +1)$ jointly occurred in expression levels of two genes.

Calculations of the conditional entropies in functions *condentropy_nominal* and *condentropy2_nominal* proceed along similar lines. For any given class label, the probability of this class is first found, followed by computing the inner sum over *i* (see the equation for individual *C*-correlation) or *i, j* (see the equation for combined *C*-correlation). After that these two quantities are multiplied, the result is accumulated and all the operations just described are repeated for another class label, etc.

```
function selected_features = RBF(train_patterns, train_targets)

% Redundancy-based feature selection (RBF stands for redundan-
cy-based
% filter)
% Original source: Lei Yu. "Feature selection for genomic data
% analysis", in Huan Liu and Hiroshi Motoda (eds.), Computa-
tional
% Methods of Feature Selection, Chapter 17, Series in Data Min-
ing and
% Knowledge Discovery, Chapman & Hall/CRC, Boca Raton, FL,
```

```
2008,
% pp.337-354.
%
% Inputs:
%    train_patterns           - DxN matrix
%                                N is the number of samples,
%                                D is the number of features
%                                (attributes, genes)
%    train_targets            - 1xN vector of class labels
%
% Output:
%    selected_features        - Indices of selected features (at-
tributes,
%                                genes)




% Get unique labels of classes
u = unique(train_targets);

% Check if there are only two different classes
if length(u) ~= 2
    error('Two-classes problems are only considered!');
end

% Get the number of features (attributes, genes)
D = size(train_patterns,1);
% Get the number of samples
N = length(train_targets);

% Perform data normalisation according to p.349
m = mean(train_patterns,2);
s = std(train_patterns,0,2);
X = zeros(D,N);
X(train_patterns < repmat((m-s/2),1,N)) = -1;
X(train_patterns > repmat((m+s/2),1,N)) = 1;
% Values within [m-s/2,m+s/2] are assigned to 0
train_patterns = X;

clear X;
```

```
% Allocate memory for individual C-correlations
isu = zeros(1,D);

% Initially, all features (attributes, genes) are candidates
for
% selection
ind = 1:D;

% Compute individual C-correlations where C-correlation is de-
fined as
% symmetrical uncertainty
hy = entropy_nominal(train_targets);
for i = 1:D
    hx = entropy_nominal(train_patterns(i,:));
    hxy = condentropy_nominal(train_patterns(i,:),train_tar-
gets);
    isu(i) = 2*((hx - hxy)/(hx + hy));
end

% Order ISU values in descending order
[isu,ind] = sort(isu,'descend');

hy = entropy_nominal(train_targets);
for i = 1:D-1
    if ind(i) > 0% only non-redundant feature (attribute, gene)
        isu_i = isu(i);
        f_i = train_patterns(ind(i),:);
        for j = i+1:D
            if ind(j) > 0% only non-redundant feature (attri-
bute,
  % gene)
                f_j = train_patterns(ind(j),:);
                % Compute combined C-correlation
                x = [f_i; f_j]'; % Nx2 matrix
                hx = entropy2_nominal(x);
                hxy = condentropy2_nominal(x,train_targets);
                csu_ij = 2*((hx - hxy)/(hx + hy));
                if isu_i >= csu_ij
                    ind(j) = -1; % feature (attribute, gene) is
```

```
    % redundant
                end
            end
        end
    end
end

% Exclude redundant features (attributes, genes)
ind(ind < 0) = [];
selected_features = ind;

return
function h = entropy_nominal(x)

% Entropy H(x) for nominal data

% Find unique nominal values
u = unique(x); % x is a 1xN vector

h = 0;
for i = 1:length(u)
    p = length(find(x == u(i)))/length(x);
    if p < eps
        p = eps;
    end
    h = h - p*log2(p);
end

return
function h = entropy2_nominal(x)

% Entropy H(x) for nominal data

% Find unique nominal values
u = unique(x,'rows'); % x is a Nxk matrix, where k > 1

h = 0;
for i = 1:length(u)
    p = sum(all(x == repmat(u(i,:),length(x),1),2))/length(x);
    if p < eps
```

```
            p = eps;
    end
    h = h - p*log2(p);
end

return
function h = condentropy_nominal(x,y)

% Conditional entropy H(x|y) for nominal data

ux = unique(x);
uy = unique(y);

h = 0;
for j = 1:length(uy)
    py = length(find(y == uy(j)))/length(y);
    s = 0;
    for i = 1:length(ux)
        pxy = length(find(x == ux(i) & y == uy(j)))/
length(find(y == uy(j)));
        if pxy < eps
            pxy = eps;
        end
        s = s + pxy*log2(pxy);
    end
    h = h - py*s;
end

return
function h = condentropy2_nominal(x,y)

% Conditional entropy H(x|y) for nominal data

ux = unique(x,'rows'); % x is a Nxk matrix, where k > 1
uy = unique(y); % y is a 1xN vector

h = 0;
for j = 1:length(uy)
    py = length(find(y == uy(j)))/length(y);
    s = 0;
```

```
    for i = 1:length(ux)
        t = all(x == repmat(ux(i,:),length(x),1),2);
        pxy = length(find(t' == 1 & y == uy(j)))/length(find(y
== uy(j)));
        if pxy < eps
            pxy = eps;
        end
        s = s + pxy*log2(pxy);
    end
    h = h - py*s;
end

return
```

REFERENCES

Braga-Neto, U. M., & Dougherty, E. R. (2004). Is cross-validation valid for small-sample microarray classification? *Bioinformatics (Oxford, England)*, *20*(3), 374–380. doi:10.1093/bioinformatics/btg419

Bramer, M. (2007). *Principles of data mining*. London: Springer-Verlag.

Cover, T. M., & Thomas, J. A. (2006). *Elements of information theory* (2nd ed.). Hoboken, NJ: John Wiley & Sons.

Ding, C., & Peng, H. (2003). Minimum redundancy feature selection from microarray gene expression data. *Proceedings of the IEEE Computer Society Bioinformatics Conference, Stanford, CA* (pp. 523-529). Los Alamitos, CA: IEEE Computer Society.

Ding, C., & Peng, H. (2005). Minimum redundancy feature selection from microarray gene expression data. *Journal of Bioinformatics and Computational Biology*, *3*(2), 185–205. doi:10.1142/S0219720005001004

Furcy, D., & Koening, S. (1995). Limited discrepancy beam search. In L. P. Kaelbling, & A. Saffiotti (Ed.), *Proceedings of the 19th International Joint Conference on Artificial Intelligence, Edinburgh, Scotland, UK* (pp. 125-131). Denver, CO: Professional Book Center.

Isaksson, A., Wallman, M., Göransson, H., & Gustafsson, M. (2008). Cross-validation and bootstrapping are unreliable in small sample classification. *Pattern Recognition Letters*, *29*(14), 1960–1965. doi:10.1016/j.patrec.2008.06.018

Jensen, D. D., & Cohen, P. R. (2000). Multiple comparisons in induction algorithms. *Machine Learning, 38*(3), 309–338. doi:10.1023/A:1007631014630

Knijnenburg, T. A., Reinders, M. J., & Wessels, L. F. (2006). Artifacts of Markov blanket filtering based on discretized features in small sample size applications. *Pattern Recognition Letters, 27*(7), 709–714. doi:10.1016/j.patrec.2005.10.019

Kohavi, R., & John, G. H. (1997). Wrappers for feature subset selection. *Artificial Intelligence, 97*(1-2), 273–324. doi:10.1016/S0004-3702(97)00043-X

Koller, D., & Sahami, M. (1996). Toward optimal feature selection. In L. Saitta (Ed.), *Proceedings of the 13th International Conference on Machine Learning, Bari, Italy* (pp. 284-292). San Francisco, CA: Morgan Kaufmann.

Liu, H., Hussain, F., Tan, C. L., & Dash, M. (2002). Discretization: an enabling technique. *Data Mining and Knowledge Discovery, 6*(4), 393–423. doi:10.1023/A:1016304305535

Pearl, J. (Ed.). (1988). *Probabilistic reasoning in intelligent systems: networks of plausible inference*. San Francisco, CA: Morgan Kaufmann.

Peng, H., Long, F., & Ding, C. (2005). Feature selection based on mutual information: criteria of max-dependency, max-relevance, and min-redundancy. *IEEE Transactions on Pattern Analysis and Machine Intelligence, 27*(8), 1226–1238. doi:10.1109/TPAMI.2005.159

Shannon, C. E. (1948). A mathematical theory of communications. *The Bell System Technical Journal, 27*, 379–423, 623–656.

Xing, E. P., Jordan, M. I., & Karp, R. M. (2001). Feature selection for high-dimensional genomic microarray data. In C. E. Brodley, & A. Pohoreckyj Danyluk (Ed.), *Proceedings of the 18th International Conference on Machine Learning, Williamstown, MA* (pp. 601-608). San Francisco, CA: Morgan Kaufmann.

Xu, Y., & Fern, A. (2007). On learning linear ranking functions for beam search. In Z. Ghahramani (Ed.), *Proceedings of the 24th International Conference on Machine Learning, Corvallis, OR* (pp. 1047-1054). Madison, WI: Omnipress.

Yu, L. (2008). Feature selection for genomic data analysis. In Liu, H., & Motoda, H. (Eds.), *Computational methods of feature selection* (pp. 337–354). Boca Raton, FL: Chapman & Hall/CRC.

Zhu, Z., Ong, Y.-S., & Dash, M. (2007). Markov blanket-embedded genetic algorithm for gene selection. *Pattern Recognition, 40*(11), 3236–3248. doi:10.1016/j.patcog.2007.02.007

ENDNOTES

[1]   MATLAB® code for the minimum-redundancy, max-relevance algorithm can be downloaded from http://www.mathworks.com/matlabcentral/fileexchange/14916. It also requires mutual information functions located at http://www.mathworks.com/matlabcentral/fileexchange/14888. If you would like to have everything packed together, then visit http://www.mathworks.com/matlabcentral/fileexchange/14608. To thrill you even more, there is possibility to run this algorithm on-line with your own data; yes, yes, you have read it correctly "ON-LINE WITH YOUR OWN DATA" at http://research.janelia.org/peng/proj/mRMR/index.htm. If you do not use MATLAB®, there are still C/C++ implementations to download for Linux and Mac OS X.

[2]   In information theory, the entropy of a random variable is a measure of the uncertainty of the random variable; it is a measure of the amount of information required to describe the random variable (Cover & Thomas, 2006), (Bramer, 2007). It was introduced for the first time by Claude E. Shannon (Shannon, 1948). Please, be aware of multiple definitions for entropy, since this measure is used in various branches of science; each of these definitions is only applied within a certain discipline.

[3]   The result of $0 \log_2 0$ in MATLAB® is NAN, which means "not a number", i.e. undefined result. In contrast, the result $\varepsilon \log_2 \varepsilon \approx -1.1546e - 14$, given $\varepsilon = 2.2204e - 16$, which correctly represents the situation. The reader must be aware of such potentially hidden cases because ignorance of them often leads to strange unexpected results. For instance, before dividing a into b, it is always advisable to check whether $b = 0$ or not.

Chapter 14
Unsupervised Feature Selection

UNSUPERVISED FEATURE FILTERING

So far we only considered supervised feature selection methods, because unsupervised feature selection methods are scarce. Varshavsky et al. (Varshavsky, Gottlieb, Linial, & Horn, 2006) proposed several variants of a feature selection algorithm which is based on singular value decomposition (SVD), where features are selected according to their contribution to the SVD-entropy, which is the entropy defined for the distribution of eigenvalues of a square data matrix. Because SVD looks for eigenvalues of a matrix, it is akin to principal component analysis.

Because wrapper models for feature selection utilize a classifier (hence, class label) for judging on feature usefulness or relevance, wrappers cannot be associated with unsupervised way for selecting features. In contrast, filter models that do not require a classifier are better suitable for unsupervised feature selection. In bioinformatics, a typical example of the unsupervised feature filtering is gene shaving (Hastie, et al., 2000).

DOI: 10.4018/978-1-60960-557-5.ch014

(Varshavsky, Gottlieb, Linial, & Horn, 2006) introduced three variants of un-supervised feature selection: simple feature ranking, forward feature selection and backward feature elimination. Due to time-consuming computations in the backward feature elimination variant, I do not describe it here. Readers are advised to consult the original article for details. Nevertheless, by learning about MATLAB® code for the forward feature selection in this chapter, interested readers will be able to easily implement the backward elimination algorithm themselves.

Singular Value Decomposition

Let A be a $D{\times}N$ data matrix, containing as its rows the expression levels of D genes, measured in N samples. The A can be written as $A = U\Sigma V^T$ where U is a $D{\times}D$ orthogonal matrix, V is an $N{\times}N$ orthogonal matrix, and Σ is a $D{\times}N$ diagonal matrix with nonnegative values. Such a decomposition of the matrix A into the product of three other matrices is called the singular value decomposition (Lay, 2003).

An algorithm to find the singular value decomposition is given in (Steeb, 2006) as follows:

Find the eigenvalues $\lambda_i, i = 1, \ldots, N$, of the $N{\times}N$ matrix A^TA and arrange the eigenvalues in descending order.

Find the number of nonzero eigenvalues of the matrix A^TA and denote this number as r.

Find the orthogonal eigenvectors v_i of the matrix A^TA corresponding to the obtained eigenvalues and arrange them in the same order to form column-vectors of the $N{\times}N$ matrix V.

Form a $D{\times}N$ matrix Σ by placing on the leading diagonal of it quantities $\sigma_i = \sqrt{\lambda_i}, i{=}1,\ldots,\min(N,D)$. These quantities are called singular values.

Find the first r column vectors of the $D{\times}D$ matrix U:

$$u_i = \frac{1}{\sigma_i} Av_i, i = 1, \ldots, r.$$

These column vectors are the left-singular (u_i) and right-singular (v_i) vectors, respectively.

Add to the matrix U the rest of the D-r vectors using the Gram-Schmidt orthogo-nalization process.

Feature Ranking by SVD-Entropy

An SVD-entropy of a dataset was first introduced in (Alter, Brown, & Botstein, 2000). It is defined as

$$E = -\frac{1}{\log N}\sum_{i=1}^{N} S_i \log S_i,$$

where $S_i = \sigma_i^2 / \sum_{j=1}^{N}\sigma_j^2$ is the normalized relative value.

This entropy varies between 0 and 1. $E = 0$ (low entropy, high order) corresponds to a dataset that can be explained by a single eigenvector associated with the only nonzero eigenvalue. $E = 1$ (high entropy, low order) corresponds to the dataset for which eigenvalues are uniformly distributed, i.e. all eigenvalues are equal in magnitude. As one can notice, there is a certain similarity between the Shannon entropy and its SDV analogue. However, instead of probability, SVD-entropy is based on the distribution of eigenvalues or singular values.

Having defined SVD-entropy, the next step is to define the contribution of the ith feature:

$$CE_i = E\left(A_{m\times N}\right) - E\left(A_{(m-1)\times N}\right),$$

where $E(A_{m\times N})$ means the SVD-entropy computed for the data contained in matrix $A_{m\times N}$ and $E(A_{(m-1)\times N})$ means the SVD-entropy computed for the data contained in matrix $A_{(m-1)\times N}$ with the ith row (feature) removed. That is, features are temporarily removed from the dataset one-by-one and the difference before and after each feature removal is calculated. Intuitively, if a feature is not important, there would be no effect on the SVD-entropy after this feature has been removed. In other words, CE for this feature will be zero. On the contrary, if a feature is significant for data characterization, its contribution will be positive.

After computing the contribution of each feature, the obtained contribution values are sorted in descending order of magnitude. Let c and d be the average of all feature contributions and their standard deviation, respectively. Then three groups of features can be distinguished:

```
CEi>c+d (features with high contribution).
c+d>CEi>c-d (features with average contribution).
CEi<c-d (features with low contribution).
```

Features of the first category are only considered relevant since removal of such feature leads to decrease in the entropy for the whole dataset. Features of other two categories are deemed to be unimportant and therefore they can be filtered out, because the second category includes neural features (entropy does not change much if one deletes them), while the third category is comprised of redundant features.

Let the number of features in the first category be m_c. This is the upper limit and the target for three feature selection algorithms described in the next section. In other words, features are selected until the number of selected features reaches m_c.

Algorithm Description

Let *Selected* and *Remaining* denote lists storing indices of already selected features and features that were not yet analyzed, respectively. In the following sections, three algorithms utilizing *CE* are given. The simplest of them is feature ranking.

Simple Ranking (SR)

In this algorithm, *CE* for each individual feature is calculated, all *CE* values are ranked and top m_c features with the highest contribution are selected. It is the fastest algorithm among all the three but its speed results from the fact that joint contributions of several features are not considered. To remedy this, (Varshavsky, Gottlieb, Linial, & Horn, 2006) also propose two variants of the forward feature selection.

Forward Selection 1 (FS1)

FS1 chooses the first feature according to the highest *CE*. Next, another feature – candidate for selection – is sought, which, together with the first feature, produces a 2-feature subset with highest entropy (D-1 features are candidates for inclusion into the list of selected features). After that, the third, fourth, etc. features are iteratively selected so that in combination with already selected features they comprise a 3-, 4-, ... m_c-feature subsets with highest SVD-entropy (see Figure 1).

Forward Selection 2 (FS2)

In this variant of forward feature selection, the first feature is selected in the same manner as in FS1. This feature is then deleted from the set of features, the contributions of the remaining D-1 features to the SVD-entropy are recalculated and the feature with the highest contribution is added to the set of selected features. The second selected feature is afterwards removed from the set of available features and the contributions of the remaining D-2 features are again recalculated in order to

Figure 1. Pseudo code of the Forward Selection 1 Algorithm

Forward Selection 1 Algorithm (FS1)

$Selected := \varnothing$;

$Remaining := \{1,\ldots,D\}$;

Let A_{Full} be the original data matrix, $A_{D\times N}$, containing all features;

For $i := 1 : D$

   Compute the contribution of the *i*th feature:

$$CE_i = E\left(A_{Full}\right) - E\left(A_{Full\backslash Remaining_i}\right);$$

End For

Let $j := \underset{i=1,\ldots,D}{\operatorname{argmax}} CE_i$;

$Selected := Selected \cup j$; /* Add feature */

$Remaining := Remaining \backslash j$; /* Remove feature */

While $|Selected| < m_c$

  For $i := 1 : |Remaining|$

---

find the third feature to be selected, etc. Such operations continue until m_c features are found. Pseudo-code of FS2 is given in Figure 2.

Brief Comments on Algorithms

In all three algorithms, the optimal number of features to find is automatically determined. Although the criterion based on which this number is determined is rather heuristic, this is nevertheless a step forward compared to the manually set number of features, which may be often far from being optimal. Thus, the algorithms in (Varshavsky, Gottlieb, Linial, & Horn, 2006) offer a certain advantage for a researcher, since it is unnecessary to guess the optimal number of features.

Figure 2. Pseudo code of the Forward Selection 2 Algorithm

Forward Selection 2 Algorithm (FS2)

$Selected := \emptyset$;

$Remaining := \{1,...,D\}$;

Let A_{Full} be the original data matrix, $A_{D\times N}$, containing all features;

For $i := 1 : D$

   Compute the contribution of the ith feature:

$$CE_i = E\left(A_{Full}\right) - E\left(A_{Full \setminus Remaining_i}\right);$$

End For

Let $j := \underset{i=1,...,D}{\operatorname{argmax}} CE_i$;

$Selected := Selected \cup j$; /* Add feature */

$Remaining := Remaining \setminus j$; /* Remove feature*/

While $\left|Selected\right| < m_c$

However, as experimental results in (Varshavsky, Gottlieb, Linial, & Horn, 2006) showed both simple ranking and forward selection algorithms find too many features (100-250). Therefore they can serve as pre-filters filling a pool of candidate features, some which might be irrelevant or weakly relevant. From this pool of features more aggressive (and often time-consuming) filters or wrappers (e.g., those based on Markov blanket (Koller & Sahami, 1996), (Knijnenburg, Reinders, & Wessels, 2006)) can further select strongly relevant features. But because these feature selection methods are applied to a smaller subset of the original set of features, this results in faster feature selection, compared to the case of the full feature set.

Given three different algorithms, one may ask which of them could be better than the others. Experiments in (Varshavsky, Gottlieb, Linial, & Horn, 2006) leave this question open, since though we could expect that FS1, taking into account feature interactions, would perform better than SR, one experiment showed almost no difference between FS1 and SR. This fact indicates that there could be problems where the simplest solution is as good as a much more computationally intensive one. However, such exceptions should not mislead a reader into thinking than FS1 or conceptually similar algorithms cannot offer significant advantages over simple ranking.

MATLAB Code

The singular value decomposition is computed in the MATLAB® function *svd*, returning N singular values for an $N \times N$ input matrix. The rest of code is straightforward and easy to understand based on the description provided in the previous section.

```
function selected_features = SVDEntropyFeatureSelection(trai
n_patterns, algorithm)


% Unsupervised SVD-entropy based feature selection
% Original source:
% Roy Varshavsky, Assaf Gottlieb, Michal Linial, and David
Horn, "Novel
% unsupervised feature filtering of biological data", Bioinfor-
matics
% 22(14), pp.e507--e513, 2006
%
% Inputs:
%    train_patterns  - Training patterns (DxN matrix, where D -
data
%                      dimensionality, N - number of patterns)
%    algorithm       - Algorithm for feature selection (string
with
%                      function name where an algorithm is
%                      implemented);
%                        valid strings: 'sr', 'fs1', and 'fs2'
%
% Outputs:
%    selected_features  - Selected features (vector of feature
indices)



% Get dataset sizes
```

```
[D,N] = size(train_patterns);

% Compute the SVD-entropy of the whole dataset
X = train_patterns;
efull = SVDEntropy(X, N);

% Allocate memory for individual feature contributions to the
SVD-
% entropy
CE = zeros(1,D);
% Compute the contribution of each feature to the SVD-entropy
for i = 1:D
    X = train_patterns;
    % Remove the ith feature before measuring its contribution
    X(i,:) = [];
    % Estimate the ith feature contribution
    CE(i) = efull - SVDEntropy(X, N);
end

% Find threshold for dividing all features into relevant and
irrelevant
thr = mean(CE) + std(CE);
% Find indices of relevant features
idx = find(CE > thr);
% Get their number
mc = length(idx);

% Now proceed according to the algorithm specified
switch upper(algorithm)
    case '',
        % No algorithm specified
        selected_features = idx;
    case 'SR',
        % Simple ranking algorithm
        selected_features = SimpleRanking(CE, mc);
    case 'FS1',
        % Forward selection 1 algorithm
        selected_features = ForwardSelection1(train_patterns,
CE, mc);
    case 'FS2',
```

```
        % Forward selection 2 algorithm
        selected_features = ForwardSelection2(train_patterns,
CE, mc);
    otherwise,
        fprintf(1, 'Algorithm %s is not supported!\n', algo-
rithm);
end

return




% Computation of the SVD-entropy
function e = SVDEntropy(X, N)

% Find singular values of an NxN matrix
s = svd(X'*X);
% s is an Nx1 vector

% Get normalized values
V = s.^2/sum(s.^2);

% Calculate the SVD-entropy
e = -(1/log(N))*V'*log(V);

return




% Simple ranking algorithm
function selected_features = SimpleRanking(CE, mc)

% Sort feature contributions to the SVD-entropy in descreasing
order of % magnitude
[dummy, idx] = sort(CE, 'descend');
% Pick up mc features with the highest contributions
selected_features = idx(1:mc);

return
```

```
% Forward Selection 1 algorithm
function selected_features = ForwardSelection1(train_patterns,
CE, mc)

% Get the initial number of features
[D,N] = size(train_patterns);

% Form indices of features
idx = 1:D;

% Find the index of the first selected feature that has the
highest
% contribution to the SVD-entropy
[dummy, i] = max(CE);
selected_features(1) = idx(i);
% Remove the selected feature from the set of features
idx(i) = [];
% Decrease the number of available features
D = D - 1;

% Incrementally add other mc-1 features one-by-one
for m = 2:mc % repeat mc-1 times
    % Compute the SVD-entropy of the "old" dataset
    e = SVDEntropy(train_patterns(selected_features,:), N);
    % Allocate memory for feature contributions
    CE = zeros(1,D);
    % Compute the contribution of each feature to the SVD-en-
tropy
    for i = 1:D
        % Estimate the ith feature contribution
        CE(i) = SVDEntropy(train_patterns([selected_features
idx(i)],:), N) - e;
    end
    % Find the index of the feature that has the highest
    % contribution to the SVD-entropy
    [dummy, i] = max(CE);
    selected_features = [selected_features idx(i)];
    % Decrease the number of available features
```

```
    D = D - 1;
    % Remove the selected feature from the set of features
    idx(i) = [];
    fprintf(1, '%d features have been added...\n', m);
end

return



% Forward Selection 2 algorithm
function selected_features = ForwardSelection2(train_patterns,
CE, mc)

% Allocate memory for selected features
selected_features = zeros(1,mc);

% Get the initial number of features
[D,N] = size(train_patterns);

% Form indices of features
idx = 1:D;

% Find the index of the first selected feature that has the
highest
% contribution to the SVD-entropy
[dummy, i] = max(CE);
selected_features(1) = idx(i);
% Remove the selected feature from the set of features
idx(i) = [];
% Decrease the number of available features
D = D - 1;

% Incrementally add other mc-1 features one-by-one
for m = 2:mc % repeat mc-1 times
    % Compute the SVD-entropy of the "whole" dataset
    efull = SVDEntropy(train_patterns(idx,:), N);
    % Allocate memory for feature contributions
    CE = zeros(1,D);
    % Compute the contribution of each feature to the SVD-en-
```

```
tropy
    for i = 1:D
        X = train_patterns(idx,:);
        % Remove the ith feature before measuring its contribu-
tion
        X(i,:) = [];
        % Estimate the ith feature contribution
        CE(i) = efull - SVDEntropy(X, N);
    end
    % Find the index of the feature that has the highest
    % contribution to the SVD-entropy
    [dummy, i] = max(CE);
    selected_features(m) = idx(i);
    % Decrease the number of available features
    D = D - 1;
    % Remove the selected feature from the set of features
    idx(i) = [];
    fprintf(1, '%d features have been added...\n', m);
end

return
```

REFERENCES

Alter, O., Brown, P. O., & Botstein, D. (2000). Singular value decomposition for genome-wide expression data processing and modeling. *Proceedings of the National Academy of Sciences of the United States of America, 97*(18), 10101–10106. doi:10.1073/pnas.97.18.10101

Hastie, T., Tibshirani, R., Eisen, M. B., Alizadeh, A., Levy, R., Staudt, L., et al. (2000). 'Gene shaving' as a method for identifying distinct sets of genes with similar expression patterns. *Genome Biology, 1*(2), research0003.1-0003.21.

Knijnenburg, T. A., Reinders, M. J., & Wessels, L. F. (2006). Artifacts of Markov blanket filtering based on discretized features in small sample size applications. *Pattern Recognition Letters, 27*(7), 709–714. doi:10.1016/j.patrec.2005.10.019

Koller, D., & Sahami, M. (1996). Toward optimal feature selection. In L. Saitta (Ed.), *Proceedings of the 13th International Conference on Machine Learning, Bari, Italy* (pp. 284-292). San Francisco, CA: Morgan Kaufmann.

Lay, D. C. (2003). *Linear algebra and its applications* (3rd ed.). Upper Saddle River, NJ: Pearson Education.

Steeb, W.-H. (2006). *Problems and solutions in introductory and advanced matrix calculus*. Singapore: World Scientific.

Varshavsky, R., Gottlieb, A., Linial, M., & Horn, D. (2006). Novel unsupervised feature filtering of biological data. *Bioinformatics (Oxford, England), 22*(14), e507–e513. doi:10.1093/bioinformatics/btl214

Chapter 15
Differential Evolution for Finding Predictive Gene Subsets

DIFFERENTIAL EVOLUTION: GLOBAL, EVOLUTION STRATEGY BASED OPTIMIZATION METHOD

This chapter introduces another evolutionary method – differential evolution – for predictive gene selection. The choice was dictated by the fact that differential evolution is one of the newest members in the family of evolutionary optimization methods.

Differential evolution is akin to the evolutionary algorithms in (Jirapech-Umpai & Aitken, 2005), (Deutsch, 2003). It is the tool for global optimization.

You can naturally suspect that finding a globally optimal solution can be much harder than getting a locally optimal solution. In many cases, the exact solution of the global optimization problem is impossible to find. In the context of combinatorial problems, such problems are called NP-hard, where the frightening 'NP-hard' stands for non-deterministic polynomial-time hard[1]. By providing this meaning and

DOI: 10.4018/978-1-60960-557-5.ch015

halting here, let me avoid of venturing further into the broad and misterious realm of computational complexity theory.

For those of you who would like to know more about global optimization, I can recommend the book of Zhigljavsky and Žilinskas (Zhigljavsky & Žilinskas, 2008).

The history and, what is more important, the development of differential evolution is well presented in (Feoktistov, 2006)[2]. I will rely on this book when explaining the basic differential evolution algorithm, despite the fact that it was invented by Kenneth Price and Rainer Storn (Price & Storn, 1997), (Storn & Price, 1997). As another view of differential evolution, the book of these authors (Price, Storn, & Lampinen, 2005) is highly recommended in addition to (Feoktistov, 2006). I preferred the book of Feoktistov, because it is thin (yes, yes, I must confess that like many of you I like thin books) and contains MATLAB® and C code that turned to be extremely useful in understanding differential evolution. I, of course, do not want to say that the inventors of differential evolution left people starving: on the contrary, the inventors turned out to be polyglots as can be found at the following web page: http://www.icsi.berkeley.edu/~storn/code.html. This web page hosts code in C, Java, MATLAB, C++, Python, R and some other programming languages. In addition, there is an advice (under 'Practical Advice' title) on how to choose values for the parameters of differential evolution. Such guidance is invaluable as we all are aware of headache caused by tedious parameter tuning.

Main Idea

According to Wolfram MathWorld http://mathworld.wolfram.com/Differentia-lEvolution.html, differential evolution is a stochastic parallel direct search based on evolution strategy optimization. It can handle non-differentiable, nonlinear and multimodal objective functions. It starts from random initialization of a population of individuals (vectors). At each iteration, called a generation, new vectors are obtained by combining vectors randomly chosen from the current population (mutation operation). These vectors are then mixed with a predetermined target vector to create a trial vector (recombination operation). Finally, the trial vector is accepted for the next generation if and only if it yields a reduction in the objective (fitness) function value (selection operation). Again, as with other evolutionary algorithms we considered, the value of the fitness function is the accuracy of a trained classifier, attained on the validation set.

As you can see, these operations resemble those of evolutionary algorithms considered earlier in our book. However, there is an important distinction that differential evolution owes its success to: the manner of the trial individual creation. I will talk more about this point below in text.

As the prototype for MATLAB® implementation of differential evolution based gene selection, I chose the algorithm from (Tasoulis, Plagianakos, & Vrahatis, 2006). However, the different encoding of individuals than in (Tasoulis, Plagianakos, & Vrahatis, 2006) is used here: if a certain gene is selected, then its value is 1; otherwise, it is 0. Real-valued numbers can be easily converted to this binary representation by means of thresholding as follows: if a gene value is greater than threshold, it is turned to 1; otherwise to 0. The binary representation has natural interpretation and is employed in other evolutionary algorithms.

Algorithm Description

Before proceeding with explanation of the main steps of differential evolution, I would like to draw to readers a general picture of a typical evolutionary algorithm. One can ask, why did not you do so in chapter introducing an evolutionary algorithm? I feel that the example of differential evolution better suits to that picture, since differential evolution contains all basic operations of a typical evolutionary algorithm whereas the evolutionary algorithm described in the previous chapter does not explicitly include crossover (recombination involving inheritance of genes from parents by children)[3].

So, a typical evolutionary algorithm encodes some basic principles of evolution in nature. Each individual is characterized by a set of features, called genes. An ensemble of individuals forms a population. Prior to optimization, an evolutionary algorithm undergoes initialization, which is often random since we do not know the optimal point location. Next, each individual is evaluated based on its fitness to optimization criterion. This is called evaluation. The initial population is thus ready and the algorithm begins its evolutionary cycle. Iterations called generations last until a chosen stopping condition is met. In each evolutionary cycle the population of individuals passes through the following steps.

- Selection of the individuals for a population.
- Variations of the selected individuals in a random manner by means crossover and mutation to produce children of the current parent individuals (or simply parents).
- Replacement by picking the best individuals among parents and children in order to form a new population.

Schematically, the typical evolutionary algorithm is represented in Figure 1 (I adopted its pseudo code from (Feoktistov, 2006) and slightly modified it).

Now it is the time to present the differential evolution algorithm. Figure 2 contains pseudo code of the differential evolution algorithm. The algorithm begins with

Figure 1. Pseudo code of the typical Evolutionary Algorithm

Typical Evolutionary Algorithm

Set generation counter i to zero;
Randomly initialize a population of individuals, *Pop*;
Evaluate fitness of this population;

While (not stopping condition)

    Increase generation counter: $i := i + 1$;
    Select individuals from *Pop* to form a set of parents;
    Vary parents to produce children via crossover, mutation, ... ;
    Evaluate fitness of children;
    Form a new population of individuals *Pop* from the best children and parents;

End While

---

random initialization of the population *Pop* of the *NP* individuals, where each individual is represented by a *D* dimensional vector of values randomly sampled from [0,1). So, basically zero and one serve as lower and upper boundary constraints. Each of the *D* values of an individual can be treated as a feature or gene. Since individuals are evaluated by a classifier, one needs to discretize each gene, since a classifier assumes that zero (one) means that gene is to be included (excluded)[4].

The simplest way to discretize is to compare a value against pre-defined threshold: if the value exceed threshold, then set it to 1, otherwise, to 0. After that, each discretized individual is evaluated by a classifier performing the role of the fitness function. It should be noted that we discretize genes only prior to calling a classifier; in all other operations, the original, real-valued genes compose each individual. In fact, discretization does not have to be carried out, because it is enough to select genes whose value exceeded threshold (see MATLAB® code below).

Gene discretization can be deemed as integer variable handling (see section 2.5 in (Feoktistov, 2006)) where discrete values of the genes are only used for fitness function evaluation.

Once initialized, the population of real-valued individuals is prepared for iterative optimization. The outer loop runs over generations whereas the inner loop runs over individuals in each generation. The number of generations, *GEN*, is used as a stopping criterion. For each individual with index j, three random members of the current generation, mutually different and different from this individual are randomly selected. Let indices of these three individuals be r_1, r_2, r_3 and $r_1 \neq r_2 \neq r_3 \neq j$. Genes of

Figure 2. Pseudo code of the Differential Evolution algorithm

---

Differential Evolution

Create the initial population *Pop* of *NP* individuals, where each individual Pop_j is assigned a vector of randomly generated values from $[0,1)$. Vector length is *D*; Make gene values discrete;
Evaluate the fitness function $Fit(Pop_j)$ for each individual Pop_j;

For $k := 1:GEN$ /* for each generation */
  For $i := 1:NP$ /* for each individual */

    Choose three random individuals from *Pop* that are mutually different and different from the *i*th individual;

    Create a trial individual *X* based on these individuals and the *i*th individual (mutation+crossover);

    Check boundary constraints for the trial individual and correct gene values if necessary;

    Compute the fitness function of the trial individual $Fit(X)$ and compare it with the fitness function of the *i*th individual $Fit(Pop_i)$:
    i.   **If** $Fit(X) \geq Fit(Pop_i)$
      a. $Pop_i := X; Fit(Pop_i) := Fit(X)$;
            Compare the fitness function of the trial individual with the fitness function of the so far best individual encountered during the search:
      b. **If** $Fit(X) \geq Fit(Pop_{ibest})$, then $ibest := i$;

  End For
End For

---

these four individuals then serve as a basis for creation of a trial individual. To guarantee that at least one gene in the trial individual will be changed, i.e. minimal mutation will occur, an index *k* between 1 and *D* is randomly selected, implying that even under very unfavorable circumstances, the value of a gene with index *k* will be different from values of the same gene in other individuals. However, one change is a small change and in order to encourage more changes, the crossover constant *CR* is enlisted (it acts like probability of acceptance of mutation). As a result, the genes of the trial individual are constructed according to the following probabilistic rule[5]:

$$x_i = \begin{cases} x_{i,r_3} + F\left(x_{i,r_1} - x_{i,r_2}\right) & if\ r < CR\ or\ i = k \\ x_{ij} & otherwise \end{cases} \qquad i = 1,\dots,D,$$

where r is a random number from $[0,1)$, F is the differentiation or mutation constant, and x_{ij} denotes the ith gene of the jth individual. As one can see, if either a randomly generated number r is less than the crossover constant or the gene index is equal to k, mutation occurs based on gene values of three randomly picked individuals with indices r_1, r_2, r_3. In all other cases, the gene of the trial individual gets the value of the current individual copied. It should be noted that there are a plenty of differentiation rules to construct a trial individual. I refer to, e.g., Chapter 3 in (Feoktistov, 2006) for their analysis.

As some gene values of the resulting trial individual may go beyond boundaries, the next step is to constrain them within boundary limits 0 and 1. This can be done simply by replacing out-of-boundary values with numbers randomly generated from $[0,1)$ while leaving all other value unchanged.

Finally, the best individual in the current generation is found. For this, first the fitness functions (classification accuracies) of the trial and the current individuals are compared. If the trial individual's function (accuracy) is greater than or equal to the current one, then the trial individual replaces the current individual in the population. If, in addition, the trial individual's accuracy is higher than or equal to that of the so far best individual with index *ibest*, then *ibest* needs to be updated, too. When the algorithm terminates upon reaching the pre-defined number of generations, the individual with index *ibest* is discretized in order to extract a subset of selected genes. This subset can be then used to classify test data.

Why Differential Evolution is Successful?

As was said above, differential evolution owes its success among different evolutionary algorithms to the way the trial individual is created. It might sound strange but the secret of success is revealed if to bring forward the formula for the trial individual creation:

$$x_i = \begin{cases} x_{i,r_3} + F\left(x_{i,r_1} - x_{i,r_2}\right) \\ x_{ij} \end{cases}.$$

Here, F serves as a scaling factor for the difference between two randomly selected individuals (hence, 'differential' in the name of differential evolution) and it determines the trade-off between exploitation and exploration of the space of genes. The product of the differentiation constant and this difference defines direction and length of the search step. We can call this product the step length. Thus obtained product is next added to the third randomly chosen individual in order to provide the trial individual.

In the beginning, when individuals are far away from each other, the step length is large. However, as the evolution goes on, the population converges and step length becomes smaller and smaller. Thus, differential evolution adapts the step length during the evolutionary process. Randomness of both search direction and individuals selected for composing a trial vector leads in many cases to the global optimum while only slightly slowing down convergence. In the words of Feoktistov (Feoktistov, 2006): the concept of differential evolution is a spontaneous self-adaptability to the function. And no pre-defined probability density function is needed to know in order to do optimization.

If in the evolutionary algorithms by (Deutsch, 2003) and (Jirapech-Umpai & Aitken, 2005) the mutation of the current individual is employed as the way to produce a new individual, then differential evolution takes a different approach where a new individual is born from recombination of the trial and current individuals. Hence, the mutation constant F might not be so important[6] in the differential evolution algorithm, because its influence can be imitated by the crossover constant CR. Thus, in differential evolution, more attention is paid to the efficient creation of a new member of a population than to the mutation of the current individual[7].

Potential diversity of the entire population is maintained by crossover guaranteeing that changes in gene values will always happen. Emphasizing the importance of crossover over that of mutation in the differential evolution algorithm can be backed by the fact that there is no disruption effect[8] for differentiation. However, it should be understood that mutation cannot be entirely ignored in the evolution process. Both mutation and crossover are necessary for an evolutionary algorithm to succeed. It is their intelligence combination that magnifies their mutual advantages, which is tricky.

Continuous improvement in the fitness function value is achieved by the elitist principle of selecting the best individual. This is the same principle that is used in many evolutionary algorithms.

Thus, keys to success of differential evolution are self-adaptability, diversity control, and continuous improvement of the fitness function.

MATLAB Code

I modified the code provided by Feoktistov in his book. However, its essence remained the same. Due to the fact that main blocks of differential evolution were extensively discussed and analyzed in the previous sections, there is not much to say about the code itself. Differential evolution employs the same fitness function as the evolutionary algorithm in the previous chapter. I copied its code for convenience. As a classifier, my favorite choices remained the same: Nearest Neighbors and Naïve Bayes. The only important question is recommended default values of parameters of the algorithm.

Population size NP should be at least 10 times larger than the number of genes per individual, i.e. $NP=10D$. It seems that this is the rule of thumb as several independent sources cite it. If population size is large, individuals densely populate the search space and therefore the mutation constant F can be small because small amplitude of their movements is sufficient.

The mutation constant F has significant impact on exploration of the search space. Feoktistov recommended $F=0.85$ while Storn (Storn, On the usage of differential evolution for function optimization, 1996) (see also http://www.icsi.berkeley.edu/~storn/code.html#prac) advised on $F=0.8$. Both values favor broad exploration in order for the algorithm not to get trapped into a local optimum.

The crossover constant CR reflects the probability with which the trial individual inherits the current individual's genes. Feoktistov reckons on $CR=0.5$ whereas Storn suggests $CR=0.9$. The larger the crossover constant is, the less probability that the trial individual will inherit properties of the current individual. Thus, the value of CR is rather problem-dependent and set up by trial-and-error.

The threshold *thr* needs to be set to a high value, say 0.9 or even higher, in order to reduce the number of genes to be evaluated by a classifier. This threshold can be adjusted to the approximate number of genes to be selected in the end. Say, if we want to select about 5% of genes, then threshold can be equal to 0.95.

The uncertainty about best parameter settings naturally leads to desire to run the algorithm multiple times, each time with distinct settings. For instance, such an approach was used in (Tvrdík, 2006) with success. Tvrdík proposed in (Tvrdík, Competitive differential evolution, 2006) to introduce a competition into differential evolution so that parameter setting can be adaptively done. It is supposed that the competition of different settings will prefer successful settings via a self-adaptive mechanism. MATLAB® code for competitive differential evolution can be downloaded from http://albert.osu.cz/oukip/optimization/. The idea of competition is explained in (Tvrdík, Differential evolution: competitive setting of control parameters, 2006) as follows.

Suppose that there are H different sets of parameter values for F and CR. Selection of a certain set h occurs with the probability q_h, $h=1,2,...,H$. The probability can vary, depending on the success rate of the associated set of parameter values. The set is successful if it generates such a trial individual that its fitness function has a higher value than that of the current individual. Given that n_h is the current number of successes of the hth set, the probability q_h is evaluated as the relative frequency:

$$q_h = \frac{n_h + n_0}{\sum_{j=1}^{H} \left(n_j + n_0 \right)},$$

where $n_0 > 0$ is a constant. Setting $n \geq 1$ prevents a sudden jump in q_h caused by a single random success. In order to avoid degeneration if a certain probability becomes too small, if any $q_h < \delta$, it is reset to the starting value $1/H$.

Besides achieving a competition, it is necessary to run differential evolution several times not only in order to discover best parameter settings, but also in order to become certain in stability of the obtained results. Like in one of the previous chapters, the z-test can be useful for this purpose.

```
function selected_features = DifferentialEvolution(train_pat-
terns, train_targets, valid_patterns, valid_targets, GEN, NP,
F, CR, thr, classifier, varargin)


% Differential Evolution based feature selection
% Original sources:
% Dimitris K. Tasoulis, Vassilis P. Plagianakos, Michael N.
Vrahatis,
% "Differential evolution algorithms for finding predictive
gene
% subsets in microarray data", In Ilias Maglogiannis, Kostas
Karpouzis, % Max Bremer (eds.), Proceedings of the 3rd IFIP
Conference on
% Artificial Intelligence Applications and Innovations, Athens,
Greece,
% pp.484-491, 2006
% Vitaliy Feoktistov, "Differential evolution: in search of so-
lutions",
% Springer Science+Business Media, New York, NY, 2006
% Vassilis P. Plagianakos and Michael N. Vrahatis, "Neural net-
work
% training with constrained integer weights", In P.J.Angeline,
Z.
```

```
% Michalewicz, M.Schoenauer, X.Yao, A.Zalzala (eds.), Proceed-
ings of
% the 1999 Congress on Evolutionary Computation, Washington,
DC,
% pp.2007-2013, 1999
%
% Inputs:
%    train_patterns  - Training patterns (DxN matrix, where D -
data
%                       dimensionality, N - number of patterns)
%    train_targets   - Training targets (1xN vector)
%    valid_patterns  - Validation patterns (DxNt matrix, where
D - data %
%                       dimensionality, Nt - number of patterns)
%    valid_targets   - Validation targets (1xNt vector)
%    GEN             - The number of generations
%    NP              - The number of predictors per generation
%    F               - Differentiation constant
%    CR              - Crossover constant
%    thr             - Threshold for feature binarization; rec-
ommended
%                        values should be at least 0.5 or closer
to 0.9%
%                       in order to dramatically reduce the num-
ber of
%                       features to feed to a fitness function
%    classifier      - Classifier (string with function name
where a
%                       classifier is implemented)
%    varargin        - Parameters of a classifier
%
% Output:
%    selected_features   - Selected features (vector of feature
indices)
%
% Examples:
% selected_features =
%   DifferentialEvolution(train_patterns, train_targets,
%   valid_patterns, valid_targets, 10000, 60, 0.9, 0.5, 0.8,
'NearestNeighbor', 3);
```

```
% selected_features =
%  DifferentialEvolution(train_patterns, train_targets,
%  valid_patterns, valid_targets, 10000, 60, 0.9, 0.5, 0.8,
'NaiveBayes');




% Get the number of genes
D = size(train_patterns,1);

% Allocate memory for a trial vector
x = zeros(D,1);
% Allocate memory for population
Pop = zeros(D,NP);
% Allocate memory for fitness of population
fit = zeros(1,NP);

% Index of the best solution
ibest = 1;

% Initialization stage
for j = 1:NP
    % Random initialization
    Pop(:,j) = rand(D,1);
    % As we need to decide which features to use, pick those
features
    % whose initial values are strictly larger than threshold
    ind = find(Pop(:,j) > thr);
    % Compute the fitness function value
    if isempty(ind)
        fit(j) = 0;
    else
        fit(j) = fitness_fcn(train_patterns(ind,:), train_tar-
gets, valid_patterns(ind,:), valid_targets, classifier, vara-
rgin{:});
    end
end

% Optimization stage
for i = 1:GEN
```

```
    fprintf(1, 'Generation %d...\n', i);
    for j = 1:NP
        % Choose three random individuals from the current
population
        % They all must be mutually different and different
from j
        % Form a list of indices of all individuals
        n = 1:NP;
        % Ecxlude index j from the list of individuals
        n(j) = [];
        % Choose indices of three distinct individuals
        ind = randsample(n,3);

        % Create a trial individual with at least one feature
changing
        % its value
        % Index of a feature to change
        k = randsample(1:D, 1);
        for ii = 1:D
            if rand < CR || ii == k
                x(ii) = Pop(ii,ind(3)) + F*(Pop(ii,ind(1)) -
Pop(ii,ind(2)));
            else
                x(ii) = Pop(ii,j);
            end
        end

        % Verify boundary constraints and apply corrections if
        % necessary
        ind = find(x < 0 | x > 1);
        x(ind) = rand(1,length(ind));

        % Find indices of features in the trial individual that
        % survived thresholding
        ind = find(x > thr);
        % Compute the fitness function value for the trial in-
dividual
        if isempty(ind)
            fit_x = 0;
        else
```

```
            fit_x = fitness_fcn(train_patterns(ind,:), train_
targets, valid_patterns(ind,:), valid_targets, classifier,
varargin{:});
        end

        % Based on the fitness function value, select the best
        % individual
        % First, try if the trial individual is better than the
current
        % one
        if fit_x >= fit(j)
            % Replace the current individual with the best one
            Pop(:,j) = x;
            fit(j) = fit_x;
            % Then check if the trial individual is better than
the
% best one
            if fit_x >= fit(ibest)
                % Update the best individual's index
                ibest = j;
            end
        end
    end % end for j
end % end for i

% Form a vector of indices of selected features
selected_features = find(Pop(:,ibest) > thr);
selected_features = selected_features';

return



% Fitness function
function fit = fitness_fcn(train_patterns, train_targets, val-
id_patterns, valid_targets, classifier, varargin)

% Inputs:
%    train_patterns  - Training patterns (DxN matrix, where D -
data
```

```
%                        dimensionality, N - number of patterns)
%    train_targets    - Training targets (1xN vector)
%    valid_patterns   - Validation patterns (DxNt matrix, where
D - data
%                        dimensionality, Nt - number of patterns)
%    valid_targets    - Validation targets (1xNt vector)
%    classifier       - Classifier (string with function name
where a
%                        classifier is implemented)
%    varargin         - Parameters of a classifier
%
% Output:
%    fit              - Fitness value (classification accuracy in
per
%                        cent)


% Get predictions of a classifier for validation data
predicted_targets = feval(classifier, train_patterns, train_
targets, valid_patterns, varargin{:});
% Compute classification accuracy as a fitness value
fit = 100 - class_error(valid_targets, predicted_targets);

return
function errprcnt = class_error(test_targets,targets)

% Calculate error percentage based on true and predicted test
labels
errprcnt = mean(test_targets ~= targets);
errprcnt = 100*errprcnt;

return
```

REFERENCES

Deutsch, J. M. (2003). Evolutionary algorithms for finding optimal gene sets in microarray prediction. *Bioinformatics (Oxford, England)*, *19*(1), 45–52. doi:10.1093/bioinformatics/19.1.45

Feoktistov, V. (2006). *Differential evolution: in search for solutions*. New York, NY: Springer Science+Business Media.

Jirapech-Umpai, T., & Aitken, S. (2005). Feature selection and classification for microarray data analysis: evolutionary methods for identifying predictive genes. *BMC Bioinformatics, 6*(148). doi:.doi:10.1186/1471-2105-6-148

Price, K., & Storn, R. (1997). Differential evolution: a simple evolution strategy for fast optimization. *Dr. Dobb's Journal of Software Tools, 22*(4), 18–24.

Price, K. V., Storn, R. M., & Lampinen, J. A. (2005). *Differential evolution: a practical approach to global optimization*. Berlin, Heidelberg: Springer-Verlag.

Spears, W. M. (1993). Crossover or mutation? In Whitley, L. D. (Ed.), *Foundations of genetic algorithms 2* (pp. 221–237). San Mateo, CA: Morgan Kaufmann.

Storn, R. (1996). On the usage of differential evolution for function optimization. *Proceedings of the Biennial Conference of the North American Fuzzy Information Processing Society, Berkeley, CA* (pp. 519-523). Los Alamitos, CA: IEEE Computer Society.

Storn, R., & Price, K. (1997). Differential evoloution - a simple and efficient heuristic for global optimization over continuous spaces. *Journal of Global Optimization, 11*(4), 341–359. doi:10.1023/A:1008202821328

Tasoulis, D. K., Plagianakos, V. P., & Vrahatis, M. N. (2006). Differential evolution algorithms for finding predictive gene subsets in microarray data. In I. Maglogiannis, K. Karpouzis, & M. Bramer (Ed.), *Proceedings of the 3rd IFIP Conference on Artificial Intelligence Applications and Innovations, Athens, Greece. IFIP International Federation for Information Processing 204* (pp. 484-491). Boston, MA: Springer.

Tvrdík, J. (2006). Competitive differential evolution. In R. Matoušek, & P. Ošmera (Ed.), *Proceedings of the 12th International Conference on Soft Computing, Brno, the Czech Republic* (pp. 7-12).

Tvrdík, J. (2006). Differential evolution: competitive setting of control parameters. In M. Ganzha, M. Paprzycki, J. Wachowicz, K. Węcel, & T. Pełech-Pilichowski (Ed.), *Proceedings of the International Multiconference on Computer Science and Information Technology, Wisła, Poland* (pp. 207-213).

Zhigljavsky, A., & Žilinskas, A. (2008). *Stochastic global optimization*. New York, NY: Springer Science+Business Media.

ENDNOTES

[1] Fans of Bruce Willis and his character, police officer John McClane in "Die hard" movies can draw a parallel. In each movie, John McClane has to solve an extremely difficult problem during a very tight period of time. So, basically, he needs to solve the 'NP-hard' problem in the time faster than polynomial (we would say 1 s faster, because the solution in the 'polynomial time' would correspond to a bomb explosion or another terrible disaster in the movie) and he does it every time! But this is a movie; mathematical reality is unfortunately harsher, that is why NP-hard problems are so hard to crack.

[2] See section 1.2 in that book.

[3] This omission, however, does not automatically make this algorithm somewhat inferior to differential evolution.

[4] The 0/1 representation is common for all evolutionary algorithms.

[5] This rule implements random strategy of search and is good to apply if one knows a little about the form of the function to be optimized.

[6] However, F is responsible for the efficiency of feature space exploration. Hence, it contribution should not be neglected.

[7] This does not imply that mutation can be absolutely ignored. Simply in differential evolution, crossover is deemed more contributing to optimization than a combination of mutation and selection.

[8] Disruption rate theory (Spears, 1993), applied to the analysis of many evolutionary strategies, estimates the likelihood that a genetic operator will disrupt a hyperplane sample. This can be also interpreted as the likelihood that individuals within a hyperplane will leave that hyperplane. It does not, however, indicate where those individuals will go. Since the trial vector is constructed as $\beta + F\delta$, where both β and δ belong to the same hyperplane, then their linear combination also belongs to this hyperplane. This implies a good survival capability of differentiation inherent to crossover.

Chapter 16
Ensembles of Classifiers

ENSEMBLE LEARNING

So far in the book, a single algorithm was assumed sufficient for data classification, which means that a single hypothesis was used to make predictions. However, no single algorithm can be superior in all possible cases. This understanding gave rise to the idea of ensemble learning, where a collection, or ensemble, of classifiers is used. Each ensemble member classifies the input data and individual predictions are then combined by a special algorithm, called a combiner, to form the prediction of the whole ensemble. For instance, several hundreds of decision trees can be generated from the same training data, each of them votes for a certain class, and all votes are then fused by a chosen combination rule to produce the final vote for a test instance. The example of a popular combination rule is the ordinary majority voting, where the winning class is determined by the simple majority. If the number of classifiers in the ensemble is 500, then for the two-class problems it is sufficient to have at least 251 votes for class C_1 in order for the ensemble to decide in favor of this class. If the number of votes for C_1 is less than 250, then it is obvious that another class, C_2, got more votes and the final classification is C_2. In case when

DOI: 10.4018/978-1-60960-557-5.ch016

votes split evenly, i.e. 250 for each class, random tie break is usually applied, which is equivalent to generating a uniform number between 0 and 1 and to deciding on C_1 if that number is less than or equal to 0.5; otherwise the decision is C_2.

The main motivation of using ensembles is to obtain smaller classification error than in the case of any single classifier. Suppose that each classifier h_i has an error p – that is, the probability that a randomly chosen instance is misclassified by h_i is p. Furthermore, suppose that the errors made by classifiers are independent. In this case, if p is small, then the probability of a large number of misclassifications is very small. For example, an ensemble of five classifiers with $p=0.1$ (1 error in 10 cases) for each classifier will have an error rate of less than 1 in 100 (Russell & Norvig, 2003)[1]. Although the independence assumption is often difficult to observe in practice, if predictions of different classifiers in an ensemble are at least a little bit different, thereby reducing the correlation between classification errors, then ensemble learning can be very useful.

Another argument in favor of ensembles is that a group of classifiers is capable of learning a more complex decision boundary separating different classes than a single classifier. This is illustrated in Figure 1, where three classifiers can correctly assign class labels to the solid black circle instances inside a triangular region, while labeling correctly all the open circle instances outside this region. None of the three classifiers can do this alone.

However, you, my ever-watchful readers, may comment: the more complex decision boundary a classifier can learn, the more such a classifier can be affected by overfitting. And you are absolutely right.

Figure 1. The example where an ensemble of three classifiers is capable of learning a more complex hypothesis than any ensemble member individually

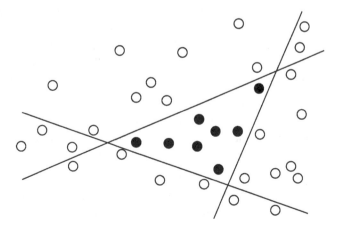

If a single classifier may be prone to data overfitting, how about an ensemble, which is a more complex model than a single classifier? Overfitting can be mitigated, though not entirely avoided, if classifiers composing an ensemble are not over-trained (Duin, 2002). If a data set is small, like in the case of microarray gene expression data sets, then the recommendation is not to use trainable combiners as simple majority voting (non-trainable combiner) may often be highly competitive with respect to more sophisticated combination schemes that require training and hence, extra data.

In order for an ensemble to deliver superior performance, classification errors made by individual classifiers in an ensemble should be as uncorrelated (diverse) as possible. Diversity of predictions among ensemble members is as important as accuracy of individual classifiers. This means that given a pool of classifiers to choose from, including only top accurate classifiers into an ensemble is a wrong strategy as the resulting ensemble is likely to deliver inferior performance because many ensemble members would produce errors on the same instances; hence, either no improvement can be obtained, compared to the best ensemble member or this improvement will be marginal. On the other hand, pursuing diversity at the expense of accuracy cannot be justified, either, as the overall result will be a poor accuracy. The exact relation between diversity and accuracy is not so far found. The only thing that is surely known is the diversity-accuracy trade-off, implying a compromise between diversity of individual predictions and their accuracy. More information about measuring diversity in practice can be found in (Kuncheva, 2004), which one of the recommended textbooks on ensemble methods.

Other exclusively ensemble-related books include (Maimon & Rokach, 2005), (Rokach, 2010) and (Seni & Elder, 2010). Proceedings of Multiple Classifier Systems Workshop published by Springer-Verlag (www.springer.com) are another valuable source of knowledge on ensembles. The surveys of Rokach (Rokach, Taxonomy for characterizing ensemble methods in classification tasks: a review and annotated bibliography, 2009) and Polikar (Polikar, 2006) are also highly recommended. Finally, do not forget to visit http://en.wikipedia.org/wiki/Ensemble_learning.

Quite often an ensemble composed of all available classifiers is not the optimal one. This means that an ensemble or ensembles including a subset of all classifiers can outperform the former. This fact leads to the necessity of classifier selection before classifiers are combined together. One of the natural criteria to incorporate into an ensemble selection process is the diversity-based selection, where selection starts from picking the most accurate classifier, which other classifiers are added to one-by-one based on the diversity of their outputs with respect to other classifiers that have been already added to the ensemble in the previous iterations.

There are many ensemble models in the literature. Out of them, I chose to describe the most wide-spread variants such as bagging and boosting, which is a typical choice of all machine learning and pattern recognition books (Bishop, 2006), (Hastie, Tibshirani, & Friedman, 2003), (Izenman, 2008), (Kuncheva, 2004), (Maimon & Rokach, 2005). One of the explanations for the success of bagging and boosting is the bias-variance decomposition (trade-off, dilemma) of the classification error. This assumes that the classification error is composed of two components[2] that one can control: bias and variance. There is the third component called noise, but it is uncontrollable and hence, it is of no interest for us, since we cannot influence on it.

The bias-variance trade-off is a term summarizing the fact that if one wants less bias in the estimate (of the true classification error, in our case), it usually costs more variance of the estimate (Everitt, 2006). If one wants to decrease variance, then it goes at the expense of increased bias. In other words, neither quantity can substantially be decreased without increasing another one. This is why it is called a dilemma or trade-off. Despite its existence, if it is somehow possible to reduce both bias and variance, then the classification error may be the smallest to attain.

Bias is a deviation of results or inferences from the truth (Everitt, 2006). More specifically, this is the extent to which a method (classifier, in our case) does not estimate the quantity (classification error, in our case) thought to be estimated. Bias is measured by the difference between an estimate (of error, in our case) and its expected value $E\left(\hat{\theta}\right)$. A biased estimator is thus the estimator such that $E\left(\hat{\theta}\right) \neq \theta$.

The motivation behind such estimators is in the potential for obtaining values that are closer, on average, to the parameter (true classification error, in our case) being estimated than would be obtained from unbiased estimators (Everitt, 2006).

Kuncheva (Kuncheva, 2004) comments, based on the talk of Dietterich (Dietterich, 2002) that bias is associated with underfitting the data, i.e. the classifier cannot deliver the optimal error rate. Put differently, a classifier with too few parameters is inaccurate because of its large bias, preventing the classifier to fit well the data. On the other hand, variance is associated with overfitting, i.e. the optimal classifier is optimal only for a given data set and it may easily cease to be optimal due to a slight change of this data set or due to the replacement of the current data set with another one. This can be interpreted as follows: a classifier with too many parameters is inaccurate because of large variance.

Naïve Bayes has a high bias for nonlinear problems because it can only model one type of decision boundary – a linear hyperplane. Like Naïve Bayes, k-nearest neighbors also have a high bias (one parameter to set up if a distance metric is fixed) and low variance. In contrast, a decision tree classifier when the tree is left unpruned after growth possesses a low bias and high variance because a slight change in data can easily result in a dramatic change in the tree structure. Bias and variance of support vector machines are heavily dependent on the kernel and its parameters as shown in (Valentini & Dietterich, 2004): a linear SVM likely has a high bias and low variance while a Gaussian or RBF SVM likely has a low bias but high variance.

One can think of bias as resulting from our domain knowledge (or the lack of it) that we build into the classifier. If we know that the true decision boundary between two classes is linear, then a linear classifier would more likely to succeed than a nonlinear classifier. However, if the decision boundary is nonlinear and we apply a linear classifier, then we are bound, on average, to incur a lot of classification errors due to our wrong bias.

As remarked in (Kuncheva, 2004), there is no general theory about how bagging and boosting effect bias and variance. Based on numerous experimental results, bagging is assumed to reduce variance without affecting the bias. Nevertheless, for some high-biased classifiers, bagging may reduce the bias, too. Boosting, on the other hand, at its early iterations primarily reduces bias while at the later stages it reduces mainly variance.

In general, the following techniques can be used to form ensembles:

- using different subsets of features for different classifiers;
- using different subsets of instances for different classifiers;
- using different types of classifiers, e.g., a decision tree and a support vector machine;
- using different parameter settings for the same classifier, e.g., 1-nearest neighbor and 5-nearest neighbors;
- using different distance metrics;
- any combination of the techniques just mentioned.

In chapters that will follow, I will utilize the first two techniques as they are most widely used for ensemble formation. Sometimes, several ensemble models are combined into a supra-ensemble (Webb, 2000), (Dettling, 2004) or (Kotsianti & Kanellopoulos, 2007).

It should be noted in the end of this chapter that the powerful idea of ensembles is not limited to machine learning, data mining and pattern recognition fields. For those who became intrigued by these words, the advice is to read (Surowiecki, 2005) and (McMahon, 2001). After that you will be able to discover ensembles everywhere (sorry, folks, no guarantee is attached though). For those who are keen to look for practical applications of ensemble methodology, the two books (Okun & Valentini, 2008), (Okun & Valentini, Applications of supervised and unsupervised ensemble methods, 2009), co-edited by Oleg Okun together with his colleague Giorgio Valentini, are humbly recommended (please, do not perceive this as self-promotion; however, if you suddenly felt an urge to buy one book or even both books, then … do not wait or hesitate).

REFERENCES

Bishop, C. M. (2006). *Pattern recognition and machine learning*. New York, NY: Springer Science-Business Media.

Dettling, M. (2004). BagBoosting for tumor classification with gene expression data. *Bioinformatics (Oxford, England)*, *20*(18), 3583–3593. doi:10.1093/bioinformatics/bth447

Dietterich, T. G. (2002, September 22-28). *Bias-variance analysis of ensemble learning*. Retrieved April 4, 2010, from The 7th Course on Ensemble Methods for Learning Machines: http://www.disi.unige.it/ person/MasulliF/ricerca/school2002/ contributions/ vietri02-lect-dietterich1.pdf

Duin, R. P. (2002). The combining classifier: to train or not to train? In *Proceedings of the 16th International Conference on Pattern Recognition, Québec, QC* (pp. 765-770). Los Alamitos, CA: IEEE Computer Society.

Everitt, B. (2006). *The Cambridge dictionary of statistics* (3rd ed.). Cambridge, UK: Cambridge University Press.

Hastie, T., Tibshirani, R., & Friedman, J. (2003). *The elements of statistical learning: data mining, inference, and prediction*. New York, NY: Springer-Verlag.

Izenman, A. J. (2008). *Modern multivariate statistical techniques: regression, classification, and manifold learning*. New York, NY: Springer Science+Business Media.

Kotsianti, S. B., & Kanellopoulos, D. (2007). Combining bagging, boosting and dagging for classification problems. In B. Apolloni, R. J. Howlett, & L. C. Jain (Ed.), *Proceedings of the 11th International Conference on Knowledge-Based Intelligent Information and Engineering Systems and the 17th Italian Workshop on Neural Networks, Vietri sul Mare, Italy. Lecture Notes in Computer Science, Vol. 4693* (pp. 493-500). Berlin/Heidelberg: Springer-Verlag.

Kuncheva, L. I. (2004). *Combining pattern classifiers: methods and algorithms*. Hoboken, NJ: John Wiley & Sons. doi:10.1002/0471660264

Maimon, O., & Rokach, L. (2005). *Decomposition methodology for knowledge discovery and data mining*. Singapore: World Scientific.

McMahon, C. (2001). *Collective rationality and collective reasoning*. Cambridge, UK: Cambridge University Press.

Okun, O., & Valentini, G. (Eds.). (2008). *Supervised and unsupervised ensemble methods and their applications*. Berlin, Heidelberg: Springer-Verlag.

Okun, O., & Valentini, G. (Eds.). (2009). *Applications of supervised and unsupervised ensemble methods*. Berlin, Heidelberg: Springer-Verlag.

Polikar, R. (2006). Ensemble based systems in decision making. *IEEE Circuits and Systems Magazine, 6*(3), 21–45. doi:10.1109/MCAS.2006.1688199

Rokach, L. (2009). Taxonomy for characterizing ensemble methods in classification tasks: a review and annotated bibliography. *Computational Statistics & Data Analysis, 53*(12), 4046–4072. doi:10.1016/j.csda.2009.07.017

Rokach, L. (2010). *Pattern classification using ensemble methods*. Singapore: World Scientific.

Russell, S., & Norvig, P. (2003). *Artificial intelligence: a modern approach* (2nd ed.). Upper Saddle River, NJ: Pearson Education.

Seni, G., & Elder, J. F. (2010). *Ensemble methods in data mining: improving accuracy through combining predictions*. San Francisco, CA: Morgan & Claypool Publishers.

Surowiecki, J. (2005). *The wisdom of crowds*. New York, NY: Anchor Books.

Valentini, G. (2003). *Ensemble methods based on bias-variance analysis*. PhD thesis. Genova, Italy: University of Genova.

Valentini, G., & Dietterich, T. G. (2004). Bias-variance analysis of support vector machines for the development of SVM-based ensemble methods. *Journal of Machine Learning Research, 5*(Jul), 725–775.

Webb, G. I. (2000). MultiBoosting: a technique for combining boosting and wagging. *Machine Learning, 40*(2), 159–196. doi:10.1023/A:1007659514849

ENDNOTES

[1]   Indeed, it is easy to calculate an ensemble error rate, provided errors made by individual ensemble members are independent and votes of individual classifiers are combined by the majority rule. We can find such an error rate based on the formula for the binomial cumulative distribution function (CDF). The ensemble of five members will make an error if there will be less than 3 correct votes out of 5. If the probability of error is 0.1, then the probability of success is 0.9. As a result, we need to find the probabilities of success in up to 2 independent votes, i.e. the number of votes resulting in error is 0, 1 or 2. This probability will be our desired ensemble error rate. Mathematically, this

fact is expressed as $\sum_{i=0}^{2} \binom{5}{i} 0.9^i 0.1^{5-i} = 0.0086$ (less than 0.01, i.e. less than 1 error in 100 cases).

[2] For mathematical details of bias-variance decomposition, you may consult, for example, (Kuncheva, 2004), (Valentini, Ensemble methods based on bias-variance analysis, 2003) or (Valentini & Dietterich, 2004). There is a plenty of information regarding this subject on the web, too, so that make use of Google to guide you in your search.

Chapter 17

Classifier Ensembles Built on Subsets of Features

SHAKING STABLE CLASSIFIERS

Out of all known techniques for building classifier ensembles, considered in the previous chapter, using different subsets of features for different classifiers in an ensemble is one of the most common. In this setting, each ensemble member is associated with its own feature subset, which can be either selected by a certain feature selection algorithm or randomly sampled from the original pool of features (Ho, 1998).

I prefer the second way because randomly chosen features do not introduce bias into the selection process and they help to avoid problems related to trapping into a local optimum while searching for a good feature subset. Stracuzzi (Stracuzzi, 2008) supports the latter argument by saying that a feature selection algorithm that chooses features at random searches for the best feature subset by sampling the space of possible subsets. Compared to the popular greedy stepwise feature selection algorithm (see, e.g., (Pudil, Novovičová, & Kittler, 1994)), which adds or removes one feature or a subset of features at a time, randomization protects against local

DOI: 10.4018/978-1-60960-557-5.ch017

minima as a random sampling is unlikely to concentrate on any particular portion of the search space.

Randomization is also very helpful when many subsets of features possess equal predictive power, i.e. they lead to equal classification performance. In such a case, random sampling of feature subsets tends to find a good solution quickly. This is a typical case with microarray gene expression data where multiple subsets of genes yield equally good classification performance (about multiplicity you may see (Díaz-Uriarte & Alvarez de Andrés, 2006) and references in that article).

There can be concerns that random subset selection does not explore many places in the search space and as such it may easily miss the target – genes potentially related to a disease under study. Such concerns are understandable and we are not going to ignore or forget about them.

Besides the abovementioned advantages, random sampling of features is fast and easy to implement. That is why it comes as no surprise that it found the application in building good ensembles of classifiers (Bay, Combining nearest neighbor classifiers through multiple feature subsets, 1998), (Bay, Nearest neighbor classification from multiple feature sets, 1999), (Altinçay, 2007), (Okun, Valentini, & Priisalu, Exploring the link between bolstered classification error and dataset complexity for gene expression-based cancer classification, 2008), (Okun & Priisalu, Dataset complexity in gene expression based cancer classification using ensembles of k-nearest neighbors, 2009).

Random feature selection is likely the only way of making stable classifiers[1] such as *k*-nearest neighbors, Naïve Bayes, support vector machines diverse. Diversity in predictions is the prerequisite for successful combination of multiple classifiers into an ensemble (Kuncheva, 2004). When each stable classifiers works with its own feature subset, which is quite different from the subsets of other ensemble members, then this classifier is likely to make errors on different instances than other classifiers in the ensemble. In this way diversity in predictions of individual ensemble members can be achieved.

Diversity can be enforced not only with random subset selection, but also with the technique called random projection (Vempala, 2004), which is based on celebrated Johnson-Lindenstrauss lemma (Johnson & Lindenstrauss, 1984) guaranteeing preservation of distances when mapping the data from the original, high dimensional space to another, low dimensional space. It is curious to notice that here again there is the word 'random', which may be perceived as rather symbolic. Random projection involves a multiplication of feature vectors representing data instances by a matrix composed of random numbers generated according to specific rules (Johnson & Lindenstrauss, 1984), (Achlioptas, 2001). If each classifier utilizes its own random matrix for mapping, then it will work with the projected data that are different from the working data of other ensemble members. Random projection has been successfully utilized in the past in ensembles of support vector machines used for the analysis of biological data (Bertoni & Folgieri, Bio-molecular cancer prediction with random subspace ensembles of support vector machines, 2005), (Folgieri, 2007), (Bertoni, Folgieri, & Valentini, Classification of DNA microarray data with random projection ensembles of polynomial SVMs, 2009).

In this chapter, I will only consider stable classifiers and will show how accurate classifier ensembles can be built based on random subset selection and by keeping in mind concerns related to randomization. With regard to schemes for combining predictions of individual classifiers, I prefer the simplest ones such as the majority vote and Naïve Bayes combination as they recommended themselves well in practice (Ré & Valentini, 2010).

Main Idea

It is known that the performance of classifiers is strongly data-dependent, i.e. it varies from data set to data set. In my early works (Okun & Priisalu, Dataset complexity and gene expression based cancer classification, 2007), (Okun, Valentini, & Priisalu, Exploring the link between bolstered classification error and dataset complexity for gene expression-based cancer classification, 2008), (Okun & Priisalu, Dataset complexity in gene expression based cancer classification using ensembles of k-nearest neighbors, 2009), I and my colleagues adopted the notion of data set complexity characterizing how well two classes of data are separated. In other words, data set complexity is a score that indicates the degree of class separability in a certain (typically 1D) space.

Given a set of original features, the data of each class are projected onto the diagonal linear discriminant axis by using only these features (for details, see (Bø & Jonassen, 2002)), i.e. they are projected onto a line. Projection coordinates then serve as input for the Wilcoxon rank sum test for equal medians[2] (Zar, 1999). Given a sample divided into two groups according to class membership, all the observations are ranked as if they were from a single sample and the rank sum statistic W is computed as the sum of the ranks, R_1, in the smaller group. Let R_2 be the sum of the ranks in the larger group. The value of the rank sum statistic, i.e. R_1, is employed as a score characterizing the separability power of a given set of features. The higher this score, the larger the overlap in projections of two classes (the closer two medians to each other), i.e. the worse separation between classes. To compare scores coming from different data sets, each score can be normalized by the sum of all ranks, i.e. if N is the data set size, then the sum of all ranks will be $\sum_{i=1}^{N} i = R_1 + R_2$. Then the normalized score is $R_1 / (R_1 + R_2)$ and it lies between 0 and 1.

The Wilcoxon rank sum test is equivalent to a Mann-Whitney U-test (Zar, 1999). In fact, Mann and Whitney and Wilcoxon independently developed two versions of the same test in the late forties of the last century. Why did we choose this test? The answer can be found in (Hanley & McNeil, 1982), where it is shown that the Mann-Whitney/Wilcoxon statistic is related to the area under the Receiver Operating Characteristic (ROC) curve (Pepe, 2004). Let us recall that this area characterizes classification performance of a learning algorithm across the entire range of false positives (Pepe, 2004), (Fawcett, 2006). As a result, the Wilcoxon rank sum statistic serves as a good indicator of class separability and its calculation does not require running a classifier, which is excellent, since one can assess the classification problem complexity prior to the classification itself. The larger R_1 is, the smaller the U-statistic is, and therefore the smaller the area under the ROC curve is.

Our choice for a dataset complexity characteristic is not accidental. Since gene expression data are very high dimensional, it is not surprising that two classes could be linearly separable in the original, high dimensional feature space (however, this does not make the classification task easy as interclass distances could be still smaller than intraclass distances, thus giving rise to classification errors). That is, the decision boundary of a classifier can be assumed to be a hyperplane. On the other hand, the complexity characteristic we employ belongs to the class of linear discriminants estimating how well a line separates two classes. As a result, we have a good match between the behavior of a classifier and the model of class separability encoded in the complexity characteristic. This is also confirmed by the extensive experiments with eight microarray gene expression data sets done in (Okun, Valentini, & Priisalu, Exploring the link between bolstered classification error and dataset complexity for gene expression-based cancer classification, 2008).

Algorithm Description

In the previous section we saw that data set complexity defined through the Wilcoxon rank sum statistic can be very useful in selecting low-complexity data sets. In this section I demonstrate how this finding can be applied to generation of ensembles of stable classifiers. Although further on we will talk about ensembles of k-nearest neighbors, everything said is equally true for other stable classifiers as well without any loss of generality. A brief survey of the literature on the ensembles of nearest neighbors can be found in (Okun & Priisalu, Multiple views in ensembles of nearest neighbor classifiers, 2005).

In a number of papers (see, e.g., (Okun, Valentini, & Priisalu, Exploring the link between bolstered classification error and dataset complexity for gene expression-based cancer classification, 2008), (Okun & Priisalu, Dataset complexity in gene expression based cancer classification using ensembles of k-nearest neighbors, 2009)) it was demonstrated that the data set complexity defined in the previous section and the bolstered resubstitution error rate (Braga-Neto & Dougherty, 2004) are related. In other words, low (high) complexity corresponds to the small (large) bolstered resubstitution error rate. When randomly sampling the original feature set, it was shown that the same complexity can correspond to several different error rates, which implies that ensemble generation based on randomly selected feature subsets will likely be composed of diverse classifiers. Hence, such an ensemble would likely to outperform a single best classifier.

The link between the data set complexity and the bolstered resubstitution error rate was established by means of the copula method (Sklar, 1959), (Joe, 2001), (Nelsen, 2006) studying dependence or concordance relations in multivariate data. As a result, selecting a low-complexity subset of genes implies an accurate classifier, which, in turn, implies an accurate classifier ensemble.

Given that it is difficult to carry out biological analysis of too many genes, it is useful to restrict the number of genes to be sampled to a small number, e.g., 50. Then each ensemble member works with 1 to 50 randomly selected (sampled with replacement) genes. This will ensure that the combined list of all genes is not too long.

Based on the abovementioned, two approaches to form ensembles consisting of L classifiers can be devised:

Randomly select L feature subsets, one subset per classifier, classify the data with each classifier and combine votes. Such an approach is utilized in (Bay, Combining nearest neighbor classifiers through multiple feature subsets, 1998), (Bay, Nearest neighbor classification from multiple feature sets, 1999).

Randomly select $M > L$ (e.g., $M = 100$) feature subsets and compute the dataset complexity for each of them. Rank subsets according to their complexity and select L least complex subsets while omitting the others. Classify the data with each classifier and combine votes. Such an approach is exercised in (Okun, Valentini, & Priisalu, Exploring the link between bolstered classification error and dataset complexity for gene expression-based cancer classification, 2008), (Okun & Priisalu, Dataset complexity in gene expression based cancer classification using ensembles of k-nearest neighbors, 2009).

As one can see, the main difference between two approaches lies in the way of choosing feature subsets: in the first (traditional) approach, subsets are chosen regardless of their classification power. As a result, one may equally expect both very good and very bad ensemble predictions. In contrast, in the second approach, subsets are chosen based on the measure *directly* related to the classification performance. As lower complexity is associated with smaller bolstered resubstitution error, selecting the subsets of smaller complexity implies more accurate classifiers included into an ensemble. Since each ensemble member works with only a small subset of all features, its individual accuracy is expected (though not always guaranteed) to be higher than or at least no worse than that of a classifier using all features. Such feature space decomposition is akin to dividing a complex problem into simpler sub-problems (Maimon & Rokach, 2005). Thus, with the second approach, both diversity and accuracy requirements for ensembles are satisfied.

Experimental results attained for ensembles of k-nearest neighbors in (Okun, Valentini, & Priisalu, Exploring the link between bolstered classification error and dataset complexity for gene expression-based cancer classification, 2008), (Okun & Priisalu, Dataset complexity in gene expression based cancer classification using ensembles of k-nearest neighbors, 2009) clearly favor the complexity-based scheme ensemble generation over 1) the traditional, complexity-ignorant scheme and 2) a single best classifier in the ensemble.

Ensemble Pruning

We already saw that it is imperative that in order for an ensemble to outperform a single best classifier, the former must be composed of diverse classifiers that are also sufficiently accurate at the same time (Kuncheva, 2004). Therefore diversity needs to be somehow measured and be incorporated into the ensemble generation process.

Although all the available classifiers are typically included in an ensemble, it is also possible to select only certain classifiers for fusion. This operation is called ensemble pruning (see a brief survey of ensemble pruning techniques in (Tsoumakas, Partalas, & Vlahavas, 2009)). Several works (see, e.g., (Cho & Won, 2003), (Park & Cho, 2003), (Peng, 2006)) have demonstrated that ensemble pruning can be beneficial for boosting the classification power of an ensemble that is especially composed of stable classifiers. In other words, a pruned ensemble may show a better performance than the full one. Below one of the ensemble pruning techniques is considered.

Prodromidis et al. (Prodromidis, Stolfo, & Chan, 1999) proposed to use diversity-based pruning before classifier combination. They defined diversity for a pair of classifiers D_i and D_j as the number of times where D_i and D_j produce different predictions. Their diversity-based algorithm is given in Figure 1 (pseudo code is adopted from (Okun & Priisalu, Ensembles of nearest neighbour classifiers and serial analysis of gene expression, 2006)), where $\mathcal{H}$ and $\mathcal{C}$ are sets containing the original and selected classifiers, respectively.

First, the diversity for all pairs of classifiers is computed. The most accurate classifier is included into $\mathcal{C}$ and excluded from $\mathcal{H}$. Next, in each round, the algorithm adds to the list of the previously selected classifiers the classifier $D_k \in \mathcal{H}$ that is most diverse to the classifiers chosen so far, i.e. the D_k that maximizes the overall diversity S over $\mathcal{C} \cup D_k, \forall k \in \left\{1, 2, \ldots, |\mathcal{H}|\right\}$, where $|\mathcal{H}|$ is the number of

Figure 1. Pseudo code of the diversity-based ensemble pruning algorithm by (Prodromidis, Stolfo, & Chan, 1999)

Diversity-based ensemble pruning algorithm by (Prodromidis, Stolfo, & Chan, 1999)

Let $\mathcal{C} := \emptyset$ and $\tilde{M}$ is the maximum number of classifiers in an ensemble;
For $i := 1 : |\mathcal{H}| - 1$
　For $j := 1 : |\mathcal{H}|$
　　Let z_{ij} is the number of times when classifiers D_i and D_j yield different predictions;
　End For
End For

Let D' is the classifier with the highest accuracy among all available classifiers;
$\mathcal{C} := \mathcal{C} \cup D'$; $\mathcal{H} := \mathcal{H} \backslash D'$;
For $i := 1 : \tilde{M} - 1$
　For $j := 1 : |\mathcal{H}|$
　　Let $S_j := \sum_{k=1}^{|\mathcal{C}|} z_{jk}$;
　End For
　Let D' is the classifier from $\mathcal{H}$ with the highest S_j;
　$\mathcal{C} := \mathcal{C} \cup D'$; $\mathcal{H} := \mathcal{H} \backslash D'$;
End For

---

classifiers in $\mathcal{H}$. The selected classifier is then deleted from $\mathcal{H}$. The selection ends when the $\tilde{M}$ most diverse classifiers are selected[3].

Prodromidis et al. (Prodromidis, Stolfo, & Chan, 1999) showed on two data sets of real credit card transactions that diversity-based ensemble pruning can lead to the ensemble with higher accuracy than that of the full ensemble. In (Okun & Priisalu, Ensembles of nearest neighbour classifiers and serial analysis of gene expression, 2006), it was shown that such an ensemble pruning scheme is advantageous to apply for gene expression data, too. Therefore, if L in the main ensemble-building algorithm was set to a large value, the ensemble-pruning algorithm in Figure 1 can be applied to eliminate the classifiers that do not significantly contribute to diversity.

Combination Schemes

The ensemble construction process would be incomplete without describing the schemes used to combine predictions of ensemble members. I adhere to simple schemes that typically do not require training. As was shown in (Ré & Valentini, 2010), such schemes could be competitive to more sophisticated schemes in the analysis of biological data. Therefore, I opted for two schemes: majority voting and

Naïve Bayes combination. Below their description is provided that was borrowed from (Kuncheva, 2004).

In the case of two classes, the majority voting is especially simple: an instance is assigned to the class that obtained 50 per cents of the votes + 1.

The Naïve Bayes combination lends its name and the principle of work to the Naïve Bayes rule. Let $s_1, \ldots, s_L$ be crisp class labels assigned to a given instance x by classifiers $D_1, \ldots, D_L$, respectively. The independence assumption means that the support for class C_k is

$$\mu_k(x) \propto P(C_k) \prod_{i=1}^{L} P(C_k | s_i),$$

where $P(C_k | s_i)$ is the probability that the true class is C_k given that D_i assigns x to class s_i, and the symbol $\propto$ means 'proportional to' (i.e. not exactly equal).

For each classifier D_i, a $C \times C$ confusion matrix CM^i is calculated by applying D_i to the training data set, where C is the number of classes (two, in our case). The (k, s) th entry of this matrix, cm_{ks}^i is the number instances whose true class label was C_k, but they were classified in class C_s by D_i. By N_k we denote the total number of instances from class C_k. Taking $cm_{ks_i}^i / N_k$ as an estimate of the probability $P(C_k | s_i)$, and N_k / N as an estimate of the prior probability for class C_k, we obtain that

$$\mu_k(x) \propto \frac{1}{N N_k^{L-1}} \prod_{i=1}^{L} cm_{ks_i}^i.$$

To avoid nullifying $\mu_k(x)$ regardless of other estimates when an estimate of $P(C_k | s_i) = 0$, the last formula can be rewritten as

$$\mu_k(x) \propto \frac{N_k}{N} \left\{ \prod_{i=1}^{L} \frac{cm_{ks_i}^i + 1/C}{N_k + 1} \right\}^B,$$

where B is a constant (typically it can be set to 1) and the Laplace correction (Kuncheva, 2004) is used to avoid zero value when $cm_{ks_i}^i = 0$.

Among all $\mu_k(x)$ (only two, in our case), the one with the highest value points to the class to be assigned to x by the ensemble.

Random Projection and Random Projection Ensembles

As was said earlier, instead of random feature selection, random projection may be employed in order to reduce data dimensionality and to provide a distinct feature set for each classifier in an ensemble. The idea of using projections in the ensemble setting is not entirely new (see, e.g., (García-Pedrajas & Ortiz-Boyer, 2009), (Bertoni & Folgieri, Bio-molecular cancer prediction with random subspace ensembles of support vector machines, 2005), (Folgieri, 2007), (Bertoni, Folgieri, & Valentini, Classification of DNA microarray data with random projection ensembles of poly-nomial SVMs, 2009), (Schclar & Rokach, 2009)). Out of all possible projection transformations, random projection represents a particular interest since it is based on the celebrated Johnson-Lindenstrauss lemma (Johnson & Lindenstrauss, 1984) saying that it is possible to project high dimensional data into low dimensional space so that interpoint distances will be approximately preserved. Put differently, this lemma concerns low-distortion embeddings of data points from high dimen-sional to low dimensional Euclidean space. If interpoint distances in both spaces are approximately the same, then two points that are close to (far from) each other in the original, high dimensional space will remain close to (far from) each other in the resulting, low dimensional space. This implies that we can safely estimate interpoint distances in the latter space by knowing that this is equivalent to do the same in the former space.

The lemma found its use in such advanced areas as compressed sensing and manifold learning (go to http://dsp.rice.edu/cs if interested in these topics). It makes sense to give this beautiful lemma here:

For any ϵ ($\frac{1}{2} > \epsilon > 0$) and any set of N points in $\mathbb{R}^D$, the following property holds with probability at least $\frac{1}{2}$ for every pair of points u and v in $\mathbb{R}^D$:

$$\left(1-\epsilon\right)u - v^2 \leq f\left(u\right) - f\left(v\right)^2 \leq \left(1+\epsilon\right)u - v^2,$$

where $f\left(u\right)$ and $f\left(v\right)$ are the projections of u and v, respectively, to a uniform random d-dimensional subspace with

$$d \geq \frac{9\ln N}{\epsilon^2 - \frac{2}{3}\epsilon^3} + 1.$$

Since $\epsilon^2 > \dfrac{2}{3}\epsilon^3$ for eligible ϵ values, it is often that

$$d \geq \frac{\ln N}{\epsilon^2}$$

is only required. In both formulas, $\ln N$ is the natural logarithm of N.

That is, the Johnson-Lindenstrauss lemma provides a lower bound on the dimensionality of the resulting low dimensional space when the projection is of form $y = R'x$ for each D-dimensional feature vector x and each d-dimensional feature vector y. Thus, all that remains is to define a $D \times d$ matrix R.

In (Vempala, 2004), two ways of defining R are presented. It can be a random orthonormal matrix or its entries can be randomly sampled from the standard normal distribution (zero mean, unit variance).

When R is a random orthonormal matrix, i.e. the columns of R are random unit vectors (their length is equal to one) and they are pairwise orthogonal (their dot product is zero), the random projection of x can be computed as

$$y = \sqrt{\frac{D}{d}}R'x,$$

where the scaling factor $\sqrt{\dfrac{D}{d}}$ is chosen so as to make the expected squared length of y equal to the squared length of x (Matoušek, 2002), (Vempala, 2004).

However, there are many more ways to define the projection matrix. For instance, Achlioptas (Achlioptas, 2001) defined entries of R as randomly generated -1's and +1's, each with probability $\dfrac{1}{2}$ (Bernoulli trials where 'success' and 'failure' are equally probable (Everitt, 2006)) . The random projection matrix is then multiplied (scaled) by $\sqrt{\dfrac{1}{d}}$ so that the final formula is

$$y = \sqrt{\frac{1}{d}}R'x.$$

Given such a powerful result and assurance that we can safely work with low dimensional data instead of high dimensional ones, it is very tempting to apply the idea of random projections to generating the ensembles of stable classifiers as was done in (Schclar & Rokach, 2009).

The idea of (Schclar & Rokach, 2009) is simple but workable. As stable classifiers are sensitive to the change in a feature set, different random projection matrices need to be associated with different ensemble members. That is, each classifier gets its own random projection matrix that is used to project both training and test data (if there are L classifiers in an ensemble, there will be L different projection matrices). After that, test data are classified in the low dimensional space by each classifier and finally, votes of all classifiers are combined with the majority vote rule.

I would like to emphasize once again that both training and test data need to be projected into the same low dimensional space; otherwise, the projection does not make sense. In other words, the same random projection matrix should be used when projecting both training and test data.

One can see a clear difference between random feature selection and random projection. The latter is, in fact, feature extraction as every derived feature is a random linear combination of the original features.

MATLAB Code

MATLAB® functions implementing the ideas and methods described in this chapter can be divided into two groups:

- those that are related to the ensembles based on random subset selection,
- those that are related to the ensembles based on random projection.

The functions of the first group are *ensemble_prediction*, *datasetcomplexity*, *diversity_based_ensemble_pruning*, *majority_vote*, and *NaiveBayes_combination*.

The functions belonging to the second group include *RandomProjection_A*, *RandomProjection_JL*, and *RandomProjectionEnsemble*.

To simplify code, I concentrated on ensembles of k-nearest neighbors as the base classifiers in *ensemble_prediction*. Given training data and their class labels as well as the number of nearest neighbors for each classifier in an ensemble, values for M and L, the number of samples per data instance in bolstered resubstitution error estimation, and the method for 1D projection, *ensemble_prediction* returns the matrix of predictions made by individual ensemble members together with indices of features employed by each classifier, the bolstered resubstitution error and an index of the best classifier in the ensemble. All returned quantities are later used by other functions, which makes *ensemble_prediction* a de-facto principal function of the first group (in other words, this function should be invoked before other functions in the group).

First, M feature subsets are randomly selected from the entire pool of the original features. Pay attention to the fact that for each subset random sampling without replacement is performed, i.e. each subset cannot include multiple copies of the same feature. However, different subsets may include common features. Feature indices of each subset are temporarily stored for further use, and the normalized data set complexity is estimated for each feature subset.

In the next step, L out of M subsets possessing the lowest complexity are chosen and each of them is associated with its own k-nearest neighbors and the bolstered resubstitution error incurred by it. All output parameters are assigned at this stage, too. Regarding nearest neighbor classification, if a data set is imbalanced, i.e. when one class dominates the other, the algorithm from (Hand & Vinciotti, 2003) could be useful to apply instead of the conventional k-nearest neighbors.

Data set complexity estimated in *ensemble_prediction* is implemented in *datasetcomplexity*, where the data can be projected onto a line with either diagonal linear discriminant or Fisher's linear discriminant method. Details of how to do the projection itself can be found in (Bø & Jonassen, 2002). After projecting the data with features specified by the input parameter *idx*, the MATLAB® function *ranksum* is called to perform the Wilcoxon rank sum test on projections. If the projections of two classes are relatively far from each other, *datasetcomplexity* returns the value of the rank sum statistic; otherwise, a very big positive real number is returned, signaling that the projections of two classes onto a discriminant axis are likely heavily overlapped.

The function *diversity_based_ensemble_pruning* is optional: it can be invoked if a user thinks that the number of classifiers, L, in an ensemble is too big and it is necessary to eliminate some classifiers before final voting. This function expects three input parameters: the matrix of base predictions, the index of the best classifier in the ensemble, and the maximum number of classifiers to retain. The first two parameters are conveniently returned by *ensemble_prediction* while the third parameter needs to be specified by a user. Given these inputs, *diversity_based_ensemble_pruning* returns indices of those classifiers that contributed most to the overall diversity of an ensemble.

Independently of whether *diversity_based_ensemble_pruning* was invoked or not, the base predictions of all available classifiers or just some classifiers are to be combined by either majority voting (function *majority_vote*) or Naïve Bayes combination (function *NaiveBayes_combination*). Both combination functions returns both 'hard' and 'soft' ensemble predictions. The latter can be then used to build a Receiver Operating Characteristic curve and to compute the area under this curve.

Random projection ensembles can be generated with the help of *RandomProjectionEnsemble* that relies on one of the two functions: *RandomProjection_A* (random projection by Achioptas (Achlioptas, 2001)) and *RandomProjection_JL* (random projection according to the Johnson-Lindenstrauss lemma). If the projection matrix is not specified, then it is computed in *RandomProjection_A* (or in *RandomProjec-*

tion_JL). Normally, this is done when the input parameter *U* points to training data. However, when this parameter is associated with test data, the random projection matrix returned by the last call of *RandomProjection_A* or *RandomProjection_JL* is given as one of the input parameters for either random projection function. This is necessary to do in order for the same projection matrix to be utilized in projecting both training and test data. *RandomProjectionEnsemble* returns both 'hard' and 'soft' predictions made by an ensemble in conformance with the output of all other classifiers and classifier ensembles. The majority vote is hard-wired in this function but nothing prevents you, dear readers, to replace it with your own fusion technique. I leave this as an exercise for you.

```
function [base_predictions, e_bc, i_bc, indices] = ensemble_
prediction(train_patterns, train_targets, params, discriminant_
method, dmax, M, L, nsamples)


% Prediction made by an ensemble of stable (k-nearest neighbor)
% classifiers.
% Each classifier employs its own subset of features randomly
sampled
% from the whole set of the original features. Feature subsets
used by
% classifiers chosen based on low data set complexity.
%
% Inputs:
%    train_patterns      - Training patterns (DxN matrix, where
D - data
%                          dimensionality, N - number of
patterns)
%    train_targets       - Training targets (1xN vector)
%    params              - Parameters of k-nearest neighbor
%    discriminant_method - Either 'DLD' (Diagonal linear dis-
criminant)
%                                        or
%                          'FLD' (Fisher's linear dis-
criminant)
%    dmax                - The maximal number of randomly se-
lected
%                          features per classifier
%    M                   - The total number of random subsets of
%                          features in search for
```

```
a good ensemble
%   L                   - The final number of feature subsets
chosen
%                               for an ensemble based
on data set
%                               complexity
%   nsamples           - The number of times to sample the
%                               multivariate normal
distribution when
%                               computing bolstered
resubstitution error of %
each classifier in an ensemble
%
% Outputs:
%
%   base_predictions    - Nbc x N matrix (Nbc - # of base clas-
sifiers)
%                               of predictions made by base classi-
fiers on %                               a test set
%   e_bc             - Bolstered resubstitution error of the
best
%                               classifier in an ensem-
ble
%   i_bc             - Index of the best classifier in an
ensemble
%                               classifier
%   indices          - Indices of features employed with each
%                               classifier in the en-
semble
%
% Example of call: [base_predictions, e_bc, i_bc, indices] =
...
% ensemble_prediction(train_patterns, train_targets, 3, 'DLD',
50, 100, 7, 10);



% Get the number of patterns
N = size(train_patterns, 2);
% Get the number of features
```

```
D = size(train_patterns, 1);
% Normalizing factor for the data set complexity estimates
sr = sum(1:N);

% Allocate memory for the data set complexity estimates
C = zeros(M,1);
% Allocate memory for indices of features in M subsets
idx = cell(1,M);

% Compute data set complexity for M subsets of features random-
ly
% sampled from the whole set of the original features
for i = 1:M
    % Set the number of features to be sampled; k is between 1
and dmax
    k = randsample(1:dmax,1);
    % Randomly sample (without replacement) k out of D original
    % features
    idx{i} = randsample(D,k,true);
    % Compute dataset complexity
    c = datasetcomplexity(train_patterns, train_targets,
idx{i}, discriminant_method);
    % Normalize
    C(i) = c/sr;
end

% Sort from lowest to highest
% Keep indices for the future use
[dummy, I] = sort(C,'ascend');

% Now we are ready to build an ensemble

% Allocate memory
indices = zeros(1,L);
e = zeros(1,L);
base_predictions = zeros(L,nsamples);

% Select L subsets of features corresponding to the lowest data
set
% complexities and include these subsets to the ensemble set
```

```
for i = 1:L
    indices{i} = idx{I(i)}; % subset of features
    j = indices{i};
    [e(i), labels] = bolstederror_Nearest_Neighbor(train_
patterns(j,:), train_targets, params, nsamples);
    base_predictions(i,:) = reshape(labels, 1, numel(labels));
end

% Get the bolstered error and an index of the best classifier
in an
% ensemble
[e_bc, i_bc] = min(e);

return
function c = datasetcomplexity(patterns, targets, idx, discrim-
inant_method)

% Data set complexity
%
% Inputs:
%    patterns            - DxN matrix of patterns (D - dimen-
sionality,
%                                  N - #patterns)
%    targets             - 1xN vector of labels
%    idx                 - Indices of the original features used
%    discriminant_method - Either 'DLD' (Diagonal linear dis-
criminant)
%                                          or
%                                'FLD' (Fisher's linear dis-
criminant)
%
% Outputs:
%    c                   - Rank sum statistic value associated
with
%                             data set complexity

% Find unique labels
u = unique(targets);
% Get the number of classes
nclasses = length(u);
```

```
% Check if there are at least two classes of data
if nclasses < 2
    error('There should be at least two classes of data!');
end
% Check if there are at most two classes of data
if nclasses > 2
    error('There should be at most two classes of data!');
end

% Find indices of patterns belonging to each of two classes
ind{1} = find(targets == u(1));
ind{2} = find(targets == u(2));

% Compute the rank sum statistic for given features by using
the
% Wilcoxon rank sum test for equal medians (it is a nonparamet-
ric
% analogue of the two-sample t-test)
% The rank sum statistic is computed using the projected data
points on % the DLD/FLD (Diagonal Linear Discriminant/Fisher's
Linear
% Discriminant) axis
% ************************************************************
******

% Compute the mean vector for each class
mu1 = mean(patterns(idx,ind{1}),2);
mu2 = mean(patterns(idx,ind{2}),2);
% Compute the covariance matrix
S = cov(patterns(idx,:)'); % Fisher's linear discriminant
if strcmpi(discriminant_method,'DLD') % Diagonal linear dis-
criminant
    S = diag(diag(S));
    S = S + eps;
end
% Find a discriminant axis
if strcmpi(discriminant_method,'FLD')
    S = S + 0.01*diag(diag(S)); % use regularization
end
warning off all;
```

```
a = S\(mu1 - mu2);
warning on all;
% Assume that medians are equal
c = realmax;
if ~any(isnan(a)) % if a is singular, the value of the rank sum
statistic
    % is forced to 0
    % Project data of each class on the discriminant axis by
using
    % a given feature
    % Notice that a projection of each point is a number!
    p1 = a'*patterns(idx,ind{1});
    p2 = a'*patterns(idx,ind{2});
    % Compute the statistic as a feature score
    warning off all;
    [p,h,stats] = ranksum(p1,p2);
    warning on all;
    if h == 1% medians are not equal
        c = stats.ranksum;
    end
end

return
function idx = diversity_based_ensemble_pruning(base_predic-
tions, i_bc, M)

% Diversity-based pruning of the base classifiers before com-
bining them
% into an ensemble
% Original source: Andreas L. Prodromidis, Salvatore J. Stolfo,
and
% Philip K. Chan, "Effective and efficient pruning of meta-
classifiers % in a distributed data mining system", Technical
Report CUCS-017-99,
% Columbia University, NY, 1999
%
% Inputs:
%   base_predictions    - Nbc x N matrix (Nbc - # of base
classifiers,
%                         N - # test patterns) of predictions
```

```
made by
%                           ensemble classifiers on a test set
%    i_bc                - Index of the best classifier in an
ensemble
%    M                   - Maximal number of base classifiers to
include
%                           into an ensemble
%
% Output:
%    idx                 - Indices of the base classifiers
selected for
%                           inclusion into an ensemble. There
could be %                                    less than M
classifiers if M is large wrt
%                                    Nbc



% Get the number of base classifiers
[Nbc, dummy] = size(base_predictions);

% Initialize
d = zeros(Nbc);

for i = 1:Nbc-1
    for j = i:Nbc
        % Retrieve predictions of the ith and jth base classi-
fiers
        pri = base_predictions(i,:);
        prj = base_predictions(j,:);
        % Find the number of instances when predictions of two
        % base classifiers disagree
        d(i,j) = length(find(pri ~= prj));
        d(j,i) = d(i,j);
    end
end

% Include the classifier with highest accuracy into the list of
% selected classifiers
idx = i_bc;
```

```
fprintf(1,'Base classifier no.%d is included ...\n', i_bc);

% Iteratively include M-1 remaining base classifiers
for i = 1:M-1
    diversity = zeros(1,Nbc);
    for j = 1:Nbc
        if isempty(find(idx == j))
            diversity(j) = sum(d(j,idx));
        end
    end
    if all(diversity == 0)
        if length(idx) < M
            fprintf(1,'Can''t collect the desired number of
base classifiers!\n');
        end
        break;
    end
    [max_diversity, k] = max(diversity);
    idx = [idx k];
    fprintf(1,'Base classifier no.%d is included ...\n', k);
    fprintf(1,'Diversity - %f\n', max_diversity);
end


return
function [ensemble_predictions, ensemble_soft_predictions] =
majority_vote(base_predictions, idx, Uc)

% Fusion of class labels generated by a group of independent
% classifiers by means of majority vote
%
% Inputs:
%   base_predictions            - Predictions of base classifi-
ers on a
%                                       test set
(Nbc x N matrix, where Nbc
%                                   the number of classifiers,
N -
%                                   the number of test pat-
terns)
%   idx                         - Indices of the classifiers
```

```
selected
%                                                for inclu-
sion into an ensemble
%    Uc                      - Unique training targets (1xNc
vector,
%                                where Nc - the number of
classes)
%
% Outputs:
%    ensemble_predictions     - Predictions (1xN vector of
class
%                                          labels),
assigned by an ensemble of
%                                            classifiers
%    ensemble_soft_predictions  - Soft predictions (similar to
%                                          probability
estimates)
%                                            (1xN vec-
tor), assigned by
%                                an ensemble of classifiers




% Sort Uc if it is unsorted like [2 3 0 1]
Uc = unique(Uc); % result is [0 1 2 3]

% Get predictions of the selected classifiers
predictions = base_predictions(idx,:);
% Build a histogram of predictions
h = hist(predictions, Uc);
% Get targets for the whole ensemble
[v,i] = max(h,[],1);
% Classify data (first regardless ties)
ensemble_predictions = Uc(i);
% Assign soft (like probability) estimates
ensemble_soft_predictions = h/length(idx);

% Now take care of tied votes ...
v = repmat(v,size(h,1),1);
% Create an indicator matrix for discovering tied votes
```

```
indicators = (h == v);
s = sum(indicators,1);
% Find tied votes if exist
idx = find(s > 1);
if ~isempty(idx)
    for j = 1:length(idx)
        % Find ties classes
        k = find(indicators(:,idx(j)) > 0);
        % Ties, if present, are broken randomly
        i = randsample(k,1);
        ensemble_predictions(idx(j)) = Uc(i);
    end
end


return
function [ensemble_predictions, ensemble_soft_predictions] =
NaiveBayes_combination(base_predictions, idx, Uc, train_base_
predictions, train_targets, B)


% Fusion of class labels generated by a group of independent
% classifiers by means of Naive Bayes combination with a cor-
rection for % zeros
% Original source: Ludmila I. Kuncheva, "Combining Pattern
Classifiers:
% Methods and Algorithms", John Wiley & Sons, Inc., NJ, 2004
%
% Inputs:
%   base_predictions          - Predictions of base classifi-
ers on a %                                      te
st set (Nbc x N matrix, where Nbc
%                                  the number of classifiers,
N -
%                                  the number of test pat-
terns)
%   idx                       - Indices of the classifiers
selected
%                                              for inclu-
sion into an ensemble
%   Uc                        - Unique training targets (1xNc
vector,
```

```
%                                    where Nc - the number of
classes)
%    train_base_predictions       - Nbc x Nt matrix (Nt - # of
training
%                                    patterns) of predictions
made by
%                                         base clas-
sifiers on a training set
%    train_targets               - Training targets (1xNt vec-
tor)
%    B                           - Constant between 0 and 1
(zero is not
%                                    included)
%
% Outputs:
%    ensemble_predictions        - Predictions (1xN vector of
class
%                                         labels),
assigned by an ensemble of
%                                         classifiers
%    ensemble_soft_predictions   - Soft predictions (similar to
%                                         probability
estimates)
%                                         (1xN vec-
tor), assigned by an
%                                    ensemble of classifiers



% Set a default value for B
if nargin < 6
    B = 1;
end

% Sort Uc if it is unsorted like [2 3 0 1]
Uc = unique(Uc); % result is [0 1 2 3]

% Get the number of different classes
c = length(Uc);
% Get the number of base classifiers
```

```
nclassifiers = length(idx);
% Get the number of test patterns
N = size(base_predictions, 2);

% Preallocate memory for confusion matrices of the base classi-
fiers
CM = cell(1,nclassifiers);

% Generate confusion matrices for base classifiers
% Base classifiers are applied to training data!
for i = 1:nclassifiers
    if isempty(train_base_predictions) % there is no a separate
training dataset
        CM{i} = confusion_matrix(base_predictions(idx(i),:),
train_targets);
    else
        CM{i} = confusion_matrix(train_base_
predictions(idx(i),:), train_targets);
    end
end

% Get the number of training patterns
Nt = length(train_targets);
% Preallocate memory for estimates of the prior probabilities
PP = zeros(1,c);
% Compute an estimate of the prior probability for each class
for i = 1:c
    PP(i) = length(find(train_targets == Uc(i)))/Nt;
end

% Create a look-up table for later use when we need to deter-
mine the
% index in the confusion matrix based on the prediction of a
base
% classifier
for i = 1:c
    LUT(Uc(i) - Uc(1) + 1) = i;
end

% Preallocate memory for the support for each class
```

```
mu = zeros(c,N);
% Compute the support
for k = 1:c
    Nk = length(find(train_targets == Uc(k)));
    % Likelihood or conditional probability of a pattern to
belong to a
    % certain class
    P = ones(1,N); % initialize
    for i = 1:nclassifiers
        % Compute likelihood with a correction for zeros
        % See section 4.4 in the original source
        bp = base_predictions(idx(i),:);
        % Consult the look-up table for the index
        j = LUT(bp - Uc(1) + 1);
        P = P.*((CM{i}(k,j) + 1/nclasses)/(Nk + 1));
    end
    % Support for class Uc(k)
    mu(k,:) = PP(k)*(P.^B);
end

% Assign soft (like probability) estimates
ensemble_soft_predictions = mu;

% Find the class yielding the maximal support
[dummy, i] = max(mu,[],1);

% Classify data
ensemble_predictions = Uc(i);

return
function [V, varargout] = RandomProjection_A(U, P, d, e, b)

% Random projection according to Achlioptas
% Original sources:
% William B. Johnson, Joram Lindenstrauss, "Extensions of
Lipschitz
% mappings into a Hilbert space", Proceedings of the Conference
in
% Modern Analysis and Probability, New Haven, CT, pp.189-206,
1984,
```

```
% AMS, Providence, RI
% Dimitris Achlioptas, "Database-friendly random projections",
% Proceedings of the 20th ACM SIGMOD-SIGACT-SIGART Annual
Symposium on % Principles of Database Systems, Santa Barbara,
CA, pp.274-281, 2001, % ACM, New York, NY
%
% Inputs:
%    U          - Input data (DxN matrix, where
%                              D - data dimensionality,
%                    N - number of patterns)
%    P          - Projection matrix (default: '' - not speci-
fied)
%    d          - Dimensionality of a low-dimensional space
%    e          - Epsilon (distance distortion allowed): [0,1)
%    b          - Positive constant
%
% Outputs:
%    V          - Projected data (dxN matrix)
%    P          - Projection dxD matrix



% Get a data dimensionality and the number of patterns
[D,N] = size(U);

% Set default values
if nargin < 5
    b = 0.2;
end
if nargin < 4
    e = 0.25;
end
if nargin < 3
    d = ceil((log(N)*(4 + 2*b))/(e^2/2 - e^3/3));
    if nargin == 1
        fprintf(1, 'Dimensionality of the resulting space is
%d\n', d);
    end
end
if nargin < 2
```

```
    P = '';
end

if d >= D
    warning('No dimensionality reduction will occur!');
end

if isempty(P)
    % Create a random matrix whose entries are either -1 or +1
with
    % the probability 0.5
    R = binornd(1, 0.5, D, d); % R is of Dxd
    R(R == 0) = -1;

    % Form a projection matrix
    P = sqrt(1/d)*R';
end

% Create the output matrix
V = P*U;
% This is how distance distortion resulting from random projec-
tion can % be seen squareform(pdist(U') - pdist(V'))
% Average absolute distance distortion
fprintf(1, 'Average absolute distance distortion is %f\n',
mean(abs(pdist(U') - pdist(V'))));

if nargout == 2
    varargout{1} = P;
end

return
function [V, varargout] = RandomProjection_JL(U, P, d, e)

% Random projection according to the Johnson-Lindenstrauss
lemma
% Original sources:
% William B. Johnson, Joram Lindenstrauss, "Extensions of
Lipschitz
% mappings into a Hilbert space", Proceedings of the Conference
in
```

```
% Modern Analysis and Probability, New Haven, CT, pp.189-206,
1984,
% AMS, Providence, RI
% Santosh S. Vempala, "The random projection method", Series in
% Discrete Mathematics and Theoretical Computer Science, Vol
65,
% The American Mathematical Society, Providence, RI, 2000
% Jirí Matoušek, "Embedding finite metric spaces into normed
spaces" in
% Jirí Matoušek (ed.), Lectures on Discrete Geometry, pp.355-
400,
% Springer-Verlag, New York, NY, 2002
%
% Inputs:
%    U          - Input data (DxN matrix, where
%                             D - data dimensionality,
%                       N - number of patterns)
%    P          - Projection matrix (default: '' - not speci-
fied)
%    d          - Dimensionality of a low-dimensional space
%    e          - Epsilon (distance distortion allowed): [0,1)
%
% Outputs:
%    V          - Projected data (dxN matrix)
%    P          - Projection dxD matrix



% Get a data dimensionality and the number of patterns
[D,N] = size(U);

% Set default values
if nargin < 4
    e = 0.1;
end
if nargin < 3
    d = ceil(log(N)/e^2);
    if nargin == 1
        fprintf(1, 'Dimensionality of the resulting space is
%d\n', d);
```

```
     end
end
if nargin < 2
    P = '';
end

if d >= D
    warning('No dimensionality reduction will occur!');
end

if isempty(P)
% Create a random orthonormal matrix, i.e., the matrix whose
% columns are unit vectors that are pairwise orthogonal
    R = orth(rand(D,d)); % R is of Dxd
    % This is how orthonormality can be seen
    % R'*R

    % Form a projection matrix
    P = sqrt(D/d)*R';
end

% Create the output matrix
V = P*U;
% This is how distance distortion resulting from random projec-
tion can % be seen squareform(pdist(U') - pdist(V'))
% Average absolute distance distortion
fprintf(1, 'Average absolute distance distortion is %f\n',
mean(abs(pdist(U') - pdist(V'))));

if nargout == 2
    varargout{1} = P;
end

return
function [ensemble_predictions, ensemble_soft_predictions] =
RandomProjectionEnsemble(train_patterns, train_targets, test_
patterns, L, RP_algorithm, classifier, varargin)

% Random Projection ensemble classifier
% Original source:
```

```
% Alon Schclar, Lior Rokach, "Random projection ensemble clas-
sifiers", % in Joaquin Filipe, José Cordeiro (eds.), Lecture
Notes in Business
% Information Processing, Vol.24 (Proceedings of the 11th
% International Conference on Enterprise Information Systems,
Milan,
% Italy, pp.309-316, 2009), Springer-Verlag, Berlin/Heidelberg
%
% Inputs:
%   train_patterns  - Training patterns (DxN matrix, where
%                                   D - data dimensionality,
%                                   N - number of patterns)
%   train_targets   - Training targets (1xN vector)
%   test_patterns   - Test patterns (DxNt matrix, where
%                                   D - data dimensionality,
%                                   Nt - number of patterns)
%   L               - The number of base classifiers in the
ensemble
%   RP_algorithm    - Random Projection algorithm (character
string
%                                   with the name of an algo-
rithm)
%   classifier      - Base classifier (character string with
the name
%                                   of an algorithm)
%   varargout       - Any parameters of a classifier (see
MATLAB
%                                   function, where a particu-
lar classifier is
%                                   implemented)
%
% Outputs:
%   ensemble_predictions - Predictions (1xN vector of class
labels),
%                                   assigned by an ensemble of clas-
sifiers
%   ensemble_soft_predictions   - Soft predictions (similar to
%                                           probability
estimates)
%                                           (1xN vec-
```

```
tor), assigned by
%                                                   an ensemble
of classifiers
%
%
% Examples of use:
%
% [ensemble_predictions, ensemble_soft_predictions] = ...
% RandomProjectionEnsemble(train_patterns, train_targets,
% test_patterns,...
% 5, 'RandomProjection_JL', 'NearestNeighbor', 3);
% [ensemble_predictions, ensemble_soft_predictions] = ...
% RandomProjectionEnsemble(train_patterns, train_targets,
% test_patterns,...
% 5, 'RandomProjection_JL', 'classification_tree_moo');
% [ensemble_predictions, ensemble_soft_predictions] = ...
% RandomProjectionEnsemble(train_patterns, train_targets,
% test_patterns,...
% 5, 'RandomProjection_A', 'NaiveBayes');



% Get unique class labels
Uc = unique(train_targets);

% Get the number of test patterns
Nt = size(test_patterns,2);

% Pre-allocate memory
base_predictions = zeros(L,Nt);

for i = 1:L
    % Projecting training data onto low-dimensional space
    [proj_train_patterns, P] = feval(RP_algorithm, train_pat-
terns);
    % Projecting test data onto low dimensional space
    proj_test_patterns = feval(RP_algorithm, test_patterns, P);
    % Notice that it is important to keep the original training
and
    % test data intact, since they are projected multiple
```

```
times!
    % Classify test data with the ith base classifier
    base_predictions(i,:) = feval(classifier, proj_train_pat-
terns, train_targets, proj_test_patterns, varargin{:});
end

% Classify test data by majority vote
[ensemble_predictions, ensemble_soft_predictions] = majority_
vote(base_predictions, 1:L, Uc);

return
```

REFERENCES

Achlioptas, D. (2001). Database-friendly random projections. In *Proceedings of the 20th ACM SIGMOD-SIGACT-SIGART Annual Symposium on Principles of Database Systems, Santa Barbara, CA* (pp. 274-281). New York, NY: ACM Press.

Altinçay, H. (2007). Ensembling evidential k-nearest neighbor classifiers through multi-modal perturbation. *Applied Soft Computing, 7*(3), 1072–1083. doi:10.1016/j.asoc.2006.10.002

Bay, S. D. (1998). Combining nearest neighbor classifiers through multiple feature subsets. In J. W. Shavlik (Ed.), *Proceedings of the 15th International Conference on Machine Learning, Madison, WI* (pp. 37-45). San Francisco, CA: Morgan Kaufmann.

Bay, S. D. (1999). Nearest neighbor classification from multiple feature sets. *Intelligent Data Analysis, 3*(3), 191–209. doi:10.1016/S1088-467X(99)00018-9

Bertoni, A., Folgieri, R., & Valentini, G. (2009). Classification of DNA microarray data with random projection ensembles of polynomial SVMs. In B. Apolloni, S. Bassis, & C. Morabito (Ed.), *Proceedings of the 18th Italian Workshop on Neural Networks, Vietri sul Mare, Salerno, Italy. Frontiers in Artificial Intelligence and Applications* (Vol.193, pp. 60-66). Amsterdam: IOS Press.

Bertoni, A., & Folgieri, R. V. (2005). Bio-molecular cancer prediction with random subspace ensembles of support vector machines. *Neurocomputing, 63*, 535–539. doi:10.1016/j.neucom.2004.07.007

Bø, T. H., & Jonassen, I. (2002). New feature subset selection procedures for classification of expression profiles. *Genome Biology, 3*, 0017.1-0017.11.

Braga-Neto, U. M., & Dougherty, E. R. (2004). Bolstered error estimation. *Pattern Recognition, 36*(7), 1267–1281. doi:10.1016/j.patcog.2003.08.017

Cho, S.-B., & Won, H.-H. (2003). Data mining for gene expression profiles from DNA microarray. *International Journal of Software Engineering and Knowledge Engineering, 13*(6), 593–608. doi:10.1142/S0218194003001469

Díaz-Uriarte, R., & Alvarez de Andrés, S. (2006). Gene selection and classification of microarray data using random forest. *BMC Bioinformatics, 7*(3).

Everitt, B. (2006). *The Cambridge dictionary of statistics* (3rd ed.). Cambridge, UK: Cambridge University Press.

Fawcett, T. (2006). An introduction to ROC analysis. *Pattern Recognition Letters, 27*(8), 861–874. doi:10.1016/j.patrec.2005.10.010

Folgieri, R. (2007). *Ensembles based on random projection for gene expression data analysis.* PhD Thesis. University of Milan, Italy.

García-Pedrajas, N., & Ortiz-Boyer, D. (2009). Boosting k-nearest neighbor classifiers by means of input space projection. *Expert Systems with Applications, 36*(7), 10570–10582. doi:10.1016/j.eswa.2009.02.065

Hand, D. J., & Vinciotti, V. (2003). Choosing k for two-class nearest neighbor classifiers with unbalanced classes. *Pattern Recognition Letters, 24*(9-10), 1555–1562. doi:10.1016/S0167-8655(02)00394-X

Hanley, J. A., & McNeil, B. J. (1982). The meaning and use of the area under a receiver operating characteristic (ROC) curve. *Radiology, 143*, 29–36.

Ho, T. K. (1998). The random subspace method for constructing decision forests. *IEEE Transactions on Pattern Analysis and Machine Intelligence, 20*(8), 832–844. doi:10.1109/34.709601

Joe, H. (2001). *Multivariate models and dependence concepts.* Boca Raton, FL: Chapman & Hall/CRC Press.

Johnson, W. B., & Lindenstrauss, J. (1984). Extensions of Lipschitz mappings into a Hilbert space. In R. Beals, A. Beck, A. Bellow, & A. Hajian (Ed.), *Proceedings of the Conference in Modern Analysis and Probability, New Haven, CT. Contemporary Mathematics* (Vol. 26, pp. 189-206). Providence, RI: The American Mathematical Society.

Kuncheva, L. I. (2004). *Combining pattern classifiers: methods and algorithms.* Hoboken, NJ: John Wiley & Sons. doi:10.1002/0471660264

Maimon, O., & Rokach, L. (2005). *Decomposition methodology for knowledge discovery and data mining*. Singapore: World Scientific.

Matoušek, J. (2002). Embedding finite metric spaces into normed spaces . In Matoušek, J. (Ed.), *Lectures on discrete geometry* (pp. 355–400). New York, NY: Springer-Verlag.

Nelsen, R. B. (2006). *An introduction to copulas* (2nd ed.). New York, NY: Springer Science+Business Media.

Okun, O., & Priisalu, H. (2005). Multiple views in ensembles of nearest neighbor classifiers. In *Proceedings of the 22nd ICML Workshop on Learning with Multiple Views, Bonn, Germany* (pp. 51-58).

Okun, O., & Priisalu, H. (2006). Ensembles of nearest neighbour classifiers and serial analysis of gene expression. In *Proceedings of the 9th Scandinavian Conference on Artificial Intelligence, Helsinki, Finland* (pp. 106-113).

Okun, O., & Priisalu, H. (2007). Dataset complexity and gene expression based cancer classification. In F. Masulli, S. Mitra, & G. Pasi (Ed.), *Proceedings of the 7th International Workshop on Fuzzy Logic and Applications, Camogli, Italy* (LNAI 4578, pp. 484-490). Berlin/Heidelberg: Springer-Verlag.

Okun, O., & Priisalu, H. (2009). Dataset complexity in gene expression based cancer classification using ensembles of k-nearest neighbors. *Artificial Intelligence in Medicine*, *45*(2-3), 151–162. doi:10.1016/j.artmed.2008.08.004

Okun, O., Valentini, G., & Priisalu, H. (2008). Exploring the link between bolstered classification error and dataset complexity for gene expression-based cancer classification . In Maeda, T. (Ed.), *New signal processing research* (pp. 249–278). New York, NY: Nova Science Publishers.

Park, C., & Cho, S.-B. (2003). Evolutionary computation for optimal ensemble classifier in lymphoma cancer classification. In N. Zhong, Z. W. Raś, S. Tsumoto, & E. Suzuki (Eds.), *Proceedings of the 14th International Symposium on Methodologies for Intelligent Systems, Maebashi City, Japan* (LNAI 2871, pp. 521-530).

Peng, Y. (2006). A novel ensemble machine learning for robust microarray data classification. *Computers in Biology and Medicine*, *36*(6), 553–573. doi:10.1016/j.compbiomed.2005.04.001

Pepe, M. S. (2004). *The statistical evaluation of medical tests for classification and prediction* (1st paperback ed.). Oxford, UK: Oxford University Press.

Prodromidis, A. L., Stolfo, S. J., & Chan, P. K. (1999). *Effective and efficient pruning of meta-classifiers in a distributed data mining system* (Technical Report CUCS-017-99). New York, NY: Columbia University.

Pudil, P., Novovičová, J., & Kittler, J. (1994). Floating search methods in feature selection. *Pattern Recognition Letters, 19*(11), 1119–1125. doi:10.1016/0167-8655(94)90127-9

Ré, M., & Valentini, G. (2010). Simple ensemble methods are competitive with state-of-the-art data integration methods for gene function prediction. *Journal of Machine Learning Research, 8*(Machine Learning in Systems Biology), 98-111.

Schclar, A., & Rokach, L. (2009). Random projection ensemble classifiers. In J. Filipe, & J. Cordeiro (Ed.), *Proceedings of the 11th International Conference on Enterprise Information Systems, Milan, Italy* (LNBIP 24, pp. 309-316).

Sklar, A. (1959). Fonctions de répartition à n dimensions et leurs marges. *Publications of the Institute of Statistics, 8*, 229–231.

Stracuzzi, D. J. (2008). Randomized feature selection . In Liu, H., & Motoda, H. (Eds.), *Computational methods of feature selection* (pp. 41–62). Boca Raton, FL: Chapman & Hall/CRC Press.

Tsoumakas, G., Partalas, I., & Vlahavas, I. (2009). An ensemble pruning primer . In Okun, O., & Valentini, G. (Eds.), *Applications of supervised and unsupervised ensemble methods* (pp. 1–13). Berlin, Heidelberg: Springer-Verlag. doi:10.1007/978-3-642-03999-7_1

Vempala, S. S. (2004). *The random projection method*. Providence, RI: The American Mathematical Society.

Zar, J. H. (1999). *Biostatistical analysis* (4th ed.). Upper Saddle River, NJ: Prentice Hall/Pearson Education International.

ENDNOTES

[1] To remind you, a stable classifier is the one whose predictions do not significantly change when the input data set is changed (perturbed).

[2] The null hypothesis of this test is two medians are equal at the 5% significance level.

[3]   As more classifiers are included into $\mathcal{C}$, the diversity among classifiers decreases. Hence, $\widetilde{M}$ serves as a threshold that prevents redundant classifiers to become members of the ensemble.

Chapter 18

Bagging and Random Forests

BOOTSTRAP AND ITS USE IN CLASSIFIER ENSEMBLES

In the previous chapter, we saw how random sampling of the original feature set can create a powerful classifier ensemble. In this chapter, we approach to ensemble generation from another direction: sampling instances instead of features. A technique for random sampling of instances is called bootstrap.

The term 'bootstrap' has a lot of meanings (http://en.wikipedia.org/wiki/Bootstrapping). The technique with such a name is said to be credited to the famous Baron Münchausen who once in a swamp, pulled himself out by either his hair or by his bootstraps (the difference is not essential, but you, dear readers, are warned not to try to reproduce his trick, since this could be dangerous for your health and even life).

The basic idea of statistical bootstrap is sampling with replacement[1] to produce random samples of size N from the original data, $x_1, x_2, \ldots, x_N$; each of these is known as a bootstrap sample and each sample is used to provide an estimate of the quantity in question (Everitt, 2006). As statistical bootstrap has nothing to do with

DOI: 10.4018/978-1-60960-557-5.ch018

Baron Münchausen (i.e. it is not his invention), it is safe to apply it in practice. If you want to learn more about statistical bootstrap, the book of Efron and Tibshirani (Efron & Tibshirani, 1998) is for you.

How bootstrap can be useful for generating classifier ensembles? An ensemble is created by drawing bootstrap samples from the original training data and fitting a classifier to each sample. A vote over the predictions of individual classifiers is used for the ensemble prediction. However, which types of classifiers are particularly suitable for this kind of scheme?

Let us recall that classifiers can be divided into stable and unstable, depending on the effect of changes in the data set on classifier predictions. The unstable classifiers, such as the classification tree, may dramatically change predictions even if the training data changed a little. In contrast, the stable classifiers, such as Naïve Bayes, k-nearest neighbors and support vector machines, are relatively unaffected by small changes in the training data.

Hence, statistical bootstrap considers the classifiers with high variance as the best suitable ones. Due to random variation in the bootstrap samples coupled with the instability of a single classifier, the ensemble will consists of diverse classifiers, i.e. ones making errors on different instances. It is known that such an ensemble when combining votes of diverse classifiers, on average, produces the correct prediction (Kuncheva, 2004).

Bootstrap lays in foundation of two ensemble methods: bagging (Breiman, Bagging predictors, 1996) and its extension, random forest (Breiman, Random forests, 2001). Both ensemble methods are in focus of this chapter. As you might already guess, the classification tree (Breiman, Friedman, Olshen, & Stone, 1984) being an unstable classifier is a good candidate to be employed in these methods. The very recently published book on classifier ensembles (Seni & Elder, 2010) exclusively concentrates on tree ensembles; hence it is highly recommended to read. Another book where tree ensembles are described in depth is (Rokach & Maimon, 2008).

Bagging

Bagging introduced by Breiman in (Breiman, Bagging predictors, 1996) stands for Boostrap AGGregatING. It is a variance-reduction technique (Breiman, Bagging predictors, 1996), (Hastie, Tibshirani, & Friedman, 2003), (Izenman, 2008), (Rokach & Maimon, 2008). The idea of bagging is very simple: boostrap the original training data, fit a separate classifier to each bootstrap sample, and combine the classifier outputs by the majority vote (Breiman, Bagging predictors, 1996). Each bootstrap sample contains the same number of instances as the original training set, i.e. it is created by sampling (N times, where N is the training set size) the training data with replacement. Once the base classifier and its parameters have been chosen or set up, the only parameter of bagging remaining to define is the number of classifier in the ensemble.

Breiman pointed out that bagging stable classifiers such as k-nearest neighbors is not a good idea as it would lead to a collection of almost identical classifiers, which would render an ensemble not being better than any its member. However, bagging unstable classifiers such as classification trees yields very good results due to the diversity of predictions of individual classifiers resulting from the tree instability. Theoretical analysis of bagging done in (Bühlmann & Yu, 2002) showed that the improvement in the prediction accuracy of ensembles is achieved by smoothing the hard cut decision boundaries created by splitting in the classification trees, which, in its turn, reduces the variance of the prediction (Strobl, Boulesteix, Kneib, Augustin, & Zeileis, 2008). (Tan & Gilbert, 2003) remarked that bagging works well in the presence of noise thanks to the smoothing effect of the majority vote rule. The success of bagging due to the majority vote combination is the likely reason for Kuncheva (Kuncheva, 2004), too, as it helps to reduce overfitting unavoidable in tree-like classifiers.

The proportions of votes for each class (the ratio of the number of votes for that class to the number of trees in the ensemble) can be viewed as the class conditional probability estimates.

Vu and Braga-Neto (Vu & Braga-Neto, 2009) by experimenting with several gene expression data sets claimed that bagged trees and random forests can easily outperform a single tree but may fail to outperform a single stable classifier such as k-nearest neighbors on small-size data. Vu and Braga-Neto applied gene filtering prior to bagging or random forest. Was feature selection used in (Vu & Braga-Neto, 2009) to blame because it was univariate, i.e. ignorant of the interactions among features? This is an open question.

Random Forest

Breiman proposed the extension of bagging, called a random forest in (Breiman, Random forests, 2001). As bagging, random forest utilizes the classification tree as the base classifier in ensemble construction and each tree casts one vote. Once all trees voted, their votes are combined with the majority rule to produce the ensemble prediction.

In addition to bootstrapping the original training data, random forest makes one step further: random feature selection carried out at each node of the tree. Namely, a subset of features is randomly chosen (sampled without replacement) from the entire set of features and the best feature from this subset is sought to split the node[2] (any splitting criterion considered in the chapter about CART can be used). Surprisingly, this randomization technique tends to improve the predictive power of the ensemble, as random selection of features reduces correlation between trees in the ensemble (hence, diversity in predictions is increased, which is the key to obtain a powerful ensemble).

A separate feature subset is selected for each node. Typical subset sizes are $0.5\sqrt{D}, \sqrt{D}, 2\sqrt{D}$, where D is the total number of features. For example, if $D = 10000$, then 50, 100 or 200 features per node are considered to be enough in order to find the best split. Such a solution is dictated by the necessity to restrict a search space in order to keep computational complexity of node splitting sufficiently low while ensuring that the search covers enough features at the same time.

Breiman suggested growing a full classification tree (no pruning), which helps to keep classifier's bias low. As averaging over all trees in the forest yields a low-variance ensemble, the result is very promising: low bias and low variance. Lin and Jeon (Lin & Jeon, 2006), however, reported that limiting the depth of the trees may be beneficial for predictive accuracy in some cases. Nevertheless, the general advice is to leave trees unpruned as it lowers the bias-component of the classification error.

The number of trees in the forest should be large (say, at least 500-1000 or even larger, say, 5000, in case of high dimensional data (Díaz-Uriarte & Alvarez de Andrés, 2006), (Izenman, 2008)). Due to randomization of the split at each node, each tree is likely to be less accurate than a tree created with the exact splits like in CART (Breiman, Friedman, Olshen, & Stone, 1984). However, by combining such seemingly non-optimal trees in the ensemble, classification accuracy can become better than that of a single tree with exact splits.

The important characteristic produced by random forests is the out-of-bag (OOB) estimate of error rate[3]. Each bootstrap sample contains approximately two-thirds of the entire training set. In other words, some instances (about one third) are left out and not used in the tree construction. These instances form an out-of-bag set and they can be used as a test set for performance evaluation as follows (Dudoit & Fridlyand, 2003). For the b th bootstrap sample, one needs to put the out-of-bag instances down this tree to get their classification. For each training instance, let the final classification of the forest be the class obtaining the most votes for the bootstrap samples in which that instance was out-of-bag. The assigned labels $C_{bag}\left(x_i\right)$ can be compared with true labels y_i and the unbiased error estimation for the OOB instances can be thus obtained (Izenman, 2008) as $N^{-1}\sum_{i=1}^{N} I\left(C_{bag}\left(x_i\right) \neq y_i\right)$, where I is an indicator function whose value is 1, if $C_{bag}\left(x_i\right) \neq y_i$ is true and 0 otherwise. This is a big plus, since one gets such an unbiased error estimate as a byproduct of the bootstrap without the need for cross-validation or test set estimation. Random forests typically do not require feature selection prior to forest generation. However, if feature selection was done beforehand, the OOB error will in general be downward biased.

How about overfitting? Breiman (Breiman, Random forests, 2001) claimed that random forests cannot overfit, because as the number of trees increases, generalization error converges to a limit. This is a very serious statement. It is safer to say that as any other classifier, the random forests may occasionally overfit on some data sets but unlike many classification models, the random forests can control and efficiently reduce overfitting. Experimental results obtained on the 22 gene expression data (see, e.g., (Statnikov, Wang, & Aliferis, 2008)) seem to support this observation, though Statnikov et al remarked that Segal (Segal, 2004) constructed examples, where the forest composed of smaller trees outperformed the forest of completely unpruned trees. In the line with the conclusions in (Vu & Braga-Neto, 2009), Statnikov et al. observed that random forests lost to such a stable classifier as support vector machine. Nevertheless, this does not mean that random forests should now be forgotten as the method giving false hope. Breathe deeply and recall no-free-lunch theorem! Did you feel relief?

As with bagging, class conditional probability estimates can be obtained as the proportions of votes for each class.

It should be noted that although the random forest has been originally defined for decision trees, this approach is applicable to any type of classifiers. The important advantage of the random forests, making this method highly competitive with respect to the state-of-the-art ensemble methods and single classifiers is its ability to handle a very large number of features (Skurichina & Duin, 2002), i.e. it does not suppose feature selection. Another important characteristic of the random forest is that it is fast to compute due to the fact that the best split at each node is only sought among a small fraction of the features instead of all features.

One of the noticeable articles about random forests for gene selection and microarray data classification is (Díaz-Uriarte & Alvarez de Andrés, 2006). It demonstrated the potential of the random forests in solving such a complicated task as microarray gene expression based cancer classification and paved the avenue for several other articles, e.g., such as (Strobl C., Boulesteix, Zeileis, & Hothorn, 2007), (Strobl C., Boulesteix, Kneib, Augustin, & Zeileis, 2008), (Statnikov, Wang, & Aliferis, 2008), (Okun & Priisalu, 2007), (Amaratunga, Cabrera, & Lee, 2008). The comprehensive research described in (Dudoit & Fridlyand, 2003) is also worth of studying.

Backward gene selection based on random forests is proposed in (Díaz-Uriarte & Alvarez de Andrés, 2006). To select genes, they iteratively fit random forests, at each iteration building a new forest after discarding genes with the smallest importance, expressed through the OOB error. In the end of each iteration, 20% of the remaining features are eliminated, which leads to fast processing. As the gene pool to select from becomes smaller, the fraction of dropped genes can become smaller, too. After fitting all forests, their OOB errors are examined and the forest with the smallest number of genes whose error is within one standard deviation of the minimum error rate of all forests is chosen as the final solution.

Amaratunga et al. (Amaratunga, Cabrera, & Lee, 2008) argued that non-informative features can degrade the classification accuracy of random forests when classifying microarray data. In microarray data, the number of noisy and irrelevant genes far exceeds the number of genes useful for disease prediction. In this case, nodes of many trees in the forest can be populated by non-informative genes. Amaratunga et al. proposed to reduce the contribution of such trees to the final vote. For this, they advocated weighted random sampling instead of simply random sampling, with the weights geared in favor of the informative features. The weights are determined by the ability of genes to separate two classes of data.

Variable Importance

Random forests could be attractive not only because of their high accuracy and the out-of-bag error estimate, but also because of variable (feature) importance that is defined in terms of the contribution to the predictive accuracy. Identifying which feature is important and which is not is, however, complicated by interactions between features: for example, a certain feature may be non-important on its own, but in combination with another feature, both features become important for class prediction.

To get variable importance, for each tree, randomly permute the values of the ith feature for the out-of-bag instances, put these new instances down the tree, and get new classifications for the forest. The motivation for random permutation is the following. If a certain feature is indeed important, then distorting its values would cause its ability to successfully classify data instances to diminish, i.e. permuted feature values would result in the increase of the OOB error. The larger this increase, the more important the feature is. The importance of the ith feature can be defined in a number of ways (see, e.g., (Izenman, 2008)), e.g., as the difference between the out-of-bag error rate for the randomly permuted ith feature and the original out-of-bag error rate, averaged over all trees in the forest. As the variable importance involves random permutation, it can, however, be a quite time-consuming procedure.

The advantage of the random forest variable importance, compared to univariate methods is that it covers the impact of each feature individually as well as in multivariate interactions with other features (Breiman, Bagging predictors, 1996), (Díaz-Uriarte & Alvarez de Andrés, 2006), (Strobl C., Boulesteix, Zeileis, & Hothorn, 2007).

Several studies demonstrated that one should treat with caution the variable importance returned by random forests (Okun & Priisalu, 2007), (Strobl C., Boulesteix, Zeileis, & Hothorn, 2007), (Strobl, Boulesteix, Kneib, Augustin, & Zeileis, 2008), especially if it is intended for gene selection.

Okun and Priisalu showed in (Okun & Priisalu, 2007) that two forests, generated with the different number of features per node split, may have very similar classification errors on the same dataset, but the respective lists of genes ranked according to feature importance can be weakly correlated. Izenman (Izenman, 2008) also warned about this contradiction to the early claims made in (Díaz-Uriarte & Alvarez de Andrés, 2006), (Svetnik, Liaw, Tong, & Wang, 2004). He wrote "note that the rankings of important features changes with the number of features randomly chosen for node splitting, the initial seed for randomization, and the number of trees in the random forest."

Strobl et al. (Strobl C., Boulesteix, Kneib, Augustin, & Zeileis, 2008) carried out another study of the variable importance and concluded that the variable importance computed by the random forests is biased towards correlated features. They discovered two reasons for this behavior: 1) inclination to select correlated features, embedded into the tree building process, and 2) the unconditional permutation scheme employed in the computation of the variable importance that gives an additional advantage to correlated features. In the words of Strobl et al. (Strobl C., Boulesteix, Kneib, Augustin, & Zeileis, 2008) "random forests show a preference for correlated predictor variables, that is also carried forward to any significance test or variable selection scheme constructed from the importance measures." Strobl et al. introduced a conditional permutation importance and they proposed to replace bootstrap sampling in the random forest with sampling without replacement[4] (Strobl C., Boulesteix, Zeileis, & Hothorn, 2007). These two modifications of the original random forest algorithm allow obtaining a reliable and unbiased variable importance estimate when features vary in their scale of measurement or number of categories. In the case of correlated features, it is important to distinguish between conditional and marginal influence of a feature. A feature can be marginally influential (i.e. the classification outcome is considered to be dependent on that feature, regardless of all other features), but when considered conditional on another feature, its influence on the classification outcome greatly diminishes or it does not influence on the classification outcome at all. In other words, different (and more sophisticated) conditional probabilities come into play in the conditional permutation importance.

MATLAB Code

Breiman and Cutler originally implemented random forests in Fortran 77 (visit http://stat-www.berkeley.edu/users/breiman/RandomForests/cc_home.htm). Its description can also be found in (Breiman & Cutler, Random forests manual,

2003). MATLAB® implementation can be downloaded, e.g., from http://shum.huji.
ac.il/~gleshem/#software0.

MATLAB® Statistics Toolbox™ 7 includes class *TreeBagger* whose methods
implement both bagging and random forests. *TreeBagger* and its methods are used
in our function *BaggingTrees_moo* that demonstrates how to utilize rich functionality offered by object-oriented MATLAB®. *BaggingTrees_moo* has six input and
three output arguments. Besides training/test data and training labels, the number
of trees in an ensemble (*ntrees*), the variable importance threshold (*fimpthr*), which
is a real number between 0 and 1, and the number of features to randomly select
for each node split (*nvar2sample*) are necessary to specify as input. The default
value for *nvar2sample* is $\sqrt{D}$, which implies that the random forest is constructed
by default. If an ensemble of bagged trees is desired instead, then *nvar2sample*
needs to be set to D (all features will be used to find the best split at each node).

The call of *TreeBagger* creates the object b, which is later used to access to the
methods of the *TreeBagger* class. *BaggingTrees_moo* consists of two parts. In the
first part, an ensemble of fifty trees is utilized to assess the variable importance and
to compute the OOB error, depending on the number of trees in an ensemble. This
is where *fimpthr* is needed in order to assign indices of important features to one of
the outputs – *selected_features*. The second part of *BaggingTrees_moo* constructs
a larger forest of *ntrees* trees, but without estimating the variable importance of
each feature. It should be noted that the features found to be important and stored
in *selected_features* are not employed here, since code in the first part of *Bagging-
Trees_moo* was only intended to demonstrate the certain capabilities of *TreeBagger*
and its methods. After an ensemble of trees has been created, it is stored in the compact form (without the information about training data) in a file *EnsembleOfTrees.
mat* in order to be used later in classifying test data that is carried out by the method
predict, which returns predicted labels of test data and associated probabilistic scores.
For each instance and each class, the score generated by each tree is the probability
of this instance originating from this class, which is computed as the fraction of
instances of this class in a tree leaf; *predict* averages these scores over all trees in
the ensemble. In this way, the remaining two outputs (*ensemble_predictions* and
ensemble_soft_predictions) of *BaggingTrees_moo* are assigned. As the *TreeBagger*
class has richer functionality than just described, my advice is to read and study
related documentation of MATLAB® Statistics Toolbox™.

I also implemented bagging with classification trees using the MATLAB® function *classregtree*. No variable importance was, however, computed as it is provided
by *TreeBagger*. The function *BaggingTrees* can be used when MATLAB® Statistics
Toolbox™ 7 is not available (older releases do not seem to include *TreeBagger*).
Instead of indices of the important features, *BaggingTrees* returns the OOB error

rate for bagging. Notice that construction of each tree is done using only in-bag instances, while tree testing is carried out with out-of-bag instances. Pay attention that matrix *oob_predictions* usually contains elements with three different values: 0, 1, and -1000000. However, when computing the OOB error rate, the elements '-1000000' are not counted, i.e. they are simply ignored, because the value of -1000000 implies that there is no vote from a given tree for a given instance. As you can see, the majority vote combines predictions of all trees and assigns the result to *ensemble_predictions*, while probabilistic predictions (*ensemble_soft_predictions*) are computed as ratios of the number of votes for a certain class to the total number of trees in the ensemble.

```
function [ensemble_predictions, ensemble_soft_predictions,
selected_features] = BaggingTrees_moo(train_patterns, train_
targets, test_patterns, ntrees, fimpthr, nvar2sample)


% Bagging (Bootstrap AGGregatING) decision trees built on
MATLAB
% Statistics Toolbox object-oriented functionality
% Original sources:
% Leo Breiman, "Bagging predictors", Machine Learning 26():
123-140,
% 1996
% Alan Julian Izenman, "Modern multivariate statistical tech-
niques:
% regression, classification, and manifold learning", Springer
% Science+Business Media, New York, NY, 2008
%
% Inputs:
%    train_patterns  - Training patterns (DxN matrix, where D
- data
%                                dimensionality, N - number
of patterns)
%    train_targets   - Training targets (1xN vector)
%    test_patterns   - Test patterns (DxNt matrix, where D -
data
%                                dimensionality, Nt - number
of patterns)
%    ntrees          - The number of trees in an ensemble
%    fimpthr         - Feature importance threshold (real value
between
```

```
%                                    0 and 1)
%    nvar2sample     - The number of features to select at
random for
%                                    each decision split. De-
fault is the square root %                                of
the original number of features. Valid
%                                    values are also 'all' or a
positive integer.
%
% Outputs:
%    ensemble_predictions      - Predictions (1xN vector of
class
%                                    labels),
assigned by an ensemble of
%                                    classifiers
%    ensemble_soft_predictions  - Soft predictions (similar to
%                                    probability
estimates) (1xN
%                                    vector),
assigned by an ensemble of
%                                    classifiers
%    selected_features        - Indices of important features
%
% Note: it requires Statistics Toolbox 7, released in 2009



% Get a version number for Statistics Toolbox
v = ver('stats');
% If MATLAB release is before 2009, generate an error message
if isempty(v)
    error('No Statistics toolbox was found!');
else
    if str2num(v.Version) < 7.1
        error('Version 7.1 (or higher) of Statistics Toolbox is
needed to run this function!');
    end
end


% Get the number of features
```

```
D = size(train_patterns, 1);

% Set default values
if nargin < 6
    nvar2sample = ceil(sqrt(D));
end

% Convert numerical labels into categorical variables
train_targets = num2cat(train_targets);

% Build an ensemble of 50 decision trees for selecting impor-
tant
% features
% Use a default value for nvar2sample
b = TreeBagger(50, train_patterns', train_targets',...
    'Method', 'classification', 'OOBVarImp', 'on');
figure(1);
% Plot classification error, depending on the number of trees
in the
% ensemble
% In-bag observations are used in computation of classification
error
plot(oobError(b));
xlabel('Number of trees');
ylabel('Out-of-bag classification error');

% In-bag observations are now excluded from computation of
% classification error
b.DefaultYfit = '';
figure(2);
% Plot classification error, depending on the number of trees
in the
% ensemble
plot(oobError(b));
xlabel('Number of trees');
ylabel('Out-of-bag classification error excluding in-bag obser-
vations');

figure(3);
% Plot feature importance for each feature
```

```
bar(b.OOBPermutedVarDeltaError);
xlabel('Feature index');
ylabel('Feature importance');
% Get indices of important features
selected_features = find(b.OOBPermutedVarDeltaError > fimpthr);

% ****************************************************************
********

% Now build a larger ensemble of decision trees for computing
% the out-of-bag classification error
% To save time, feature importance is not computed
% Out-of-bag predictions for training data are computed
b = TreeBagger(ntrees, train_patterns', train_targets',...
    'Method', 'classification', 'OOBPred','on', 'OOBVarImp',
'off',...
    'NVarToSample', nvar2sample, 'DefaultYfit', 'MostPopular');
figure(4);
plot(oobError(b));
xlabel('Number of trees');
ylabel('Out-of-bag classification error');

% Build a lightweight version of the ensemble in order to save
memory
% It can be used to predict class labels of test data
c = compact(b);

% Save the ensemble in a.mat file for later use
save EnsembleOfTrees c;

% Classify test data
[predictions, scores] = predict(c, test_patterns');
% As predictions is a cell array of strings, we need to convert
it into % a numeric array
ensemble_predictions = cat2num(predictions', unique(train_tar-
gets));

ensemble_soft_predictions = scores';

return
```

```
function [ensemble_predictions, ensemble_soft_predictions,
oobError] = BaggingTrees(train_patterns, train_targets, test_
patterns, ntrees)

% Bagging (Bootstrap AGGregatING) decision trees built on
MATLAB
% function classregtree.m
% Original sources:
% Leo Breiman, "Bagging predictors", Machine Learning 26():
123-140,
% 1996
% Alan Julian Izenman, "Modern multivariate statistical tech-
niques:
% regression, classification, and manifold learning", Springer
% Science+Business Media, New York, NY, 2008
%
% Inputs:
%    train_patterns   - Training patterns (DxN matrix, where D
- data
%                                   dimensionality, N - number
of patterns)
%    train_targets   - Training targets (1xN vector)
%    test_patterns   - Test patterns (DxNt matrix, where D -
data
%                                   dimensionality, Nt - number
of patterns)
%    ntrees          - The number of trees in an ensemble
%
% Outputs:
%    ensemble_predictions        - Predictions (1xNt vector of
class
%                                              labels),
assigned by an ensemble of
%                                              classifiers
%    ensemble_soft_predictions   - Soft predictions (similar to
%                                              probability
estimates) (CxNt
%                                              matrix,
where C - the number of
%                                              classes),
```

```
assigned by an ensemble
%                                                     of classi-
fiers
%   oobError                  - Out-of-bag misclassification
error



% Get the number of training patterns
N = length(train_targets);

% Set default value
if nargin < 4
    ntrees = 500;
end

% Pre-allocate memory
idx = zeros(ntrees,N);
base_predictions = zeros(ntrees,size(test_patterns,2));
oob_predictions = -1000000*ones(ntrees,N);
% It is assumed that there are no real labels equal to -1000000
among
% the data

for i = 1:ntrees
    % Set indices of all patterns to 'on'
    all_idx = ones(1,N);

    % Randomly sample the training set with replacement
    idx = randsample(N, N, true);

    % Set indices of in-bag patterns to 'off'
    all_idx(idx) = 0;
    % Get indices of out-of-bag patterns
    oob_idx = find(all_idx);

    % Create a tree from the in-bag training data
    % Don't prune the tree
    t = classregtree(train_patterns(:,idx)', train_
targets(idx)',...
```

```
        'method', 'classification', 'prune', 'off');

    fprintf(1, 'Tree %d has been constructed...\n', i);

    % Classify out-of-bag data
    targets = eval(t, train_patterns(:,oob_idx)');
    % Targets is a cell array of character strings now
    % Convert categorical labels into numerical values
    oob_predictions(i,oob_idx) = cat2num(targets, unique(train_
targets));

    % Classify test data by the ith tree
    test_targets = eval(t, test_patterns');
    % Convert categorical labels back into the numerical repre-
sentation
    test_targets = cat2num(test_targets, unique(train_tar-
gets));
    % Accumulate predictions
    base_predictions(i,:) = test_targets;
end

% Get unique class labels
% If necessary, convert first to the numerical representation
Uc = unique(cat2num(train_targets, unique(train_targets)));

% Compute the out-of-bag misclassification rate which is an
unbiased
% estimate of ensemble preformance
h = histc(oob_predictions, Uc, 1);
[dummy,i] = max(h, [], 1);
predictions = Uc(i);
oobError = class_error(predictions, cat2num(train_targets,
unique(train_targets)));

% Classify test data by majority vote among all predictors
h = histc(base_predictions, Uc, 1);
[dummy,i] = max(h, [], 1);
ensemble_predictions = Uc(i);
ensemble_soft_predictions = h/ntrees;
```

```
return
function errprcnt = class_error(test_targets,targets)

% Calculate error percentage based on true and predicted test
labels
errprcnt = mean(test_targets ~= targets);
errprcnt = 100*errprcnt;

return
```

REFERENCES

Amaratunga, D., Cabrera, J., & Lee, Y.-S. (2008). Enriched random forests. *Bioinformatics (Oxford, England)*, *24*(18), 2010–2014. doi:10.1093/bioinformatics/btn356

Breiman, L. (1996). Bagging predictors. *Machine Learning*, *24*(2), 123–140. doi:10.1007/BF00058655

Breiman, L. (2001). Random forests. *Machine Learning*, *45*(1), 5–32. doi:10.1023/A:1010933404324

Breiman, L., & Cutler, A. (2003). *Random forests manual. Technical Report, Department of Statistics*. Berkeley, CA: University of California.

Breiman, L., Friedman, J. H., Olshen, R. A., & Stone, C. J. (1984). *Classification and regression trees*. New York, NY: Wadsworth.

Bühlmann, P., & Yu, B. (2002). Analyzing bagging. *Annals of Statistics*, *30*(4), 927–961. doi:10.1214/aos/1031689014

Díaz-Uriarte, R., & Alvarez de Andrés, S. (2006). Gene selection and classification of microarray data using random forest. *BMC Bioinformatics*, *7*(3).

Dudoit, S., & Fridlyand, J. (2003). Classification in microarray experiments. In Speed, T. (Ed.), *Statistical analysis of gene expression microarray data* (pp. 93–158). Boca Raton, FL: Chapman & Hall/CRC Press. doi:10.1201/9780203011232.ch3

Efron, B., & Tibshirani, R. J. (1998). *An introduction to the bootstrap*. Boca Raton, FL: Chapman & Hall/CRC Press.

Everitt, B. (2006). *The Cambridge dictionary of statistics* (3rd ed.). Cambridge, UK: Cambridge University Press.

Hastie, T., Tibshirani, R., & Friedman, J. (2003). *The elements of statistical learning: data mining, inference, and prediction*. New York, NY: Springer-Verlag.

Izenman, A. J. (2008). *Modern multivariate statistical techniques: regression, classification, and manifold learning*. New York, NY: Springer Science+Business Media.

Kuncheva, L. I. (2004). *Combining pattern classifiers: methods and algorithms*. Hoboken, NJ: John Wiley & Sons. doi:10.1002/0471660264

Lin, Y., & Jeon, Y. (2006). Random forests and adaptive nearest neighbors. *Journal of the American Statistical Association, 101*(474), 578–590. doi:10.1198/016214505000001230

Okun, O., & Priisalu, H. (2007). Random forest for gene expression based cancer classification: overlooked issues. In J. Marti, J.-M. Benedi, A. M. Mendonça, & J. Serrat (Ed.), *Proceedings of the 3rd Iberian Conference on Pattern Recognition and Image Analysis, Girona, Spain. Lecture Notes in Computer Science, Vol.4478*, pp. 483-490. Berlin/Heidelberg: Springer-Verlag.

Rokach, L., & Maimon, O. (2008). *Data mining with decision trees: theory and applications*. Singapore: World Scientific.

Segal, M. R. (2004). *Machine learning benchmarks and random forest regression. Technical Report. Center for Bioinformatics and Molecular Biostatistics*. San Francisco, CA: University of California.

Seni, G., & Elder, J. F. (2010). *Ensemble methods in data mining: improving accuracy through combining predictions*. San Francisco, CA: Morgan & Claypool Publishers.

Skurichina, M., & Duin, R. P. (2002). Bagging, boosting and the random subspace method for linear classifiers. *Pattern Analysis & Applications, 5*(2), 121–135. doi:10.1007/s100440200011

Statnikov, A., Wang, L., & Aliferis, C. F. (2008). A comprehensive comparison of random forests and support vector machines for microarray-based cancer classification. *BMC Bioinformatics, 9*(319).

Strobl, C., Boulesteix, A.-L., Kneib, T., Augustin, T., & Zeileis, A. (2008). Conditional variable importance for random forests. *BMC Bioinformatics, 9*(307).

Strobl, C., Boulesteix, A.-L., Zeileis, A., & Hothorn, T. (2007). Bias in random forest variable importance measures: illustrations, sources and a solution. *BMC Bioinformatics, 8*(25).

Svetnik, V., Liaw, A., Tong, C., & Wang, T. (2004). Application of Breiman's random forest to modeling structure-activity relationships of pharmaceutical molecules. In F. Roli, J. Kittler, & W. Terry (Ed.), *Proceedings of the 5th International Workshop on Multiple Classifier Systems, Cagliari, Italy* (LNCS 3077, pp. 334-343). Berlin/Heidelberg: Springer-Verlag.

Tan, A. C., & Gilbert, D. (2003). Ensemble machine learning on gene expression data for cancer classification. *Applied Bioinformatics*, *2*(3Suppl), S75–S83.

Vu, T. T., & Braga-Neto, U. M. (2009). Is bagging effective in the classification of small-sample genomic and proteomic data? *EURASIP Journal on Bioinformatics & Systems Biology*.

ENDNOTES

[1]   To remind you, sampling with replacement is a technique for taking samples from a finite population when each element is replaced before the next one is drawn (Everitt, 2006).

[2]   Recall that in CART (Breiman, Friedman, Olshen, & Stone, 1984), the best split is sought among all available features. Thus, bagging represents a special case of random forests when the number of features sampled without replacement is equal to the total number of features (Rokach & Maimon, 2008).

[3]   As pointed out by Izenman in (Izenman, 2008), the OOB error can be computed for bagging, too, since bagging is a special case of random forests.

[4]   In order to mimic bootstrap sampling, sample size for sampling without replacement was set to $0.632N$, because in bootstrap sampling about 63.2% of the data instances end up in the bootstrap sample. Sampling without replacement leads to a faster random forest and it does not introduce an artificial association between features as bootstrap sampling does (Strobl C., Boulesteix, Zeileis, & Hothorn, 2007).

Chapter 19
Boosting and AdaBoost

WEIGHTED LEARNING, BOOSTING AND ADABOOST

In this chapter, one of the most widely used ensemble method, called boosting, will be considered. Boosting is based on the idea of a weighted learning, where each instance in a training set gets a nonnegative weight. The higher the weight is, the higher is the importance of an instance. As a result of weight assignment during training, the training set becomes a weighted training set.

Boosting starts from all weights equal. From this set, boosting generates the first hypothesis (classification result), h_1 that classifies some instances correctly and some of them incorrectly. As we would like the next hypothesis to do better on the misclassified instances, the weights of misclassified instances increase while those of correctly classified decrease[1]. From this new weighted training set, a new hypothesis, h_2, is generated and so on until we have generated M hypotheses, where M is pre-defined.

After that, all hypotheses are combined with a weighted majority vote, where each hypothesis is weighted according to how well it performed on the training set.

A typical representative of the boosting algorithms is AdaBoost (short for Adaptive Boosting) that will be described below. AdaBoost was proposed by Freund and Schapire in (Freund & Schapire, 1996) and analyzed theoretically by Schapire

DOI: 10.4018/978-1-60960-557-5.ch019

(Schapire, 1999). AdaBoost was named in (Wu & Kumar, 2009) among the top ten algorithms in data mining.

The key factor attributing to the success of AdaBoost is the selection of the base algorithm used to learn hypotheses (Russell & Norvig, 2003). It needs to be a weak learner, which implies that it needs to perform only slightly better than a randomly guessing classifier (50% accurate), in order to provide very good ensemble performance. In contrast, a strong learner is a classifier that has a much better accuracy than the weak one.

An example of the weak learner is a decision stump, which is a decision tree with just one split (therefore one test at the root) and two terminal nodes[2]. Moreover, as M grows larger, AdaBoost tends to perfectly (i.e. with zero error) classify the training data. Thus, the aggregated classifier boosts the accuracy of the original weak classifier; hence, the name. This is true regardless of how complex the decision boundary between different data classes is.

One may wonder why AdaBoost works so well. One of the answers is that it belongs to the family of probably approximately correct (PAC) learning algorithms (Valiant, 1984). These algorithms return hypotheses that are probably approximately correct, which means that these hypotheses are consistent with a sufficiently large set of training instances, i.e. they are unlikely to be seriously wrong (Russell & Norvig, 2003). An approximately correct hypothesis is therefore the hypothesis that has the probability of error bounded by a small positive constant ε. In the absence of knowledge about the true function describing the data, an approximately correct hypothesis is said to lie (with probability at least $1 - \delta$, where $0 < \delta \leq 0.5$) inside the ε-ball around the true function (Russell & Norvig, 2003).

Based on such reasoning, PAC theory arrives at the sample complexity, i.e. the number of training instances needed for a learning algorithm to be probably approximately correct. It says that the lower limit on the sample complexity is huge if one does not restrict the hypothesis space. However, if one insists that a learning algorithm returns only simple consistent hypotheses (for example, like it is done in decision trees or decision stumps), then the sample complexity results (and hence, the learning process) become generally better.

Another possible explanation for remarkable performance is given in (Friedman, Hastie, & Tibshirani, 2000). Friedman, Hastie, and Tibshirani showed that AdaBoost is equivalent to running a gradient-descent algorithm to fit a logistic regression model. I omit their proof as it is presented in many books (see, for example, (Hastie, Tibshirani, & Friedman, 2003), (Izenman, 2008)). However, results in (Mease & Wyner, 2008) show that AdaBoost is not very suitable for estimating class-conditional probabilities, which implies that there is a doubt that a similarity of AdaBoost to logistic regression plays a key role in the success of AdaBoost.

Numerous experiments with AdaBoost demonstrated a surprising outcome: test error continues to decrease after the training set error reached zero. In other words, adding more weak learners after observing zero training error will benefit the classification performance. This contradicts to a typical classification scenario where test error first decreases for a while and then increases due overfitting as a classifier become more and more complex. In other words, AdaBoost tends to be quite resistant to overfitting.

Nevertheless, for small data sets like microarray gene expression data, one needs to always remember about overfitting. Moreover, AdaBoost is not absolutely immune from this harmful effect. One possible cure might be to have the number of classifiers in the ensemble quite small, e.g., 10, which would prevent the ensemble to become too complicated. However, the work (Mease & Wyner, 2008) demonstrates that overfitting may happen very early and running AdaBoost for a much larger number of iterations surprisingly reduces the amount of overfitting.

As the subject of this book is microarray gene expression data, we will consider in this chapter how to strike a good balance between the excellent classification performance of AdaBoost and the necessity to avoid overfitting.

Algorithm Description

Almost every book on machine learning, data mining or pattern recognition treats AdaBoost in detail (Bishop, 2006), (Hastie, Tibshirani, & Friedman, 2003), (Izenman, 2008), (Kuncheva, 2004), (Maimon & Rokach, 2005), (Dudoit & Fridlyand, 2003), (Russell & Norvig, 2003), (Wu & Kumar, 2009). My description of AdaBoost is mainly based on (Bishop, 2006), (Friedman, Hastie, & Tibshirani, 2000) and is presented in Figure 1. It should be noted that unlike our convention about $\{0,1\}$ -class labels, AdaBoost assumes that labels can be either -1 or +1 in order to facilitate a decision about class membership as can be seen below.

In the training or learning phase, weak learners are trained in order to find the values of weights reflecting each learner's classification accuracy. Once these weights are learned, the trained learners classify test data and the obtained predictions are then combined into a weighted sum. In order to obtain a binary decision, such a weighted output is finally thresholded as followed:

- zero and negative output is assigned 0-class,
- positive output is assigned to 1-class.

In the beginning, all instances get equal weights ($1/N$, where N is the number of training instances). Given M classifiers to combine, weight related to the ith

Figure 1. Pseudo code of AdaBoost

AdaBoost

Initialize weights of data points by setting $w_i^{(1)}$ to $1/N$, where $i = 1, \ldots, N$;

For $m := 1 : M$
    Fit a classifier $y_m(x)$ to the training data and evaluate two quantities:

$$\varnothing_m = \frac{\sum_{i=1}^{N} w_i^{(m)} I\left(y_m(x_i) \neq t_i\right)}{\sum_{i=1}^{N} w_i^{(m)}},$$

$$\alpha_m = \ln\left(\frac{1 - \varnothing_m}{\varnothing_m}\right),$$

where $I\left(y_m(x_i) \neq t_i\right)$ is the indicator function and equals 1 if $y_m(x_i) \neq t_i$ and 0 otherwise;

    Update weights for all instances as follows:

$$w_i^{(m+1)} = w_i^{(m)} exp\left(\alpha_m I(y_m(x_i) \neq t_i)\right);$$

    Normalize weights:

$$w_i^{(m+1)} = \frac{w_i^{(m+1)}}{\sum_{i=1}^{N} w_i^{(m+1)}};$$

End For

Classify test data with each of the trained classifiers.
Combine the obtained predictions $y_m(x_{test})$ as follows:

$$Y_M(x_{test}) = sign\left(\sum_{m=1}^{M} \alpha_m \left(2 y_m(x_{test}) - 1\right)\right).$$

---

instance and the *m*th classifier is denoted by $w_i^{(m)}$. Let the *m*th hypothesis (classifier's output) for the input *x* be $y_m(x)$ and the true class label of *x* be t_i.

It is assumed that classifiers can cope with weighted instances. If it is not the case, an unweighted data set is generated from the weighted data by resampling. In other words, instances are chosen with probability according to their weights until the data set becomes as large as the original training set. This option is chosen in the AdaBoost implementation below.

A decision tree or its extreme variant, decision stump, is the most widely selected as the base classifier in AdaBoost, since both classifiers meet the condition of a weak learner. However, Meade and Wyner (Mease & Wyner, 2008) found out that stumps are not necessarily the best classifiers for AdaBoost and that larger trees can actually be more effective. Hence, each tree is fully grown and left unpruned.

The quantity ϵ_m represents a weighted measure of the classification error rate attained by the mth classifier on the weighted training data set. Thus, another weight, α_m, gives a weight to the mth classifier when all M classifiers are combined. The larger α_m is, the more accurate the associated classifier is. After computing α_m and having the mth hypothesis, $I\left(y_m\left(x_i\right) \neq t_i\right)$, for all instances, weights $w_i^{(m)}$ are updated to $w_i^{(m+1)}$ in order to incorporate new information into the learning process at the next iteration. In order to prevent some weight to run to infinity, all weights are then normalized so that their sum is equal to 1. Once all classifiers had a chance to generate hypotheses, classifiers' predictions are finally combined using weights attached to classifiers.

According to our convention on class labels, the latter can be either 0 or 1. On the other hand, AdaBoost supposes $\{-1,+1\}$-labels. Therefore, a simple transformation, $2t - 1$, delivers the desired outcome. In other words, 0 becomes -1 while 1 remained 1. Having weighted outputs of the ensemble, the natural partitioning data into two classes is to assign positive outputs to one class and zero and negative outputs to another class. This is exactly what the sign function does: it decides which class a given instance is to be assigned to: $sign(F)=1$, if $F > 0$ and it is 0 otherwise.

AdaBoost as described above seems to be sensitive to noise and outliers, especially for small data sets (Breiman, 1998), (Bauer & Kohavi, 1999). One solution targeting noisy class labels, i.e. when some labels are incorrect, is proposed in (Krieger, Long, & Wyner, 2001). It consists of bagging followed by boosting. One has to generate $B = \rho N$ ($0 < \rho < 1$) bootstrap samples from the training data, compute a boosted classifier from each bootstrap sample using M iterations, combine the B resulting classifiers into an ensemble, and then average.

In several works (Dudoit & Fridlyand, 2003), (Giarratana, Pizzera, Masseroli, Medico, & Lanzi, 2009), (Long & Vega, 2003), (Tan & Gilbert, 2003) it is claimed that the classical AdaBoost algorithm is not fit to microarray data analysis as it does not yield good generalization (i.e. its classification performance on the training data is excellent, but it degrades on the test data). With no exception, the known algorithms perform boosting in conjunction with decision trees (mostly with decision stumps but also with C4.5 (Quinlan, 1993) – technique very similar to the classification tree).

The remedies proposed include either replacing AdaBoost with one of its extensions, e.g., such as LogitBoost[3] (Dettling & Bühlmann, 2003), that can better cope with the presence of misclassified instances and training data inhomogeneities (Dudoit & Fridlyand, 2003), or modifying the classical AdaBoost (Long & Vega, 2003) in order to overcome the poor generalization capability of AdaBoost. It is my guess after reading (Mease & Wyner, 2008) that one reason for poor generalization might be a decision stump itself. Mease and Wyner argued that a larger tree may be more optimal in terms of performance than such a simple classifier as the decision stump.

Long and Vega's modifications of AdaBoost with decision stumps resulted in AdaBoost-VC[4] employing each feature only once in the ensemble and computing ϵ_m as follows:

$$\varepsilon_m^{new} = \varepsilon_m + \frac{d}{N}\left[\ln N + \sqrt{1 + \frac{\varepsilon_m N}{d}}\right],$$

where $d \in \{1, 2, 3\}$ is an adjustable parameter and N is the training set size.

Dettling and Bühlmann proposed to run feature selection prior to LogitBoost with stumps in order to enhance robustness of AdaBoost to noise and training set imbalance.

Tan and Gilbert (Tan & Gilbert, 2003) compared C4.5, Bagging C4.5 and Boosting C4.5 on seven microarray gene expression data sets. Their results show that, in general, bagging C4.5 outperforms both a single C4.5 and AdaBoost with C4.5 as the weak learner. Even though AdaBoost was preceded by gene filtering, it did not seem to help. Tan and Gilbert see the reason of bad AdaBoost performance in overfitting due to noise, still remaining in a data set after filtering out redundant genes. This implies that the small sample size of gene expression data prompts a very dramatic reduction in the number of genes in order for AdaBoost to regain its property of resistance to overfitting.

It seems that the classical AdaBoost alone (i.e. without aggressive dimensionality reduction) is incapable of dealing with such high dimensional data as microarray gene expressions (Lausser, Buchholtz, & Kestler, 2008). Hence, a general recommendation could be to carry out drastic gene filtering before running any noise-resistant variant of AdaBoost.

Mease and Wyner (Mease & Wyner, 2008) advocated to replace stumps in favor of a larger decision trees as according to them "boosting with larger trees actually often overfits less than boosting with smaller trees in practice since the larger trees are more orthogonal and a self-averaging process prevents overfitting." However, Mease and Wyner did not find any evidence in favor of LogitBoost over AdaBoost. Thus, our combined (Molotov) cocktail for battling the AdaBoost overfitting would include 1) gene selection, 2) either the original AdaBoost or its robust version (but not LogitBoost) with a large (possibly unpruned) tree as the weak learner (AdaBoost should be first tried before experimenting with any of its modifications), and 3) the number of trees in the ensemble should also be large, possibly as large as 500-1000 (like in Random Forests (Breiman, Random forests, 2001)).

Finally, it should be mentioned that AdaBoost cannot boost performance just any type of classifiers. This classifier must be a weak, i.e. its error rate must be slightly above 0.5. However, I think that a classifier is also desired (though not required!) to be unstable in the sense that small changes in the training set can cause large changes in classifier's predictions (Breiman, Bagging predictors, 1996). Out of all classifier types considered in this book, only the classification tree satisfies these requirements. All the other classifiers (Naïve Bayes, k-nearest neighbors, support vector machines (SVMs)) appear to be much more stable (they have a low variance) than the classification tree. However, in the literature, some examples can be found, where certain modifications led to promising results with stable classifiers (see, e.g., (Ting & Zhu, 2009) for boosting SVMs and (Athitsos & Sclaroff, 2005) for boosting nearest neighbors). Nevertheless, given that these results are obtained for other than microarray gene expression data, it is impossible to say anything about their extrapolation to our case.

MATLAB Code

There are many implementations of AdaBoost due to its popularity and the easy-to-implement factor. Check, for example, Wikipedia (http://en.wikipedia.org/wiki/AdaBoost), Classification Toolbox for MATLAB® (Stork & Yom-Tov, 2004) or the Spider for MATLAB® (http://www.kyb.mpg.de/bs/people/spider). Pseudo code of AdaBoost and AdaBoost-VC with decision stumps can be found in (Long & Vega, 2003). Pseudo code of LogitBoost can be found in (Dettling & Bühlmann, 2003). MATLAB® code for AdaBoost and LogitBoost can be downloaded from http://shum.huji.ac.il/~gleshem/#software0.

I also implemented AdaBoost in function *BoostingTrees*. As its name says, the weak learner is the classification tree (Breiman, Friedman, Olshen, & Stone, 1984), considered earlier in this book. This selection was motivated by the fact that the classification tree is the most widely used weak learner in AdaBoost.

BoostingTrees expects four input parameters related to the training and test data and the number of trees, *ntrees*, to combine. Both numerical and categorical class labels can be processed by *BoostingTrees* thanks to using the function *cat2num*, which was introduced in the chapter on classification trees. This function converts categorical values into a numerical representation. If its input is already a real number, nothing is done and this input is simply returned by *cat2num*.

In order to make code more compact, the training and test instances are combined into a single matrix *full_patterns*. As a result, when each tree performs classification, we obtain class labels needed in both training and test phases.

Although it was said in the previous section that AdaBoost assumes that individual classifiers can cope with weighted instances, there is another option which was chosen here: resampling the original training set according to specified weights. Sampling is done with MATLAB® function *randsample* that randomly selects with replacement N indices of the training instances, where N is the training set size. Each instance is picked with probability associated with that instance (vector w of weights consists of N elements that are probabilities). Thus, each weighted set is of the same size as the original training set.

For classification with trees, the function *classification_tree_moo* with default input parameters was selected, which, in its turn, relies on the MATLAB® function *classregtree* and the methods related to a tree object. Having indices of sampled instances in vector *indices*, the weighted training set just contains these instances, while the test set includes all original unweighted training and test instances in the matrix *full_patterns*. The length of the returned vector *targets* is $N + N_t$, where N_t is the number of test instances. Due to the composition of *full_patterns*, the first N elements of *targets* are the training labels, whereas other N_t elements are the test labels. Generated trees are not displayed as the number of trees can be large.

After the current tree completed classification of the data in *full_patterns*, the indicator vector I and error ϵ_m are computed. If ϵ_m attained the zero value (no errors), the current classifier is not a weak learner; hence, iterations stop and *BoostingTrees* halts. If this weighted error is non-zero, calculations continue in order to compute α_m (importance weight of a given classifier) and to update weights w.

Also ensemble's hard and soft predictions are updated at this stage. This part is a bit tricky so a few words are to be said about it. First, test labels are extracted from *targets* and converted to the numerical format if necessary. The obtained real-valued labels may not necessarily be 0 or 1: in general, they can be expected to be a and b, where a and b can be any real numbers. Thus, the test labels need to be converted to our conventional $\{0,1\}$ representation of class labels.

Soft ensemble predictions are uncalibrated scores (i.e. they do not sum up to 1) rather than calibrated probabilities. As was found in (Mease & Wyner, 2008), AdaBoost provides a poor estimation of class-conditional probabilities, especially if the sample size is small. Hence, I decided on scores instead of probabilities. Although (Niculescu-Mizil & Caruana, 2005) proposed several techniques of how to calibrate AdaBoost outputs, some of them requires an extra training set in addition to the original training set, while others suffers from the drawback noticed by Mease and Wyner in (Mease & Wyner, 2008).

In order to be useful for such purposes as Receiver Operating Characteristic (ROC) curve generation and area under the ROC curve (AUC) computation, there should be two-element vector associated with each soft prediction. As AdaBoost does not offer anything like this, I was resorted to invent my own scoring scheme. I noticed that each prediction of the entire classifier ensemble is a sum of two components: negative score and positive score. If the absolute value of the negative score is larger than the positive score, the outcome is negative, while if the positive score is larger than the negative score, the sum will be positive. Furthermore, the ensemble predictions are updated according to:

$$new\_value = old\_value + \alpha_m \left(2t - 1\right)$$

where t is either 0 or 1. If $t = 0$, then the score is decreased by $-\alpha_m$. If $t = 1$, the score is increased by $+\alpha_m$. Thus, if an instance got the label 0 (1), the negative (positive) score is adjusted. According to our conventions, soft predictions (either scores or probabilities) for class 0 are stored in the first row of matrix *ensemble_ soft_predictions*, whereas soft predictions for class 1 are stored in the second row of this matrix.

After *ntrees* iterations are finished, all ensemble predictions are thresholded, so that positive predictions are replaced with 1, while zero and negative predictions are replaced with 0. This is how the hard ensemble predictions are added to the soft ones.

```
function [ensemble_predictions, ensemble_soft_predictions] =
BoostingTrees(train_patterns, train_targets, test_patterns,
ntrees)

% Two-class classification using AdaBoost with the classifica-
tion tree % as a weak learner
% Original sources:
% Yoav Freund, Robert E. Schapire, "Experiments with a new
boosting
% algorithm": In L. Saitta (Ed.), Proceedings of the 13th In-
ternational
% Conference on Machine Learning, Bari, Italy, pp.148-156, Mor-
gan
% Kaufmann, San Francisco, CA, 1996
% Christopher M. Bishop,"Pattern recognition and machine learn-
ing",
% Springer Science+Business Media, New York, NY, 2006
```

```
%
% Inputs:
%   train_patterns  - Training patterns (DxN matrix, where D - data
%     dimensionality, N - number of patterns)
%   train_targets   - Training targets (1xN vector)
%   test_patterns   - Test patterns (DxNt matrix, where D - data
%     dimensionality, Nt - number of patterns)
%   ntrees          - The number of trees in an ensemble
%
% Outputs
%   ensemble_predictions      - Predictions (1xNt vector of class
%                                labels), assigned by an ensemble of
%                                classifiers
%   ensemble_soft_predictions - Uncalibrated scores (CxNt matrix,
%                                where C - the number of classes)
%                                assigned by an ensemble of
%                                classifiers
%
% Attention: this function is only suitable for two classes
% Default parameters of a classification tree can be changed in
% classification_tree_moo.m


% Get the number of training patterns
N = size(train_patterns,2);

% Get unique training labels
Uc = unique(train_targets);
% Get the number of classes
c = length(Uc);

% Check if the number of classes is two
if length(Uc) > 2
    error('This function assumes only two classes of data!');
end
if length(Uc) < 2
    error('There is just one class of the data!');
end

% Class labels must be either 0 or 1!
```

```
i = find(train_targets == Uc(1));
train_targets(i) = 0;
i = find(train_targets == Uc(2));
train_targets(i) = 1;

% Get unique training labels again
Uc = unique(train_targets);

% Set initial values for weights
w = ones(1,N)/N;

% Combine training and test data together
full_patterns = [train_patterns test_patterns];

% Pre-allocate memory for the outputs
ensemble_predictions = zeros(1,size(test_patterns,2));
ensemble_soft_predictions = zeros(c,size(test_patterns,2));

for i = 1:ntrees
   % Get a weighted sample using a vector of positive weights w.
   % The probability that the integer i is selected is w(i)/sum(w)
   indices = randsample(1:N, N, true, w);

   % Train a classification tree with default parameters and classify
   % both training and test patterns
   targets = classification_tree_moo(train_patterns(:,indices),
train_targets(indices), full_patterns, 'off');

   fprintf(1, 'Tree %d has been constructed ... \n', i);

   % Get an indicator vector returning either 0 or 1, depending on
   % whether a pattern is misclassified or not
   I = xor(cat2num(targets(1:N),Uc),cat2num(train_targets,Uc));
   % Get weighted training error
   e = sum(w.*I)/sum(w);

   % Break iterations if error is zero
   if e == 0
      break;
   end
```

```
    % Get alpha
    alpha = log((1 - e)/e);

    % Update weights w
    w = w.*exp(alpha*I);
    % Normalize weights w
    w = w./sum(w);

    % Extract test labels that are appended after the first N elements
    % in targets
    % Convert test labels into numerical representation if necessary
    t = cat2num(targets(N+1:end),Uc);
    % Test labels need to be either 0 or 1 so we need to convert them if
    % required
    ut = cat2num(Uc,Uc);
    t(t == ut(1)) = 0;
    t(t == ut(2)) = 1;

    % Assign uncalibrated scores to the data of the "negative" class
    j = find(t == 0);
    ensemble_soft_predictions(1,j) = ensemble_soft_
predictions(1,j) - alpha;
    % Assign uncalibrated scores to the data of the "positive" class
    j = find(t == 1);
    ensemble_soft_predictions(2,j) = ensemble_soft_
predictions(2,j) + alpha;
    % Uncalibrated scores are assigned based on the formula below

    % Update test targets via accumulation of predictions made by the
    % current tree
    ensemble_predictions = ensemble_predictions + alpha*(2*t-1);
end

% Get binary test targets by converting logical type to double
ensemble_predictions = double(ensemble_predictions > 0);

return
```

REFERENCES

Athitsos, V., & Sclaroff, S. (2005). Boosting nearest neighbor classifiers for multiclass recognition. *Proceedings of the 2005 IEEE Computer Society Conference on Computer Vision and Pattern Recognition, San Diego, CA* (p. 45). Los Alamitos, CA: IEEE Computer Society Press.

Bauer, E., & Kohavi, R. (1999). An empirical comparison of voting classification algorithms: bagging, boosting, and variants. *Machine Learning, 36*(1-2), 105–142. doi:10.1023/A:1007515423169

Bishop, C. M. (2006). *Pattern recognition and machine learning.* New York, NY: Springer Science+Business Media.

Breiman, L. (1996). Bagging predictors. *Machine Learning, 24*(2), 123–140. doi:10.1007/BF00058655

Breiman, L. (1998). Arcing classifiers. *Annals of Statistics, 26*(3), 801–849.

Breiman, L. (2001). Random forests. *Machine Learning, 45*(1), 5–32. doi:10.1023/A:1010933404324

Breiman, L., Friedman, J. H., Olshen, R. A., & Stone, C. J. (1984). *Classification and regression trees.* New York, NY: Wadsworth.

Dettling, M., & Bühlmann, P. (2003). Boosting for tumor classification with gene expression data. *Bioinformatics (Oxford, England), 19*(9), 1061–1069. doi:10.1093/bioinformatics/btf867

Dudoit, S., & Fridlyand, J. (2003). Classification in microarray experiments . In Speed, T. (Ed.), *Statistical analysis of gene expression microarray data* (pp. 93–158). Boca Raton, FL: Chapman & Hall/CRC Press. doi:10.1201/9780203011232.ch3

Freund, Y., & Schapire, R. E. (1996). Experiments with a new boosting algorithm. In L. Saitta (Ed.), *Proceedings of the 13th International Conference on Machine Learning, Bari, Italy* (pp. 148-156). San Francisco, CA: Morgan Kaufmann.

Friedman, J. H., Hastie, T., & Tibshirani, R. (2000). Additive logistic regression: a statistical view of boosting. *Annals of Statistics, 28*(2), 337–407. doi:10.1214/aos/1016218223

Giarratana, G., Pizzera, M., Masseroli, M., Medico, E., & Lanzi, P. L. (2009). Data mining techniques for the identification of genes with expression levels related to breast cancer prognosis. In J. J. Tsai, P. C.-Y. Sheu, & H. C. Hsiao (Ed.), *Proceedings of the 9th IEEE International Conference on Bioinformatics and Bioengineering, Taichung, Taiwan* (pp. 295-300). Los Alamitos, CA: IEEE Computer Society Press.

Hastie, T., Tibshirani, R., & Friedman, J. (2003). *The elements of statistical learning: data mining, inference, and prediction*. New York, NY: Springer-Verlag.

Izenman, A. J. (2008). *Modern multivariate statistical techniques: regression, classification, and manifold learning*. New York, NY: Springer Science+Business Media.

Krieger, A., Long, C., & Wyner, A. (2001). Boosting noisy data. In C. E. Brodley, & A. Pohoreckyj Danyluk (Eds.), *Proceedings of the 18th International Conference on Machine Learning, Williamstown, MA* (pp. 274-281). San Francisco, CA: Morgan Kaufmann.

Kuncheva, L. I. (2004). *Combining pattern classifiers: methods and algorithms*. Hoboken, NJ: John Wiley & Sons. doi:10.1002/0471660264

Lausser, L., Buchholtz, M., & Kestler, H. A. (2008). Boosting threshold classifiers for high-dimensional data in functional genomics. In L. Prevost, S. Marinai, & F. Schwenker (Ed.), *Proceedings of the 3rd IAPR Workshop on Artificial Neural Networks in Pattern Recognition, Paris, France* (LNCS 5064, pp. 147-156). Berlin/Heidelberg: Springer-Verlag.

Long, P. M., & Vega, V. B. (2003). Boosting and microarray data. *Machine Learning*, *52*(1-2), 31–44. doi:10.1023/A:1023937123600

Maimon, O., & Rokach, L. (2005). *Decomposition methodology for knowledge discovery and data mining*. Singapore: World Scientific.

Mease, D., & Wyner, A. (2008). Evidence contrary to the statistical view of boosting. *Journal of Machine Learning Research*, *9*, 131–156.

Niculescu-Mizil, A., & Caruana, R. (2005). Obtaining calibrated probabilities from Boosting. *Proceedings of the 21th Confrence in Uncertainty in Artificial Intelligence, Edinburgh, Scotland* (pp. 413-420). Arlington, VA: AUAI Press.

Quinlan, J. R. (1993). *C4.5: programs for machine learning*. San Mateo, CA: Morgan Kaufmann.

Russell, S., & Norvig, P. (2003). *Artificial intelligence: a modern approach* (2nd ed.). Upper Saddle River, NJ: Pearson Education.

Schapire, R. E. (1999). Theoretical views of boosting and applications. In O. Watanabe, & T. Yokomori (Ed.), *Proceedings of the 10th International Conference on Algorithmic Learning Theory, Tokyo, Japan* (LNCS 1720, pp. 13-25). Berlin/Heidelberg: Springer-Verlag.

Stork, D. G., & Yom-Tov, E. (2004). *Computer manual in MATLAB to accompany Pattern Classification* (2nd ed.). Hoboken, NJ: John Wiley & Sons.

Tan, A. C., & Gilbert, D. (2003). Ensemble machine learning on gene expression data for cancer classification. *Applied Bioinformatics*, *2*(3Suppl), S75–S83.

Ting, K. M., & Zhu, L. (2009). Boosting support vector machine successfully. In J. A. Benediktsson, J. Kittler, & F. Roli (Ed.), *Proceedings of the 8th International Workshop on Multiple Classifier Systems, Reykjavik, Iceland* (LNCS 5519, pp. 509-518). Berlin/Heidelberg: Springer-Verlag.

Valiant, L. G. (1984). A theory of the learnable. In *Proceedings of the 16th Annual ACM Symposium on Theory of Computing, Washington, D.C.* (pp. 436-445). New York, NY: ACM Press.

Vapnik, V., & Chervonenkis, A. (1971). On the uniform convergence of relative frequencies of events to their probabilities. *Theory of Probability and Its Applications*, *16*(2), 264–280. doi:10.1137/1116025

Vapnik, V. N. (1995). *The nature of statistical learning theory*. New York, NY: Springer-Verlag.

Vapnik, V. N. (1998). *Statistical learning theory*. New York, NY: John Wiley & Sons.

Wu, X., & Kumar, V. (Eds.). (2009). *The top ten algorithms in data mining*. Boca Raton, FL: Chapman & Hall/CRC Press.

ENDNOTES

[1]   By assigning a larger weight to misclassified instances, we force the next classifier to concentrate on hard to classify instances.

[2]   A decision stump bases its prediction on whether a single level of expression of a given gene is above or below a threshold.

[3]   LogitBoost was proposed in (Friedman, Hastie, & Tibshirani, 2000).

[4]   VC stands for Vapnik-Chervonenkis and came from the fact that the new formula for the weighted training error came from statistical learning theory (Vapnik & Chervonenkis, 1971), (Vapnik V. N., The nature of statistical learning theory, 1995), (Vapnik V. N., Statistical learning theory, 1998), which Vapnik and Chervonenkis greatly contributed to.

Chapter 20
Ensemble Gene Selection

GETTING IMPORTANT GENES OUT OF A POOL

Note that when we bag a model, any simple structure in the model is lost. For instance, a bagged tree is no longer a tree. For interpretation of the model this is clearly a drawback (Hastie, Tibshirani, & Friedman, 2003). The same is true for boosted trees and ensembles of stable classifiers. In fact, classifier ensembles bring more genes to consideration than any single classifier. We are sure that biologists would not like this, but machine learning researchers would defend ensembles due to their superior classification performance. So, how to marry interests of two groups of researchers? It looks like a daunting task. However, this is not the case.

In fact, many researchers agree that if a certain gene is indeed important, it will be selected by different feature selection algorithms. Because different feature selection algorithms are built on different principles, they have different biases. Hence, if despite of such a difference, the same gene appears multiple times in the pool of selected genes used in accurate classifiers, there is something special about that gene. This argument can be straightforwardly extended to the case when the same feature selection algorithm is applied to random samples taken from the original data set. And, of course, feature selection also includes random feature selection used in ensembles of stable classifiers[1].

DOI: 10.4018/978-1-60960-557-5.ch020

For example, Chan et al. (Chan, Bridges, & Burgess, 2008) proposed the following procedure for ensemble gene selection:

- run feature selection followed by classification based on selected features;
- use a voting procedure for features to reward features that occur in many accurate classifiers;
- build and test classifier with the features that got the most votes.

Voting can be as simple as counting the number of times a given gene has been selected or as was done in (Chan, Bridges, & Burgess, 2008), it can be more sophisticated.

In (Chan, Bridges, & Burgess, 2008), two gene scoring techniques were suggested. According to the first technique, the score of a gene f_j is computed as

$$s\left(f_j\right) = \sum_{i=1}^{L} e_{ij} a_i,$$

where L is the number of classifiers in an ensemble, $e_{ij} = 1$ if f_j is a gene selected for the ith classifier and zero otherwise, and a_i is the accuracy of the ith classifier. Thus, a gene score is the sum of accuracies for all classifiers that include that gene. Such a scoring scheme favors the genes that lead to more accurate classification.

The second technique utilizes the weighted score:

$$ws\left(f_j\right) = \sum_{i=1}^{L} \frac{1}{F_i} e_{ij} a_i,$$

where F_i is the number of genes used by the ith classifier. This scheme rewards the genes that are members of small subsets that yield accurate classifiers as it takes into account not only accuracies of individual classifiers but also the number of genes they employ to achieve high accuracy. In contrast, the simple scoring scheme relies exclusively on classification accuracies of ensemble members.

In experiments done in (Chan, Bridges, & Burgess, 2008), the weighted scoring turned out to be more advantageous to apply than the simple scoring.

Applied to classifier ensembles considered so far, this procedure, called ensemble gene selection, means that one needs to

- collect together the genes used with each classifier in an ensemble;
- count the number of times each gene appeared;

- pick up only those genes that got the number of votes exceeding a certain threshold;
- build a classifier with the genes that passed the last step.

For ensembles of stable classifiers such as *k*-nearest neighbors, Naïve Bayes, support vector machines, all gene subsets are merged into a single list. For ensembles of bagged/boosted trees and random forests, all genes from all tree nodes are collected together.

Having a single list of genes instead of multiple subsets, gene appearance counting and selection of most frequent genes are straightforward to do. It is important to note that in ensemble gene selection, genes rather than classifiers get votes.

The final classifier could be a single classifier like a classification tree that allows easy interpretability of how the classification was done. This final classifier employs only the genes accrued the most votes. In principle, nothing prevents us from choosing another ensemble as a final classifier. It is likely that this decision would yield more accurate results but at the expense of interpretability. So, together with Chan and her coauthors (Chan, Bridges, & Burgess, 2008) I feel that if interpretability is more important than prediction accuracy, then one needs to pick up a single classifier as the final classifier.

To ensure that the most frequently selected genes do not include any gene accrued its votes by a pure chance, different ensemble models may be constructed and the procedure above can be repeated for each ensemble model either separately from other ensembles or it can be carried out for the subsets of genes coming from all available ensemble models.

If individual gene subsets are too large, then the approach exercised in a number papers (see, e.g., (Peng, 2006), (Liu, Liu, & Zhang, 2010)) may work. It consists in gene clustering so that each subset is reduced to the small number of representatives before further analysis.

Due to simplicity of the overall idea borrowed from (Chan, Bridges, & Burgess, 2008), I deviate from common practice and do not provide MATLAB® code implementing ensemble gene selection. I encourage you, my patient readers, to do it yourselves.

Meta-Analysis or Integration of Several Microarrays

Meta-analysis or integration of several microarrays, i.e. several independent gene expression data sets that have the same experimental objectives, may loosely be considered as ensemble gene selection. Its essence is a combination of several data sets into a single data set (Rhodes, Barrette, Rubin, Ghosh, & Chinnaiyan, 2002), which is then subjected to a typical analysis path. The purpose of meta-analysis is to enlarge the number of instances available for analysis so as to mitigate the effect of overfitting and to make analysis results more trustworthy.

It should be noted that integrating multiple microarray data sets poses a number of challenges. Difficulties can arise from differences in gene sets, differences in microarray platforms, differences in the measurement scale. This implies that direct fusion of all data sets is not an appropriate solution.

One interesting integration method proposed in (Yoon, Lee, Park, Bien, Chung, & Rha, 2008) is described below. It starts from the extraction of genes common to all microarray data sets. The expression values of each data set are then transformed into ranks, independently of other data sets. Such a transformation is done across expression values of each instance and for each class separately. In other words, if a certain instance belonging to class 'normal' is characterized by the expression values 300, 100, 580, and 78, then these values are replaced with the following ranks: 3, 2, 4, and 1, respectively[2]. Once the expression values are replaced with ranks, the integration of instances from different data sets becomes feasible as long as the gene order in all data set is maintained the same (i.e. a certain gene needs to be in the same position in all lists of genes).

The obtained ranks are sorted from the smallest to the largest for each gene. Each sorted rank is changed to 0 if the instance associated with it is 'normal' and to 1 if the instance associated with it is 'tumor'. Thus, a matrix of sorted ranks is replaced with a 0-1 matrix.

After that, a score measuring how differently a gene is expressed in two classes is calculated for each gene. A gene score is the number of swaps between adjacent 0 and 1 needed to obtain the perfect partitioning into two classes. For example, if there are six instances, three of which are normal and the other three are tumor, then the perfect partitioning is encoded with the sequence 000111[3], i.e. two classes do not overlap. For the sequence 001011, one needs only one swap, while for the sequence 111000 requires nine swaps to turn it into 000111.

Given d genes to be selected (d is a user-specified parameter), $d / 2$ genes with a very large score and $d / 2$ genes with a very small score are chosen[4]. These genes are considered to be informative for two-class classification. Finally, a classifier (called k-GeneTriple), operating on these d genes, is built that consists of k ($k \geq 5$) decision rules, with each rule encoding a relation between three genes and a class label. The rules allow easy interpretation of a decision made. Decisions of all rules are combined by the majority vote in order to assign the final class label to each test instance.

Experiments performed in (Yoon, Lee, Park, Bien, Chung, & Rha, 2008) on the integration three prostate cancer data sets and three colon cancer data sets demonstrated the feasibility of the proposed approach and its advantages over other methods.

REFERENCES

Chan, D., Bridges, S. M., & Burgess, S. C. (2008). An ensemble method for identifying robust features for biomarker discovery. In Liu, H., & Motoda, H. (Eds.), *Computational methods of feature selection* (pp. 377–392). Boca Raton, FL: Chapman & Hall/CRC Press.

Hastie, T., Tibshirani, R., & Friedman, J. (2003). *The elements of statistical learning: data mining, inference, and prediction*. New York, NY: Springer-Verlag.

Liu, H., Liu, L., & Zhang, H. (2010). Ensemble gene selection by grouping for microarray data classification. *Journal of Biomedical Informatics*, *43*(1), 81–87. doi:10.1016/j.jbi.2009.08.010

Peng, Y. (2006). A novel ensemble machine learning for robust microarray data classification. *Computers in Biology and Medicine*, *36*(6), 553–573. doi:10.1016/j.compbiomed.2005.04.001

Rhodes, D. R., Barrette, T. R., Rubin, M. A., Ghosh, D., & Chinnaiyan, A. M. (2002). Meta-analysis of microarrays: interstudy validation of gene expression profiles reveals pathway dysregulation in prostate cancer. *Cancer Research*, *62*, 4427–4433.

Yoon, Y., Lee, J., Park, S., Bien, S., Chung, H. C., & Rha, S. Y. (2008). Direct integration of microarrays for selecting informative genes and phenotype classification. *Information Sciences*, *178*(1), 88–105. doi:10.1016/j.ins.2007.08.013

ENDNOTES

[1] Recall that classification error is related to data set complexity and low-complexity feature subsets are selected for stable classifiers such as k-nearest neighbors. As a result, random feature selection may not be so random after all.

[2] The smallest value gets rank 1; the second smallest value gets rank 2, etc.

[3] A gene tends to be over-expressed in tumor instances, compared to normal instances where that gene has normal expression. Hence, normally expressed values get smaller ranks while over-expressed values are associated with higher ranks. As a result, one obtains the perfect partitioning when all zeroes in the sequence are followed by all ones.

[4] Gene selection rules can be more sophisticated than the one just described.

Chapter 21
Introduction to Classification Error Estimation

PROBLEM OF CLASSIFICATION ERROR ESTIMATION

Given a gene expression data set and the multitude of classification algorithms, one is often tempted to ask three questions:

- What is the minimum error rate that can be attained on a given data set?
- Given the training error rate, what could be the future classification performance?
- How effectively (i.e. without over-optimistic or over-pessimistic bias) the classification error can be estimated?

The answer to the *first* question is the Bayes error rate or simply the Bayes error. It is named so after Thomas Bayes – the reverend who lived in the 18th century and was interested in mathematics; he is principally remembered for the version of what is today known as Bayes' Theorem, which is one of the important theorems in probability theory and statistical decision theory.

So what is this error? It is a *lower bound* on classification performance of *any* algorithm on a given data set. Pay attention to the words 'any algorithm'. They mean that the Bayes error concerns a particular data set rather than a classifier. Put

DOI: 10.4018/978-1-60960-557-5.ch021

differently, the Bayes error rate for a classification problem is the minimum achievable error rate, i.e. the error rate of the best possible classifier. The best possible classifier always makes the optimal decision by picking the class with the highest posterior probability[1]. In other words, the optimal decision is to assign the observation x to the class C_j, where $j = \arg\max_i P(C_i|x)$ and $P(C_i|x)$ is the posterior probability. As we assume two classes, i takes values of 1 and 2. The Bayes error is nonzero if the classes overlap.

Any data set consists of individual observations. Thus, the Bayes error is defined by taking this fact into account. It is the average error rate of the optimal decision rule (best classifier) in the sense above.

Formally, for two classes the Bayes error is defined in (Ripley, 2007) as

$$E^* = E\left[1 - \max\left(P(C_1|x), P(C_2|x)\right)\right] = E\left[\min\left(P(C_1|x), P(C_2|x)\right)\right],$$

where $E[.]$ is the expectation, i.e. the averaging over the minimum error probabilities of all observations.

The last formula can be further reformulated as

$$E^* = \int_{x \in H_1} P(C_2|x)\, p(x)\, dx + \int_{x \in H_2} P(C_1|x)\, p(x)\, dx,$$

where $p(x)$ is the prior distribution of x, H_1 is the area where a classifier assigns x to C_1 (probability of misclassification of x as C_1 though it belongs to C_2) and H_2 is the area where a classifier assigns x to C_2 (probability of misclassification of x as C_2 though it belongs to C_1).

To clarify the last paragraph, let us turn to Figure 1 that shows two Gaussian probability density functions $P(C_1|x)$ and $P(C_2|x)$ that overlap (MATLAB® code to reproduce this figure can be found in *BayesErrorGraphicalExplanation*). The area under each curve is equal to 1. The overlap area of two curves is the probability of misclassification, i.e. the Bayes error.

However, there is one small problem with the Bayes error: it cannot be computed analytically, except for some really toy examples. Oops! I see your surprised face, dear reader: What? After so many words and graphical explanation about this error, you said that I cannot compute it? Why in this case anyone needs such an estimator? Do not worry, there is still hope for the desperate (not in the sense of Spanish 'desperado' (furioso, violento)): lower and upper bounds for this error can be calculated.

Figure 1. Two overlapping posterior probability density functions and the Bayes error

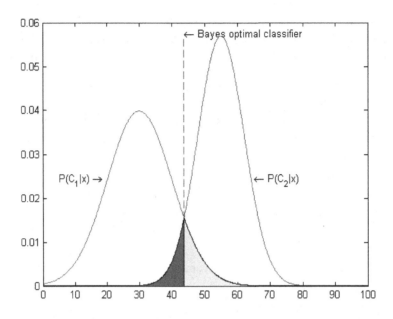

The bounds for the Bayes error are determined by means of the Bhattacharyya bound (Bhattacharyya, 1943). For two normal distributions with the means μ_1 and μ_2 and covariance matrices Σ_1 and Σ_2 it is calculated as (Ripley, 2007)

$$J_B = \frac{1}{8}(\mu_1 - \mu_2)' \left[\frac{1}{2}(\Sigma_1 + \Sigma_2)\right]^{-1} (\mu_1 - \mu_2) + \frac{1}{2} \log \frac{\left|\frac{1}{2}(\Sigma_1 + \Sigma_2)\right|}{\sqrt{|\Sigma_1||\Sigma_2|}}.$$

Let us look carefully at this formula. Is everything Ok? How about high dimensionality and the covariance matrix (and especially, its inverse)? Now we hope you got it: special care must be taken when computing these matrices and their inverses. If you ignore this warning, then you will likely obtain either very poor estimates due to small sample size and high data dimensionality or your program will likely crash because of the covariance matrix singularity caused by high data dimensionality[2] (hence, the found J_B will stand for 'just bad' in this case). For the correct approaches to the covariance matrix estimation for high dimensional data, consult, for example, (Cao & Bouman, 2008), (Rütimann & Bühlmann, 2009),

(Friedman, Hastie, & Tibshirani, 2008), (Scheinberg & Rish, 2009), (Pavlenko & Björkström, 2010). And you should remember that you will get approximates rather than exact estimates.

If necessary precautions in the covariance matrix estimation are observed, then the Bhattacharyya bound can serve as a class separability indicator that can be useful for gene/feature selection (Saon & Padmanabhan, 2000), (Zhang & Deng, 2007), (Yang & Hu, 2008) or even for classifier selection. Of course, it is not the only separability characteristic.

What if you occasionally got an error smaller than the Bayes error? The article (Fu, Mallick, & Caroll, 2009) gives some clues to this problem and circumstances where it occurs.

As you already know, zero error rate on the training data does not mean a classifier would deliver similar news for the test data. That is, the training error cannot be a reliable predictor of the future classification performance. But what can be? The answer is the VC dimension, where VC stands for Vapnik-Chervonenkis – two Soviet scientists, Vladimir Vapnik and Alexey Chervonenkis, who have invented it (Vapnik & Chervonenkis, 1971). The VC dimension is the answer to the *second* question.

The VC dimension provides the *probabilistic upper bound* on the test error made by a classifier.

Let E_{train} and E_{test} be the training and expected test[3] error, respectively. Then with probability $1 - \eta$ $(0 \leq \eta \leq 1)$, the following bound holds:

$$E_{test} \leq E_{train} + \sqrt{\frac{h\left(\log\left(2N \,/\, h\right) + 1\right) - \log\left(\eta \,/\, 4\right)}{N}},$$

where N is the training set size and h (nonnegative integer) is the VC dimension that does not depend on the choice of training data.

Knowing h straightforwardly yields the upper bound on the expected test error. Computing the VC dimension also helps to understand which learning algorithms are more powerful. Power is measured by the number of points that can be shattered[4] by a learning algorithm. In this regard, the VC dimension is the maximum number of points that can be shattered by a certain learning algorithm (machine). Two tutorials (Burges, 1998 and http://www.autonlab.org/tutorials/vcdim08.pdf) give examples of how VC dimension can be found in simple cases. You can notice that the minimization of h yields a tighter bound on the expected test error. Thus, out of two learning algorithms, having the same training error, we would prefer the algorithm with a smaller h. However, a better performance may be associated with another algorithm, since knowing h gives only the upper bound but not the error itself.

What if a learning algorithm has the infinite VC dimension? For instance, a 1-NN (nearest neighbor) classifier has such a dimension and zero training error (Burges, 1998). Does this mean that a 1-NN is a poor classifier? No, as numerous studies and experiments revealed. Hence, the word of caution: infinite h does not imply poor performance.

The answer to the *third* question is the bolstered error rate (Braga-Neto & Dougherty, 2004), which, in my opinion, is ideally suitable for classification error estimation in the small-sample size setting. In this technique, a test set is randomly generated, based on the training examples. Thus, not only splitting (common in cross-validation) a small training set is avoided, but also the bolstered error turns out to have low variance of error estimate and it is much faster to compute than bootstrap error estimators (Braga-Neto & Dougherty, Bolstered error estimation, 2004), (Braga-Neto, Fads and fallacies in the name of small-sample microarray classification - A highlight of misunderstanding and erroneous usage in the applications of genomic signal processing, 2007). Critics of cross-validation and bootstrap that are even now dominant error estimation techniques can be found in (Isaksson, Wallman, Göransson, & Gustafsson, 2008), where the authors emphasize the importance of Bayesian confidence (credible) interval estimation in addition to a single point error estimation. Of course, all the mentioned does not diminish the value of the Area Under the Receiver Operation Characteristic (ROC) Curve (AUC) (Cortes & Mohri, 2004), (Pepe, 2004), (Huang & Ling, 2005), (Fawcett, 2006).

REFERENCES

Bhattacharyya, A. (1943). On a measure of divergence between two statistical populations defined by their probability distributions. *Bulletin of the Calcutta Mathematics Society, 35*, 99–110.

Braga-Neto, U. M. (2007). Fads and fallacies in the name of small-sample microarray classification - A highlight of misunderstanding and erroneous usage in the applications of genomic signal processing. *IEEE Signal Processing Magazine, 24*(1), 91–99. doi:10.1109/MSP.2007.273062

Braga-Neto, U. M., & Dougherty, E. R. (2004). Bolstered error estimation. *Pattern Recognition, 36*(7), 1267–1281. doi:10.1016/j.patcog.2003.08.017

Burges, C. J. (1998). A tutorial on support vector machines for pattern recognition. *Data Mining and Knowledge Discovery, 2*(2), 955–974. doi:10.1023/A:1009715923555

Cao, G., & Bouman, C. A. (2008). *Covariance estimation for high dimensional data vectors using the sparse matrix transform. Technical Report ECE 08-05.* West Lafayette, IN: Purdue University.

Cortes, C., & Mohri, M. (2004). AUC optimization vs. error minimization. In Thrun, S., Saul, L. K., & Schölkopf, B. (Eds.), *Advances in Neural Information Processing Systems (Vol. 16)*. Cambridge, MA: MIT Press.

Fawcett, T. (2006). An introduction to ROC analysis. *Pattern Recognition Letters, 27*(8), 861–874. doi:10.1016/j.patrec.2005.10.010

Friedman, J., Hastie, T., & Tibshirani, R. (2008). Sparse inverse covariance estimation with the graphical Lasso. *Biostatistics (Oxford, England), 9*(3), 432–441. doi:10.1093/biostatistics/kxm045

Fu, W. J., Mallick, B., & Caroll, R. J. (2009). Why do we observe misclassification errors smaller than the Bayes error? *Journal of Statistical Computation and Simulation, 79*(5), 717–722. doi:10.1080/00949650801905221

Huang, J., & Ling, C. X. (2005). Using AUC and accuracy in evaluating learning algorithms. *IEEE Transactions on Knowledge and Data Engineering, 17*(3), 299–310. doi:10.1109/TKDE.2005.50

Isaksson, A., Wallman, M., Göransson, H., & Gustafsson, M. (2008). Cross-validation and bootstrapping are unreliable in small sample classification. *Pattern Recognition Letters, 29*(14), 1960–1965. doi:10.1016/j.patrec.2008.06.018

Pavlenko, T., & Björkström, A. (2010). Sparse inverse covariance estimation in the supervised classification of high-dimensional data. In *Proceedings of the International Conference on Trends and Perspectives in Linear Statistical Inference, Tomar, Portugal*.

Pepe, M. S. (2004). *The statistical evaluation of medical tests for classification and prediction* (1st paperback ed.). Oxford, UK: Oxford University Press.

Ripley, B. (2007). *Pattern recognition and neural networks* (1st paperback ed.). Cambridge, UK: Cambridge University Press.

Rütimann, P., & Bühlmann, P. (2009). High dimensional sparse covariance estimation viar directed acyclic graphs. *Electronic Journal of Statistics, 3*, 1133–1160. doi:10.1214/09-EJS534

Saon, G., & Padmanabhan, M. (2000). Minimum Bayes error feature selection. In *Proceedings of the 6th International Conference on Spoken Language Processing, Beijing, China* (pp. 75-78).

Scheinberg, K., & Rish, I. (2009). SINCO - an efficient greedy method for learning Sparse INverse COvariance matrix. In *Proceedings of the 2nd NIPS Workshop on Optimization for Machine Learning, Whistler, BC, Canada.*

Vapnik, V., & Chervonenkis, A. (1971). On the uniform convergence of relative frequencies of events to their probabilities. *Theory of Probability and Its Applications, 16*(2), 264–280. doi:10.1137/1116025

Yang, S.-H., & Hu, B.-G. (2008). Feature selection by nonparametric Bayes error minimization. In T. Washio, E. Suzuki, K. M. Ting, & A. Inokuchi (Ed.), *Proceedings of the 12th Pacific-Asia Conference on Knowledge Discovery and Data Mining, Osaka, Japan* (LNCS 5012, pp. 417-428). Berlin/Heidelberg: Springer-Verlag.

Zhang, J.-G., & Deng, H.-W. (2007). Gene selection for classification of microarray data based on the Bayes error. *BMC Bioinformatics, 8*(1), 370. doi:10.1186/1471-2105-8-370

ENDNOTES

[1] The optimal prediction of any observation x is the class label that has highest probability given x.

[2] This is the most probable case in practice.

[3] It should be kept in mind that E_{test} is the expected value of the test error rather than the error attained on a particular test set. Hence, this quantity cannot be computed.

[4] For example, a line can shatter (separate) any three points, i.e. for a linear classifier, $h = 3$. See also http://en.wikipedia.org/wiki/VC_dimension.

Chapter 22

ROC Curve, Area under it, Other Classification Performance Characteristics and Statistical Tests

CLASSIFICATION PERFORMANCE EVALUATION

It is not uncommon to become confused about performance evaluation of classification algorithms run on such small-sample size data as microarray gene expression data sets. In fact, I think that before one starts running experiments with these algorithms, it is good practice to think of how the results of these experimental runs will be evaluated.

First, it is reasonable to choose one or several numerical performance characteristics such as classification accuracy, true positive rate, false positive rate, positive predictive value, etc. After estimating one or several such characteristics for multiple algorithm runs and several different algorithms, one collects this information in the form of a matrix whose columns are algorithms and each column contains performance statistics attained during runs by a particular algorithm. In order to create the same conditions for all algorithms under test (this is necessary in order to

DOI: 10.4018/978-1-60960-557-5.ch022

eliminate any accidental bias), each run needs to contain the data that are the same for all algorithms participating in that run. One can have two scenarios:

- Two algorithms to compare.
- More than two algorithms to compare.

Although nobody prevents me from treating the first scenario as a special case of the second scenario, I prefer to keep them separated due to the different tests to be conducted in each case. Thus, in the first scenario, the McNemar's test can be enough in order to establish whether there is a significant difference in classification performance of two algorithms. In the second scenario, things become more complicated: one needs to run the one-way Analysis of Variance (either parametric or nonparametric), followed a multiple comparison test. The one-way Analysis of Variance tests whether all means (parametric test) or medians (nonparametric test) are equal[1]. If the result of the test is that there is a statistically significant difference between algorithms compared, the multiple comparison procedure then determines, based on results of the one-way Analysis of Variance, which algorithms demonstrate significantly different performance than the others.

If you met many unfamiliar words while reading the previous paragraphs, do not worry because all necessary definitions will be given in sections that follow below.

Classification Performance Characteristics

In this chapter, I will heavily borrow (not money but knowledge, which makes my loan especially attractive in these hard times) from my favorite source – Dictionary of Statistics by Everitt (Everitt, 2006) in order to standardize all necessary definitions.

Let us assume that a set of possible class labels consists of positive (p) and negative (n) labels. Let us also assume that if an instance is assigned to the positive class, a diagnosis is a disease (e.g., cancer), while if an instance is assigned to the negative class, it implies 'no disease', i.e. healthy. Such classification corresponds to that of microarray gene expression data sets. It is important for discussion in this chapter that a classifier produces not only discrete labels but also a continuous output – an estimate of an instance's class membership probability. All the classifiers introduced earlier in the book are capable of estimating the class probability.

Given a classifier and an instance which label is to be predicted, there are four possible outcomes:

- If the instance is positive and it is classified as positive, then this is counted as a true positive.

Figure 1. Confusion matrix

		True class	
		Positive	**Negative**
Predicted class	**Positive**	True Positive	False Positive
	Negative	False Negative	True Negative

- If the instance is positive and it is classified as negative, then this is counted as a false negative.
- If the instance is negative and it is classified as positive, then this is counted as a false positive.
- If the instance is negative and it is classified as negative, then this is counted as a true negative.

These four cases can be represented by a confusion matrix shown in Figure 1.

From this matrix several useful characteristics of classification performance can be derived. For example, numbers along the main diagonal represent correct decisions made when classifying positive and negative instances, while numbers off this diagonal are classification errors.

The True Positive (TP) rate (also called recall) is estimated as the number of correctly classified positives (TP), divided by the total number of positives (P) in a data set.

The True Negative (TN) rate is estimated as the number of correctly classified negatives (TN), divided by the total number of negatives (N) in a data set.

The False Positive (FP) rate (also known as the false alarm rate) is estimated as the number of negatives incorrectly classified as positives (FP), divided by the total number of negatives (N) in a data set. In other words, the FP rate is the proportion of cases in which a diagnostic test indicates disease present in healthy patients (Everitt, 2006).

The False Negative (FN) rate is estimated as the number of positives incorrectly classified as negatives (FN), divided by the total number of positives (P) in a data set. Put differently, the FN rate is the proportion of cases in which a diagnostic test indicates disease absence in patients who have the disease (Everitt, 2006).

For better memorizing these notions, these can be represented by the following formulas:

$$TP\,rate = \frac{TP}{P}, TN\,rate = \frac{TN}{N}, FP\,rate = \frac{FP}{N}, FN\,rate = \frac{FN}{P}.$$

All rates take values between 0 and 1. An ideal classifier should have $TP\,rate = 1$ (all disease cases are correctly predicted) and $FP\,rate = 0$ (no healthy patient is classified as having a disease). On the other hand, a useless classifier has $TP\,rate = FP\,rate$ (random guess in disease prediction).

Additional performance characteristics include but are not limited to

$$accuracy = \frac{TP + TN}{P + N},$$

$$sensitivity = recall = TP\,rate = \frac{TP}{P},$$

$$specificity = 1 - FP\,rate = 1 - \frac{FP}{N} = \frac{TN + FP - FP}{TN + FP} = \frac{TN}{TN + FP},$$

$$positive\,predictive\,value = precision = \frac{TP}{TP + FP},$$

$$negative\,predictive\,value = \frac{TN}{TN + FN},$$

$$F - measure = \frac{2}{1\,/\,precision + 1\,/\,recall}.$$

Sensitivity is an index of the performance of a diagnostic test, calculated as the percentage of individuals with a disease who are correctly classified as having the disease, i.e. the conditional probability of having a positive test result given having the disease (Everitt, 2006). A test is good if it is sensitive to the disease, i.e. it is positive for most individuals having the disease.

Specificity is an index of the performance of a diagnostic test, calculated as the percentage of individuals without the disease who are classified as not having the disease, i.e. the conditional probability of a negative test result given the disease is absent (Everitt, 2006). A test is good if it is specific to the disease, i.e. it is positive for only a small percentage of those without the disease.

A Positive Predictive Value (PPV) is the probability that a person having a positive result on a diagnostic test actually has a particular disease (Everitt, 2006). A Negative Predictive Value (NPV) is the probability that a person having a negative result on a diagnostic test does not have the disease (Everitt, 2006). Both predictive values can also be expressed via sensitivity and specificity as follows:

$$PPV = \frac{\rho \, sensitivity}{\rho \, sensitivity + (1 - \rho)(1 - specificity)},$$

$$NPV = \frac{(1 - \rho) \, specificity}{(1 - \rho) \, specificity + \rho(1 - sensitivity)}.$$

where ρ is the prevalence of a disease in a population (the total number of cases of this disease in the population at a given time).

Predictive values are often used in medicine in order to evaluate the usefulness of a diagnostic test. Both predictive values depend on the disease prevalence, ρ. To learn more about predictive values, consult the book (Pepe, 2004).

Receiver Operating Characteristic Curve and Area under It

One of the characteristics of a classifier, frequently used in medical decision making, is a Receiver Operating Characteristic (ROC) curve. It is a visual characteristic allowing for visualizing classification performance of one or several algorithms. When there are several ROC graphs on the same plot, it is straightforward to determine which algorithms perform better than the others. ROC curves provide much more information about classification performance than the number of correctly classified cancer and healthy patients. The article of Swets et al. (Swets, Dawes, & Monahan, 2000) brought the public attention to ROC curves.

A Receiver Operating Characteristic curve is a plot of the sensitivity (or the TP rate) of a diagnostic test against one minus its specificity (or the FP rate), as the cut-off criterion for indicating a positive test is varied (Everitt, 2006). This plot depicts relative trade-offs between true positives and false positives. The FP rate is plotted on the X axis while the TP rate is plotted on the Y axis. The range of values on each axis is from 0 to 1.

ROC curves have one nice property: they are insensitive to changes in class distribution (Fawcett, 2006). In other words, if the proportion of positive to negative instances changes in a test set, the ROC curve will not change. This is in contrast to such performance characteristics as classification accuracy, precision or F-measure that do not stay stable when a class distribution changes. As a result, a ROC curve remains reliable when the disease prevalence changes over time.

Typical ROC curves are shown in Figure 2. Here, there are ROC curves for three classifiers: A, B, and C. In the case of a discrete classifier that outputs only class labels a ROC curve would degenerate to a point in ROC space. However, for classifiers in Figure 2 with the probability estimates as outputs a ROC graph is indeed a curve. Classifier A represents random guessing a class label. Classifier B is better than A, since for any fixed FP rate, the TP rate of B is higher than that of A. By following the same logic, classifier C is better than B. The ROC graph for the ideal classifier would represent a curve composed of two lines: a vertical line along Y axis from point (0,0) to point (0,1), followed by a horizontal line from point (0,1) to point (1,1). As it is difficult to attain the perfect performance in practice, the closer a ROC curve to the upper-left corner of ROC space, the better a classifier is.

But how to judge on the case when two ROC curves cross each other? The answer is neither classifier dominates another one on the entire range of the false positive rate. Each classifier is only superior to the other on an interval of the whole range. For example, one classifier is superior to another classifier on $[0, 0.3]$, while it is inferior to another classifier on $]0, 3, 1]$. Overlapping ROC curves is a visual sign that classifiers cannot be absolutely superior. This is an indication that they are rather relatively superior or inferior, depending on which interval of the false positive rates is considered.

However, how to build a ROC curve like those curves in Figure 1? First of all, a classifier needs to produce either a probability estimate or a score (a numeric value that represents the degree to which an instance is a member of a class) (Fawcett, 2006). Probability estimates or scores can be strict probabilities (they adhere to the probability theorems) or they can be uncalibrated scores, with higher score implying higher probability. Although there known techniques for converting uncalibrated scores into proper probabilities (see, e.g., (Zadrozny & Elkan, 2001), (Drish, 2001), (Niculescu-Mizil & Caruana, 2005)), this conversion is not required for ROC curve generation. A classifier just needs to produce good relative scores that allow for reliable discrimination of positive and negative instances.

The continuous output of such a ranking or scoring classifier can be thresholded as follows: if the classifier output is above the threshold t, the classifier produces a positive label; otherwise, the label is negative. Thus, each threshold value leads to a point in ROC space. By varying t from 0 to 1, it is possible to trace the entire ROC curve. However, this straightforward way of generating ROC curves is computationally inefficient, though it may be good to follow if one wants to construct a ROC curve for a small set of instances as done, e.g., in Figure 3 in (Fawcett, 2006). It should be noted that any ROC curve generated from a finite set of instances is actually a step function, approaching a smooth curve as the number of instances

Figure 2. A typical ROC plot showing ROC curves for three classifiers: A, B, and C

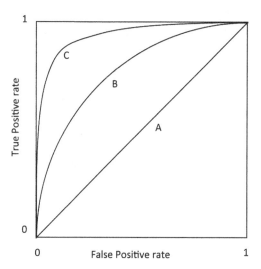

Figure 3. ROC curve when the number of instances in a data set is finite and rather small

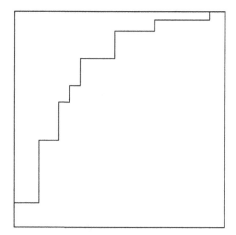

goes to infinity. That is, do not be surprised if your real ROC curve will look like a set of "steps" shown in Figure 3.

A better way to generate a ROC curve adopted from (Fawcett, 2006) is shown in Figure 3. Suppose that we are given a test set, class labels of test instances and probability estimates or scores attached to the test instances. The efficient algorithm exploits the monotonicity of thresholded classifications: any instance that is classified positive with respect to a given threshold will be classified positive for all lower thresholds as well. Hence, if we sort probabilistic outputs of a classifier in descending order of magnitude, move along the sorted list and update counters of true and false positives, then we can quickly construct a ROC curve during a single linear pass over the data. If a test set consists of N instances, then this algorithm requires an $O(N \log N)$ sort, followed by an $O(N)$ scan of the list, thus resulting in the total complexity $O(N \log N)$.

The algorithm in Figure 2 starts from setting the true and false positive counters to zero. When moving along the sorted list of probabilistic outputs, the true positive counter is increased if the current output belongs to a positive instance and the false positive counter is increased if the current output belongs to a negative instance. A stack of ROC points is maintained that receives a new point each time when we meet the element in the sorted list that has a smaller value than its predecessor. Such a procedure is necessary in order not to emit an ROC point until all instances of the equal probabilistic output values have been processed. Instances scored equally are not uncommon for some classifiers.

Area Under Curve (AUC) is a way of summarizing the ROC curve (Everitt, 2006). An AUC is calculated by adding the areas under the curve between each pair of consecutive observations, using, for example, the trapezium (trapezoidal) rule[2]. The AUC is a scalar representation of a 2D ROC curve. Since the AUC is a portion of the area of the unit square, its value will range from 0.0 to 1.0. The AUC of a randomly guessing classifier is equal to 0.5, which means that in order for a classifier to be useful, its AUC must be larger than 0.5.

The AUC has an important statistical property: it is equivalent to the probability that a classifier will rank a randomly chosen positive instance higher than a randomly chosen negative instance (Fawcett, 2006). This is equivalent to the Wilcoxon test of ranks[3] (Hanley & McNeil, 1982). As demonstrated in (Breiman, Friedman, Olshen, & Stone, 1984), the AUC is also closely related to the Gini index, which is twice the area between the diagonal and the ROC curve. Hand and Till (Hand & Till, 2001) pointed out that Gini + 1 = 2AUC.

The AUC can be computed with a slight modification of the algorithm for ROC generation as shown in Figure 4. The principle is the same: sort probabilistic outputs and then move down along the sorted list of values. However, instead of gathering ROC points, now we need to add successive areas of trapezoids.

As one may be interested in a sub-range of false positive rates instead of the entire range from 0 to 1, the partial AUC needs to be used instead of the conventional AUC. As follows from its name, the partial AUC estimates the area under a part of the ROC curve, typically in the range of false positive rates from 0 to a certain FP rate. The algorithm in Figure 5 can be easily modified to incorporate the

Figure 4. Pseudo code for efficient ROC curve generation

Algorithm for ROC curve generation

Let T be class labels of test instances, S be the probabilistic estimates that instances are positive, P and N be the number of positive and negative instances in a data set, TP and FP be the true positive and false positive counters, and R be a list of ROC points;

$R := \varnothing$; $TP := 0$; $FP := 0$; $s_{prev} := -\infty$ (very large negative value);

Sort S in descending order of magnitude;
$i := 1$;
While $i \le |S|$

   If $S_i \ne s_{prev}$

$$R := R \cup \left(\frac{FP}{N}, \frac{TP}{P} \right);$$

     $s_{prev} := S_i$;
   End If
   If T_i is positive
     $TP := TP + 1$;
   Else
     $FP := FP + 1$;
   End If
   $i := i + 1$;
End While
$R := R \cup \left(\frac{FP}{N}, \frac{TP}{P} \right)$; /* This is point (1,1) */

---

partial AUC calculation. More information about the partial AUC can be found in (Pepe, 2004).

Useful Statistical Tests

As you already learned, a statistical test involves formulation of a null hypothesis H_0 and an alternative hypothesis H_1. A null hypothesis is the 'no difference' or 'no association' hypothesis to be tested (usually by means of a significance test) (Everitt, 2006). An alternative hypothesis is the hypothesis that postulates non-zero difference or association (Everitt, 2006). A statistical test tries to reject the null hypothesis in favor of the alternative one. If it cannot reject the null hypothesis, then it is assumed that there is not enough evidence to do so. For more information about statistics and statistical tests, the following books are recommended: (Sheskin,

Figure 5. Pseudo code for calculating the area under the ROC curve

Algorithm for calculating the AUC

Let T be class labels of test instances, S be the probabilistic estimates that instances are positive, P and N be the number of positive and negative instances in a data set, TP and FP be the true positive and false positive counters, and A be the AUC;

$A := 0$; $TP := 0$; $FP := 0$; $TP_{prev} := 0$; $FP_{prev} := 0$;

$s_{prev} := -\infty$ (very large negative value);

Sort S in descending order of magnitude;
$i := 1$;
While $i \leq |S|$
   If $S_i \neq s_{prev}$
      $A := A + trapezoid\_area\left(FP, FP_{prev}, TP, TP_{prev}\right)$;
      $s_{prev} := S_i$;
      $FP_{prev} := FP$; $TP_{prev} := TP$;
   End If
   If T_i is positive
      $TP := TP + 1$;
   Else
      $FP := FP + 1$;
   End If
   $i := i + 1$;
End While
$A := A + trapezoid\_area(N, FP_{prev}, P, TP_{prev})$;
$A := A / \left(P \times N\right)$; /* scale to [0,1] */

2004), (Zar, 1999), (Good & Hardin, 2006), (Drăghici, 2003), (Kranzler, Moursund, & Kranzler, 2006), and (van Emden, 2008) (especially, the last two for those who are 'terrified').

Rejection of H_0 is decided by comparing the p-value of a test and the significance level α.

The p-value of a test is the probability, under the null hypothesis, of obtaining a value of the test statistic as extreme or more extreme than the value computed from the sample. The p-value is neither the probability of the null hypothesis, nor the probability of the data having arisen by chance (Everitt, 2006).

The significance level of a test is a threshold of probability α agreed before the test is conducted. A typical value of α is 0.05. If the *p*-value of a test is less than

α, H_0 is rejected. If the p-value is greater than α, there is no sufficient evidence to reject the null hypothesis. Note that lack of evidence for rejecting the null hypothesis is not evidence for accepting the null hypothesis.

As a result of a statistical test, one may commit the following types of errors. Type I error is the error that results in falsely rejecting the null hypothesis (Everitt, 2006). The significance level is often interpreted as the probability of rejecting the null hypothesis when it is actually true. As you may guess, there is also type II error, which is the error that results in falsely accepting the null hypothesis (Everitt, 2006). And there is also type III error (Everitt, 2006), but it is of no interest for us, though nobody is immune from committing it. Since we want to avoid both type I and type II errors, ideally both errors should be small. However, this is not easy to achieve in practice as these errors are interrelated, i.e. lowering one of them increases the other. For example, we could drive the type I error to zero, but this will result in the large type II error.

If α determines the probability of rejecting H_0 when it is true, then β determines the probability of accepting H_0 when it is false. The quantity $1 - \beta$, which is the probability of correctly rejecting H_0 when it is false, is called the power of a test (Everitt, 2006). Naturally, the test with higher power is preferred.

Rejection of the null hypothesis can also be decided based on the critical value of the test statistic. A test statistic is the statistic used to assess a particular hypothesis in relation to some population (Everitt, 2006). The essential requirement of a test statistic is a known distribution when the null hypothesis is true. The critical value is the value with which a test statistic is compared in order to decide whether H_0 should be rejected (Everitt, 2006). For example, if the statistic calculated during the test is larger than or equal to the critical value for $\alpha = 0.05$, then H_0 is rejected. As you can see the critical value is related to the particular significance level chosen (Everitt, 2006), i.e. for each significance level there is its own critical value. The values of a test statistic that lead to rejection of H_0 form a critical region whose size is equal to α.

When statistical tests are usually applied in the context of topics of this book? Of course, after we collected classification performance quantitative indicators and need to get the answer to a burning question: which of classification algorithms[4] provides the best results? Equivalently, we could ask if there is any difference in performance of several algorithms or not?

As quantitative indicators, different characteristics introduced in the previous section can be used: classification accuracy, PPV, TP rate, FP rate, AUC or any other performance characteristic. That is, the same performance characteristic is estimated for several competing algorithms. This is to be done multiple times, each time for a different sample of the original data set, in order to gather enough data to run statistical tests and, which is more important, to make conclusions derived from these tests statistically meaningful. As a result, a matrix of numbers is formed whose rows are samples and columns are classification algorithms.

If you have only a pair of classification algorithms, then the McNemar's test can be useful to test the performance difference. The McNemar's test is a statistical test for comparing proportions in data involving paired samples (e.g., samples obtained from people who got two treatments/drugs, one on each of two separate occasions) (Everitt, 2006). The test statistic is given by

$$X^2 = \frac{(b-c)^2}{b+c},$$

where b is the number of pairs for which the individual receiving treatment A has a positive response and the individual receiving treatment B does not, and c is the number of pairs for which the reverse is the case (Everitt, 2006). If the probability of a positive response is the same in each group, then X^2 has a chi-squared distribution with a single degree of freedom.

You may wonder how this test can fit our purpose. For the answer let us look in (Kuncheva, 2004). Let b be the number of times when classifier A gives correct prediction while classifier B makes an error. Let c be the number of times when classifier A is erroneous while classifier B is correct. That is, the performance characteristic is the classification accuracy.

The null hypothesis is $b = c$, i.e. there is no difference in classification accuracies of two classifiers. The alternative hypothesis is hence $b \neq c$. The alternative hypothesis is two-tailed, i.e. non-directional. It is surprising to see that the McNemar's test does not take into account how many times each classifier was correct; only misclassifications play critical role.

Then one gets the same statistic X^2 as before with the null hypothesis that there is no difference between the classification accuracies of two classifiers. Dietterich in (Dietterich, 1998) in the line with early sources recommended introducing '-1' in the numerator for a continuity correction[5] with small sample sizes:

$$X^2 = \frac{(|b-c|-1)^2}{b+c}.$$

Thus, one needs to compute X^2 and compare it with the tabulated chi-square value for a given significance level[6] α (typical value for α is 0.05, and the chi-square value for this significance level is 3.8415 that can be obtained by running

MATLAB® function $chi2inv(0.95,1)$, where the first parameter $0.95 = 1 - 0.05$ and the second parameter is the number of degrees of freedom). In other words, if $X^2 \geq 3.8415$, then we reject the null hypothesis and consider that two classifiers have significantly different classification accuracies (Sheskin, 2004).

But what if you have and would like to compare more than a pair of classifiers? The tempting solution could be to run multiple McNemar's tests for all desired pairs. Do not rush! The explanation for this will shortly follow. The recommended action consists of running two tests: the one-way ANOVA followed by one of the multiple comparison procedures.

Analysis of Variance (ANOVA) is the separation of variance attributable to one cause from the variance attributable to others (Everitt, 2006). By partitioning the total variance of a set of observations into parts due to within-groups and between-groups variance and comparing variances (mean squares) by way of F-tests[7], differences between means can be assessed. If variance within groups is smaller than variance between groups, there is a clear difference in group means. If two variances are of the same order of magnitude, all means are not significantly different from each other. The ratio of between-groups variance to within-group variance, denoted by F, is compared with the critical value for this test, determined by the chosen significance level, the number of algorithms compared, and the sum of the number of performance values in all groups (Zar, 1999). If the calculated F is at least as large as the critical value, we reject the null hypothesis that all means are equal. But one should remember that all we can conclude in such a case is that all means are not equal and we do not know which means are different and which are not.

There are two types of one-way ANOVA tests: parametric or classical one-way ANOVA and nonparametric (distribution free) Kruskal-Wallis test. The parametric test relies on the assumption of a Gaussian data distribution[8]. In contrast, the Kruskal-Wallis test does not necessarily expect that the data came from a normal distribution. The Kruskal-Wallis test is based on an analysis of variance using the ranks of the data values, not the data values. The Kruskal-Wallis test is 95% as powerful as the parametric one-way ANOVA test if assumptions of the latter hold. It may be more powerful than the parametric one-way ANOVA test when group data are not Gaussian distributed.

The classical one-way ANOVA tests whether the groups to be compared have the same population mean (e.g., the average AUC or TP rate), while its nonparametric cousin tests whether the groups have the same medians (e.g., the median AUC or TP rate).

The test statistic of the Kruskal-Wallis test is derived by ranking all the N observations from 1 to N regardless of which group they belong to, and the calculating

$$H = \frac{12 \sum_{i=1}^{k} n_i \left(\bar{R}_i - \bar{R} \right)^2}{N \left(N - 1 \right)},$$

where k is the number of different groups of data, n_i is the number of observations in the ith group, $N = \sum_{i=1}^{k} n_i$, $\bar{R}_i$ is the average rank in the ith group, and $\bar{R} = (N + 1) / 2$ is the average of all ranks (Everitt, 2006).

When the null hypothesis is true the test statistic is chi-squared distributed with k-1 degrees of freedom. That is, the calculated test statistic H is compared with the critical value of the chi-square distribution at the specified α and k-1 degrees of freedom. If $H \geq \chi^2_{\alpha, k-1}$, then H_0 is rejected.

The rule of thumb is to apply the classical test when it is known that data in groups are normally distributed and to conduct the Kruskal-Wallis test if there is uncertainty or doubts about normality. In the latter case, the Kruskal-Wallis test may be capable of detecting a significant difference while the classical test may fail to do so.

When one faces with multiple comparisons, it is tempting to apply two-sample tests (like the McNemar's test) to all possible pairs of classifiers. For example, if there are three classifiers with the mean TP rates μ_1, μ_2, μ_3, then one might test each of the following null hypotheses: $H_0 : \mu_1 = \mu_2$, $H_0 : \mu_1 = \mu_3$, $H_0 : \mu_2 = \mu_3$. But such a procedure would be invalid (Toothaker, 1993), (Zar, 1999), (Sheskin, 2004).

If for each two-sample test $\alpha = 0.05$, then there is a 95% probability that H_0 will not be rejected when H_0 is true. For three hypotheses the probability of correctly declining to reject all of them is only $0.95^3 = 0.86$ according to the product rule of probabilities. This implies that the probability of incorrectly rejecting at least one of these hypotheses is equal to $1\text{-}0.95^3 = 0.14$ (Zar, 1999). In a general case, if there are k classifiers and $C = k(k-1) / 2$ possible pairwise comparisons, the probability of making the type I error is given by $1 - (1 - \alpha)^C$, where α is chosen for a single comparison (Toothaker, 1993). Thus, for $\alpha = 0.05, k = 4$ and $C = 6$, this probability is 0.26. If $k = 10$ and $C = 45$, then this probability jumps to 0.9! That is, if we do not take care of the error control, we are destined to incorrectly reject the null hypothesis much more often than we would wish to. The main conclusion is that two-sample tests should not be applied to multisample hypotheses. The right way to proceed is multiple comparison tests.

Multiple comparisons or multiple comparison tests are procedures for detailed examination of the differences between a set of means or medians, usually after a general hypothesis that they are all equal has been rejected (Everitt, 2006). If the one-way ANOVA test (either parametric or distribution free) led to rejection of the null hypothesis, all that we know is that some groups are different than the others. However, we do not know which groups are different and which are not. Multiple comparisons allow us to answer to this question.

The name 'multiple comparisons' hints that many (but not necessarily all) pairs of means/medians are to be compared, but a care is to be taken to compensate for multiple comparisons when making a decision of whether to reject H_0 or not. Usually some correction of the probability of a type I error for a single comparison is done in order for the probability of a type I error for all comparisons to be not larger than α (Drăghici, 2003). It is this correction that makes multiple comparison tests less powerful than the one-way ANOVA tests as it is well known that the larger α is, the more power a test has.

There are several multiple comparisons tests differing in the way how they control the possible inflation of the type I error (Bonferroni correction, Scheffé's test, Dunnett's test) (Toothaker, 1993). I leave out the description of these tests here and instead recommend to look in (Toothaker, 1993), (Drăghici, 2003), (Sheskin, 2004) for details. As we rely on MATLAB® implementation of these tests, additional information about their outputs and their interpretation will be given in Section "MATLAB Code".

It is sometimes possible that $H_0 : \mu_1 = \mu_2 = \ldots \mu_k$ is rejected by an analysis of variance but the subsequent multiple comparison test fails to detect differences between any pair of means. Although this is not common, one should not be surprised as the analysis of variance is more powerful than the multiple comparison test. In other words, type II errors are more likely to happen in multiple comparison tests than in ANOVA tests (Zar, 1999). Usually repeating the experiment with a larger number of runs per algorithm would make a multiple comparison analysis capable of detecting differences among means or medians.

MATLAB Code

MATLAB® has a rich arsenal of own functions for performance evaluation, though they are spread over several toolboxes and thus can sometimes be difficult to find. I found the following functions that can be useful for the purpose of this chapter:

roc (Neural Network Toolbox™);
plotroc (Neural Network Toolbox™);
confusion (Neural Network Toolbox™);
plotconfusion (Neural Network Toolbox™);
confusionmat (Statistics Toolbox™);

classperf (Bioinformatics Toolbox™);
boxplot (Statistics Toolbox™);
perfcurve (Statistics Toolbox™).

Almost all these function require both class labels assigned to the instances of the test set and scores/probability estimates.

Function *roc* generates ROC points together with thresholds corresponding to them so that the *i*th point is represented by a triple (TP rate, FP rate, threshold).

Function *plotroc* plots the ROC curve. Multiple ROC curves on the same plot are possible, too, where each plot is constructed based on its own class labels and probabilistic outputs.

Function *confusion* returns the classification confusion matrix as well as the percentage of false positives, false negatives, and true positives for each class of data. Function *confusionmat* only outputs the confusion matrix itself. Function *plotconfusion* plots a confusion matrix for a given classifier.

Function *classperf* compares true class labels and classifier outputs and based on this comparison it returns a classification performance object containing such fields as accuracy, error rate, sensitivity, specificity, positive and negative predictive values, positive (sensitivity/(1-specificity)) and negative ((1-sensitivity)/specificity) likelihoods, prevalence, confusion matrix. Given the fact that this function is from Bioinformatics Toolbox™, it is certainly recommended for performance evaluation purposes related to bioinformatics tasks.

Function *boxplot* produces a box and whiskers plot for each column of an input matrix. Box and its orientation as well as whiskers length can be set up by the parameters of the function.

Function *perfcurve* is yet another function that is able to report a comprehensive summary of classification performance. For input, *perfcurve* takes true class labels for given data and scores assigned by a classifier to these data. By default, this function computes a ROC curve and returns values of false positive rate, in X and true positive rate, in Y.

One can choose other criteria for X and Y by selecting one out of several provided criteria (e.g., false positive rate, accuracy, specificity, sensitivity, predictive values, etc.) or specifying an arbitrary criterion through an anonymous function. One can then display the computed performance curve using *plot*(X, Y). Function *perfcurve* can compute values for various criteria to plot either on the X- or the Y-axis. All such criteria are described by a 2-by-2 confusion matrix, a 2-by-2 cost (of misclassification) matrix, and a 2-by-1 vector of scales (based on the prior information on class probabilities) applied to class counts. Except for X and Y, *perfcurve* can also return an AUC, the optimal operating point of the ROC curve, and thresholds on classifier scores for the computed values of X and Y.

By default, *perfcurve* computes values of the X and Y criteria for all possible score thresholds. Alternatively, it can compute a reduced number of specific X values supplied as an input argument. Thus, it is possible to compute the partial AUC with *perfcurve*.

As an example, I also implemented the own classification performance function *classperfchr* that computes error rate, true and false positive rates, and F-measure. As a by-product, the confusion matrix is also computed in *confusion_matrix*.

```
function [err, TPR, FPR, TP_FP_spread, F_measure] =
classperfchr(predictions, targets)

% Classification performance characteristics
%
% Inputs:
%    predictions              - Predictions (1xN vector,
%                                where N - the number of patterns)
%    targets                  - Targets (1xN vector)
%
% Outputs:
%    err                    - Error rate (not in percent!)
%   TPR                     - True positive rate (not in per cent!)
%   FPR                     - False positive rate (not in per
%                              cent!)
%    TP_FP_spread           - TP-FP spread (not in per cent!)
%    F_measure                - F_measure (not in per cent!)


% Compute the error rate
err = class_error(predictions, targets);
err = err/100;

% Compute the confusion matrix
%              Predictions
%                 P    N
%              * * * * * * * * * *
% True       P  *  TP  *  FN  *
% labels     N  *  FP  *  TN  *
%              * * * * * * * * * *
CM = confusion_matrix(predictions, targets);
% Get the number of true positives
```

```
TP = CM(1,1);
% Get the number of true negatives
TN = CM(2,2);
% Get the number of false positives
FP = CM(2,1);
% Get the number of false negatives
FN = CM(1,2);

% Compute the true positive rate (sensitivity) or the ratio of the
% number of true positives to the number of positives
TPR = TP/(TP + FN);
% Compute the false positive rate (1 - specificity) or the ratio of the
% number of false positives to the number of negatives
% Specificity: TN/(FP + TN)
FPR = FP/(FP + TN);
% Compute the TP - FP spread
% It denotes the ability of a classifier to make correct predictions
% while minimizing false alarms
TP_FP_spread = TPR - FPR;
% Compute the F-measure combining TP and FP
% For this we need to compute precision and recall
% Compute precision
Precision = TP/(TP + FP);
% Compute recall
Recall = TPR;
% Finally, compute the F-measure
F_measure = (2*Precision*Recall)/(Precision + Recall);

return
function CM = confusion_matrix(predictions, targets)

% Confusion matrix
% Inputs:
%    predictions          - Predictions
%    targets              - True labels
%
% Output:
%    CM                   - confusion matrix
% Check if sizes of two vectors are equal
if ~isequal(size(predictions),size(targets))
```

```
        error('Predictions and targets must have equal length!');
end


% Get the number of classes
Uc = unique(targets);
c = length(Uc);


% Allocate memory
CM = zeros(c);
% Construct the confusion matrix
for i = 1:c
    for j = 1:c
        CM(i,j) = length(find(targets == Uc(i) & predictions == Uc(j)));
    end
end
% Elements of CM are not per cents!!!


return
```

I also provided the own code for ROC generation and AUC computation in functions *ROC* and *AUC*, respectively. Pay attention that though for two-class problems the probabilistic estimate for each instance is a 2D vector, we only need the estimate of the degree to belong to the positive class in both functions. If no output is given in the calling function, *ROC* plots a ROC graph where line style and color are specified by the parameters, *linestyle* and *color*, respectively.

```
function R = ROC(targets, prob, p, n, linestyle, color)

% ROC curve generation
% Original source: Tom Fawcett, "An introduction to ROC analysis",
% Pattern Recognition Letters 27(8), pp.861--874, 2006
%
% Inputs:
%    targets        - Labels (1xN vector, where N - the number of
%                         patterns)
%    prob           - Probability estimates (1xN vector)
%                         (ith estimates is the probability that
%                         pattern i is "positive")
%    p, n           - Number of "positive" and "negative" patterns
%    color          - Color of a line connecting the points on the ROC
```

```
%                       curve
%     linestyle         - Line style
%
% Output:
%    R                  - Mx2 matrix of (x,y) coordinates of M points on
%                           the ROC curve
%                           x - false positive rate,
%                           y - true positive rate
% Check the number of classes since ROC analysis requires only two
% classes
if length(unique(targets)) ~= 2
    error('Only two classes are allowed!');
end

% Check if prob is a vector
if size(prob,1) ~= 1
    prob = prob';
    if size(prob,1) ~= 1
        error('Prob must be a vector!');
    end
end

% Check if the number of probability estimates and that of labels is
% the same
if size(prob,2) ~= length(targets)
    error('Unequal number of probability estimates and class
labels!');
end

% Sort probability estimates in descending order
[dummy, idx] = sort(prob, 'descend');
% Initialise false positive and true positive counts
fp = 0; tp = 0;
probprev = -1;
R = [];
i = 1;
while i <= length(prob)
    if prob(idx(i)) ~= probprev
        R = [R; [fp/n tp/p]];
        probprev = prob(idx(i));
```

```
    end
    if targets(idx(i)) == 0
        tp = tp + 1;
    else
        fp = fp + 1;
    end
    i = i + 1;
end
% Record the last point (this is (1,1))
R = [R; [fp/n tp/p]];

% Plot the ROC curve if no output argument is provided
if nargout == 0
    plot(R(:,1), R(:,2), 'Color', color, 'LineStyle', line-
style, 'LineWidth', 2);
    xlabel('False positive rate');
    ylabel('True positive rate');
end


return
function area = AUC(targets, prob, p, n)

% Area under an ROC curve
% Original source: Tom Fawcett, "An introduction to ROC analysis",
% Pattern Recognition Letters 27(8), pp.861--874, 2006
%
% Inputs:
%   targets         - Labels (1xN vector, where N - the number of
%                                   patterns)
%   prob            - Probability estimates (1xN vector)
%                         (ith estimates is the probability that
%                         pattern i is "positive")
%   p, n            - Number of "positive" and "negative" patterns
%
% Output:
%   area            - Area under the ROC curve
% Check the number of classes since ROC analysis requires only two
% classes
if length(unique(targets)) ~= 2
    error('Only two classes are allowed!');
```

```
end

% Check if prob is a vector
if size(prob,1) ~= 1
    prob = prob';
    if size(prob,1) ~= 1
        error('Prob must be a vector!');
    end
end

% Check if the number of probability estimates and that of labels is
% the same
if size(prob,2) ~= length(targets)
    error('Unequal number of probability estimates and class
labels!');
end

% Sort probability estimates in descending order
[dummy, idx] = sort(prob, 'descend');
% Initialise false positive and true positive counts
fp = 0; tp = 0;
fpprev = 0; tpprev = 0;
probprev = -1;
R = [];
i = 1;
area = 0;
while i <= length(prob)
    if prob(idx(i)) ~= probprev
        area = area + trapezoid_area(fp, fpprev, tp, tpprev);
        fpprev = fp;
        tpprev = tp;
        probprev = prob(idx(i));
    end
    if targets(idx(i)) == 0
        tp = tp + 1;
    else
        fp = fp + 1;
    end
    i = i + 1;
end
```

```
area = area + trapezoid_area(n, fpprev, p, tpprev);
area = area/(p*n); % scale from pxn onto the unit square


return



function a = trapezoid_area(x1, x2, y1, y2)

% Compute trapezoid area given by x1, x2, y1, y2

base = abs(x1 - x2);
average_height = (y1 + y2)/2;
a = average_height*base;

return
```

Finally, let us turn to the code implementing statistical test (see *stattests*). The *stattests* function is heavily commented; I intentionally brought comments from MATLAB® documentation into a single place in order to efficiently present ANOVA and multiple comparison tests to readership. These comments are advised to be carefully read in order to gain better understanding of the material. Not only they explain the meaning of the textual and graphical output of each function, but also provide the prior assumptions regarding each test.

In order to get acquainted with the graphical output *stattests* produces, let us process the following artificial example, where each column of a matrix X represents AUCs computed during seven runs of one of the three classifiers:

$$X = \begin{bmatrix} 0.60 & 0.65 & 0.55 \\ 0.58 & 0.70 & 0.60 \\ 0.62 & 0.64 & 0.57 \\ 0.59 & 0.58 & 0.59 \\ 0.63 & 0.68 & 0.53 \\ 0.59 & 0.57 & 0.59 \\ 0.61 & 0.66 & 0.54 \end{bmatrix}$$

Let us apply the nonparametric Kruskal-Wallis test ($\alpha = 0.05$) to X in order to find out whether all classifiers have equal median AUC values or not. The output provided by MATLAB® is shown in Figure 6.

Figure 6. Kruskal-Wallis ANOVA table

Source	SS	df	MS	Chi-sq	Prob>Chi-sq
Columns	300.2857	2	150.1429	7.866	0.019585
Error	463.2143	18	25.7341		
Total	763.5	20			

Besides the table, *kruskalwallis* displays a box plot shown in Figure 7. A box plot or box-and-whisker plot is a graphical method of displaying the important characteristics of a set of observations (Everitt, 2006). The display is based on the five-number summary[9] (minimum value, lower quartile Q_L, median, upper quartile Q_U and maximum value) of the data with the 'box' part covering the inter-quartile range[10] (the difference between the third and first quartiles of the data, $Q_U - Q_L$), the 'whiskers' extending to include all but outside observations[11] (that is, the observations falling outside the interval $\left[Q_L - 1.5\left(Q_U - Q_L\right), Q_U + 1.5\left(Q_U - Q_L\right)\right]$), these being indicated separately (Everitt, 2006). Quartiles are the values that divide a frequency or probability distribution into four equal parts (Everitt, 2006).

The five-number summary for a vector of values x is returned by MATLAB® function $quantile(x, [0.025\ 0.25\ 0.50\ 0.75\ 0.975])$ as a vector, the second and fourth elements of which are the first and third quartiles.

A box plot is generated by MATLAB® function *boxplot* that produces a box and whiskers for each column of an input matrix. The box has lines at the lower quartile, median, and upper quartile values. Whiskers extend from each end of the box to the most extreme values within 1.5 times the inter-quartile range from the ends of the box. Outliers (data with values beyond the ends of the whiskers) are displayed with a red plus sign.

From Figure 7 it can be seen that the medians are clear different (and the p-value is about 0.02, which is less than the significance level; hence, the null hypothesis that all medians are equal is rejected), but to find out which medians are different, we need to run *multcompare* using the structure *stats* returned by *kruskalwallis*. Graphical results of *multcompare* are given in Figure 8, Figure 9 and Figure 10. Each column of X is called a group in the MATLAB® terminology. Each figure represents a graphical plot when one of the groups was highlighted, i.e. the cursor was moved over a line associated with a particular group. By clicking on that line as suggested in the plot title, one can get the outcome of multiple comparison tests with regard to the group of interest. In particular, if the intervals of the ranks for groups i and j overlap, this means that the null hypothesis that the mean ranks of two groups are equal was not rejected.

Figure 7. Kruskal-Wallis ANOVA plot

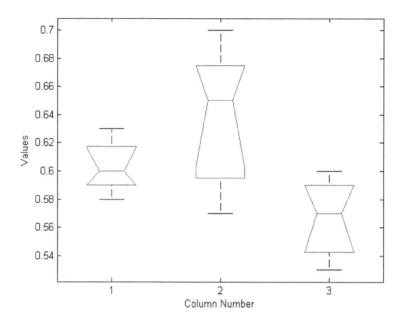

For example, in Figure 8 the interval of ranks for group 1 overlaps the intervals of the other groups. Hence, the conclusion is that no groups have mean ranks significantly different from group 1. In all figures related to the multiple comparison test, open circles show the mean rank of a given group.

By clicking on the interval of ranks corresponding to group 2 (see Figure 9), we can see that the mean ranks of groups 2 and 3 are significantly different. The same conclusion is drawn when looking at Figure 10. Thus, the result is the classification performance of the second and third classifiers are significantly different, with the second classifier significantly outperforming the third one as demonstrated by its interval of ranks.

You, dear readers, may wonder why we seemingly ignored the conclusion from Figure 8. Did we get a contradiction? The answer is no; everything is correct.

Let us carefully read the conclusion regarding Figure 8: no groups have mean ranks significantly different from group 1. That is, we had tested the following null hypotheses: $H_0 : \mu_1 = \mu_2$ and $H_0 : \mu_1 = \mu_3$. However, we did not test $H_0 : \mu_2 = \mu_3$, therefore we cannot conclude anything about it from the results of other two tests.

On the other hand, comparison tests related to group 2 resulted in the statements that $\mu_2 = \mu_1$ and $\mu_2 \neq \mu_3$, while tests for group 3 led to the conclusions that $\mu_3 = \mu_1$ and $\mu_3 \neq \mu_2$. By gathering all conclusions together, we came to $\mu_2 \neq \mu_3$.

Figure 8. Multiple comparison plot when group 1 is highlighted

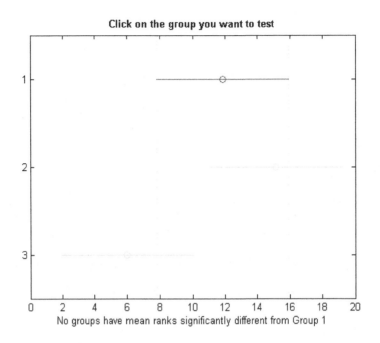

Figure 9. Multiple comparison plot when group 2 is highlighted

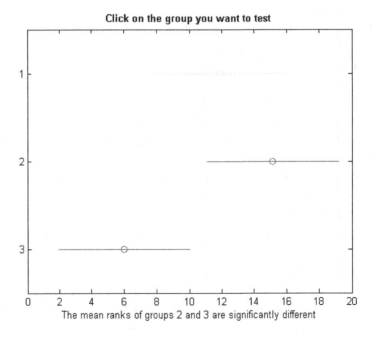

Figure 10. Multiple comparison plot when group 3 is highlighted

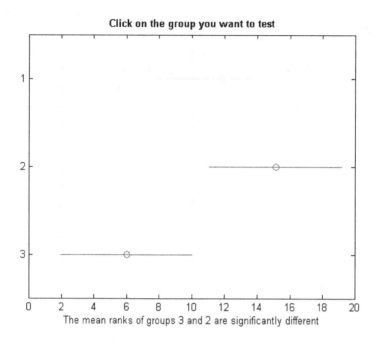

Figure 11. Results of multiple comparisons for our artificial example

	Gr1	Gr2	LowBnd	Diff.	UpBnd	Outcome
1	1	2	-11.3696	-3.2857	4.7982	1
2	1	3	-2.2268	5.8571	13.9411	1
3	2	3	1.0589	9.1429	17.2268	0

In fact, one does not need to test all possible pairs and it is not usually done in practice. For example, let us look at the text output written in file *TestResults_c.txt* and shown in Figure 11.

As you can see only the following comparisons were done: μ_1 versus μ_2, μ_1 versus μ_3, and μ_2 versus μ_3. The last column in Figure 11 indicates 1 if a null hypothesis was not rejected and 0 otherwise. Another way to see whether a null hypothesis is to be rejected is to look at the values in columns 'UpBnd' and 'LowBnd'. If zero is included in the interval between lower and upper bounds, then the decision is not to reject the null hypothesis. If the interval does not contain zero, then the null hypothesis is rejected.

In addition to *TestResults_c.txt*, another text file *TestResults_m.txt* is generated by *stattests* that contains the mean ranks and their standard deviations for each group.

I performed multiple comparison procedures with Scheffé correction for the significance level of a single comparison. However, other α-control techniques are available as the options of *multcompare*, e.g., Bonferroni correction.

Before performing a test, it is good to know its power, especially if a null hypothesis is not rejected. To remind you, the power of a test is the probability of correctly rejecting H_0 when it is false. It determines how good test is in detecting a false null hypothesis.

If H_0 is true for an analysis of variance, then the variance ratio follows the F distribution that is characterized by the numerator and denominator degrees of freedom (ν_1 and ν_2, respectively) (Zar, 1999). If H_0 is false, then the ratio between-groups variance (groups mean square or groups MS) to within-groups variance (error mean square or error MS) follows instead the noncentral F distribution, which is defined by ν_1, ν_2, and a third parameter known as the noncentrality parameter. The power of ANOVA thus depends on this parameter (Zar, 1999).

In the case of ANOVA, power can be calculated based on the estimates of within-groups and between-groups variances and the number of groups, k. From this information, one can calculate a quantity ϕ, which is related to the noncentrality parameter, as follows (see p.193 in (Zar, 1999))

$$\phi = \sqrt{\frac{(k-1)\left(groups\,MS - s^2\right)}{ks^2}},$$

where $groups\,MS$ and s^2 are the between-groups variance and within-group variance, respectively. All the information needed to calculate ϕ can be retrieved from a cell array *table* returned by MATLAB® function *anova1* as shown in our function *stattests*. Except for ϕ, the values of ν_1 and ν_2 are returned by *stattests*. All three values are then needed in order to find the power of ANOVA by consulting Appendix Fig. B.1 in (Zar, 1999).

It can be seen from Appendix Fig. B.1 that, for a given significance level, ν_1, and ν_2, greater values of ϕ are associated with greater power. The value of ϕ will also increase with

- increased sample size;
- increased between-groups variance;
- decreased between-groups variance;
- a fewer number of groups.

For the artificial example above, the returned $\phi \approx 1.8$, $\nu_1 = 2$, $\nu_2 = 18$. Hence, power is approximately 0.73 as found in Appendix Fig. B1, which means that there was a 27% chance of having committed a type II error, i.e. to falsely accept the fact that all three algorithms have the same mean AUC.

```
function [phi,nu1,nu2] = stattests(fcn, X, group, alpha)

% Perform either
% 1) parametric one-way ANOVA test, followed by parametric
multiple
% comparison test (anova1+multcompare)
% or
% 2) nonparametric one-way ANOVA, followed by nonparametric
multiple
% comparison test (kruskalwallis+multcompare)
%
% Original source:
% Jerrold H. Zar, "Biostatistical Analysis", Upper Saddle
River, NJ:
% Prentice Hall/Pearson Education International, 4th Edition,
1999.
%
% Inputs:
%   fcn        - Either 'anova1' or 'kruskalwallis'
%   X          - Data matrix whose columns are samples and
rows are
%                observations
%   group      - Character array or a cell array of strings,
with one
%                   row per column of X, containing
group names. Enter
%                an empty array ([]) if you do not want to
specify
%                      group names
%   alpha      - Significance level (e.g., 0.05 or 0.01)
%
% Outputs:
%   phi        - Variability among columns of X (groups)
%   nu1        - Groups degrees of freedom (see Zar's book,
pp.181-
```

```
%                          182)
%    nu2         - Error degrees of freedom = Total degrees of
freedom
%                          minus groups degrees of freedom (see
Zar's book,
%                          pp.181-182)
%
%
% Comments in this function may in part duplicate those in the
MATLAB
% documentation. Duplication is intended to gather all useful
% information in one source.
%
%
%
% Perform a balanced one-way ANOVA for comparing the means of
two or
% more columns of data in the m-by-n matrix X, where each
column
% represents an independent sample containing m mutually inde-
pendent
% observations. The function returns the p-value for the null
% hypothesis that all samples in X are drawn from the same
population
% (or from different populations with the same mean).
%
% If the p-value is less than the significance level alpha,
this casts
% doubt on the null hypothesis and suggests that at least one
sample
% mean is significantly different than the other sample means.
%
% The anova1 function displays two figures. The first figure is
the
% standard ANOVA table, which divides the variability of the
data in
% X into two parts:
% 1) Variability due to the differences among the column means
% (variability between groups)
% 2) Variability due to the differences between the data in
```

each column

% and the column mean (variability within groups)

%

% The ANOVA table has six columns:

% The first shows the source of the variability.

% The second shows the Sum of Squares (SS) due to each source.

% The third shows the degrees of freedom (df) associated with each

% source.

% The fourth shows the Mean Squares (MS) for each source, which is

% the ratio SS/df.

% The fifth shows the F statistic, which is the ratio of the MS's.

% The sixth shows the p-value, which is derived from the cdf of F.

% As F increases, the p-value decreases.

%

% The second figure displays box plots of each column of X. Large

% differences in the center lines of the box plots correspond to

% large values of F and correspondingly small p-values.

%

% p = anova1(X,group) uses the values in group (a character array or

% cell array) as labels for the box plot of the samples in X, when X

% is a matrix. Each row of group contains the label for the data in

% the corresponding column of X, so group must have length equal to

% the number of columns in X.

%

% [p,table] = anova1(...) returns the ANOVA table (including column

% and row labels) in cell array table.

%

% [p,table,stats] = anova1(...) returns a stats structure that you

```
% can use to perform a follow-up multiple comparison test.
%
% The anova1 test evaluates the hypothesis that the samples all
have
% the same mean against the alternative that the means are not
all
% the same. Sometimes it is preferable to perform a test to
determine
% which pairs of means are significantly different, and which
are
% not. Use the multcompare function to perform such tests by
supplying
% the stats structure as input.
%
% Assumptions:
%
% The ANOVA test makes the following assumptions about the data
in X:
%
% All sample populations are normally distributed.
% All sample populations have equal variance.
% All observations are mutually independent.
%
% The ANOVA test is known to be robust to modest violations of
the
% first two assumptions.
%
%
%
% Perform a Kruskal-Wallis test to compare samples from two or
more
% groups. Each column of the m-by-n matrix X represents an
independent
% sample containing m mutually independent observations. The
function
% compares the medians of the samples in X, and returns the
p-value for
% the null hypothesis that all samples are drawn from the same
% population (or equivalently, from different populations with
the same % distribution).
```

```
%
% Note that the Kruskal-Wallis test is a nonparametric version
of the
% classical one-way ANOVA, and an extension of the Wilcoxon
rank sum
% test to more than two groups.
%
% If the p-value is less than the significance level alpha,
this casts
% doubt on the null hypothesis and suggests that at least one
sample
% median is significantly different from the others.
%
% The kruskalwallis function displays two figures. The first
figure is
% a standard ANOVA table, calculated using the ranks of the
data rather
% than their numeric values. Ranks are found by ordering the
data from
% smallest to largest across all groups, and taking the numeric
index
% of this ordering. The rank for a tied observation is equal to
the
% average rank of all observations tied with it.
%
% The entries in the ANOVA table are the usual sums of squares,
degrees
% of freedom, and other quantities calculated on the ranks. The
usual
% F statistic is replaced by a chi-square statistic. The p-val-
ue
% measures the significance of the chi-square statistic.
%
% The second figure displays box plots of each column of X (not
the
% ranks of X)
%
% p = kruskalwallis(X,group) uses the values in group (a char-
acter
% array or cell array) as labels for the box plot of the sam-
```

```
ples in X,
% when X is a matrix. Each row of group contains the label for
the data
% in the corresponding column of X, so group must have length
equal to
% the number of columns in X.
%
% [p,table] = kruskalwallis(...) returns the ANOVA table (in-
cluding
% column and row labels) in cell array table.
%
% [p,table,stats] = kruskalwallis(...) returns a stats struc-
ture that
% you can use to perform a follow-up multiple comparison test.
% The kruskalwallis test evaluates the hypothesis that all
samples come
% from populations that have the same median, against the
alternative
% that the medians are not all the same. Sometimes it is pref-
erable to
% perform a test to determine which pairs are significantly
different,
% and which are not. You can use the multcompare function to
perform
% such tests by supplying the stats structure as input.
%
% Assumptions:
%
% The Kruskal-Wallis test makes the following assumptions about
the
% data in X:
% All samples come from populations having the same continuous
% distribution, apart from possibly different locations due to
group
% effects.
% All observations are mutually independent.
%
% The classical parametric one-way ANOVA test replaces the
first
% assumption with the stronger assumption that the populations
```

```
have
% normal distributions.
%
% Sometimes, a statistically significance difference between
% medians can be detected by the Kruskal-Wallis test when its
% parametric counterpart failed to do so. However, if it is
surely
% known that data in columns of X are normally distributed,
then
% running the parametric one-way ANOVA is enough in order to
detect a
% significant difference.




% Check if the function name is correct
if ~strcmpi(fcn,'anova1') && ~strcmpi(fcn,'kruskalwallis')
    error('Either ANOVA1 or KRUSKALWALLIS function must be
used!');
end

% Run the test on the equality of all means/medians
[p, table, stats] = feval(fcn, X, group);

fprintf(1,'P-value of one-way ANOVA test using function %s:
%e\n', upper(fcn), p);

% Estimate the power of the ANOVA after it has been performed
% See Zar's book, pp.82-83,193
% The power of a statistical test is defined as 1-beta, i.e.
the power % is the probability of rejecting the null hypothesis
when it is in
% fact false and should be rejected, where beta is the prob-
ability of
% not rejecting the null hypothesis when it is in fact false
k = size(X,2); % no. of groups
s2 = table{3,4}; % error MS
groupsMS = table{2,4}; % groups (columns) MS
phi = real(sqrt((k-1)*(groupsMS-s2)/(k*s2))); % phi
nu1 = table{2,3}; % nu1
```

```
nu2 = table{3,3}; % nu2
% Concult Appendix Fig.B1 of Zar's book for estimation of the
power
% of the ANOVA, given nu1, nu2 and phi;
% beta = (1-power) determines the probability of comitting a
Type II
% error, i.e. probability of not rejecting the null hypothesis
when it
% is in fact false

if p < alpha % difference is statistically significant
    % At least one mean (median) is significantly different
from others
    % Apply multiple comparison test of means or medians
    % Perform a multiple comparison test using the information
in the
    % stats structure, and return a matrix c of pairwise com-
parison
    % results. It also displays an interactive graph of the
estimates
    % with comparison intervals around them.
    %
    % In a one-way analysis of variance, you compare the means
or
    % medians of several groups to test the hypothesis that
they are
    % all the same, against the general alternative that they
are not
    % all the same. Sometimes this alternative may be too
general. You
    % may need information about which pairs of means are
significantly
    % different, and which are not. A test that can provide
such
    % information is called a "multiple comparison procedure."
    %
    % When you perform a simple t-test of one group mean
against
    % another, you specify a significance level that determines
the
```

```
    % cutoff value of the t statistic. For example, you can
specify the
    % value alpha = 0.05 to ensure that when there is no real
    % difference, you will incorrectly find a significant
difference no
    % more than 5% of the time. When there are many group
means, there
    % are also many pairs to compare. If you applied an ordi-
nary t-test
    % in this situation, the alpha value would apply to each
    % comparison, so the chance of incorrectly finding a sig-
nificant
    % difference would increase with the number of comparisons.
    % Multiple comparison procedures are designed to provide an
upper
    % bound on the probability that any comparison will be
incorrectly
    % found significant.
    %
    % The output c contains the results of the test in the form
of a
    % five-column matrix. Each row of the matrix represents one
test,
    % and there is one row for each pair of groups. The entries
in the
    % row indicate the means being compared, the estimated
difference
    % in means, and a confidence interval for the difference.
    %
    % For example, suppose one row contains the following
entries.
    % 2.0000    5.0000    1.9442    8.2206    14.4971
    % These numbers indicate that the mean of group 2 minus the
mean
    % of group 5 is estimated to be 8.2206, and a 95% confi-
dence
    % interval for the true mean is [1.9442, 14.4971].
    %
    % In this example the confidence interval does not contain
0.0,
```

```
    % so the difference is significant at the 0.05 level.
    % If the confidence interval did contain 0.0, the differ-
ence would
    % not be significant at the 0.05 level.
    %
    % The multcompare function also displays a graph with each
group
    % mean represented by a symbol and an interval around the
symbol.
    % Two means are significantly different if their intervals
are
    % disjoint, and are not significantly different if their
intervals
    % overlap. You can use the mouse to select any group, and
the graph
    % will highlight any other groups that are significantly
different
    % from it.
    %
    % [c,m] = multcompare(...) returns an additional matrix m.
    % The first column of m contains the estimated values of
the means
    % (or whatever statistics are being compared) for each
group, and
    % the second column contains their standard errors.
    %
    % [c,m,h] = multcompare(...) returns a handle h to the
comparison
    % graph. Note that the title of this graph contains in-
structions
    % for interacting with the graph, and the x-axis label
contains
    % information about which means are significantly different
from
    % the selected mean.
    %
    % [c,m,h,gnames] = multcompare(...) returns gnames, a cell
array
    % with one row for each group, containing the names of the
groups.
```

```
    figure,
    [c, m, h, gnames] = multcompare(stats, 'ctype', 'scheffe');
    % Write m to a text file
    varnames = {'Mean'; 'Std. Error'};
    casenames = num2str((1:size(m,1))');
    tblwrite(m, varnames, casenames, 'TestResults_m.txt',
'tab');
    % Write the augmented c to a text file
    c(:,6) = (c(:,3) <= 0 & 0 <= c(:,5));
    % If c(i,6) = 1, H0 is not rejected
    % If c(i,6) = 0, H0 is rejected
    varnames = {'Gr1'; 'Gr2'; 'LowBnd'; 'Diff.'; 'UpBnd';
'Outcome'};
    casenames = num2str((1:size(c,1))');
    tblwrite(c, varnames, casenames, 'TestResults_c.txt',
'tab');
end

% When using nonparametric one-way ANOVA and nonparametric
multiple
% comparison tests, means/std. errors are rank sums!!!
%
% For example, for 160x24 matrix the total number of ranks is
equal to
% 3,840!

return
```

REFERENCES

Breiman, L., Friedman, J. H., Olshen, R. A., & Stone, C. J. (1984). *Classification and regression trees*. New York, NY: Wadsworth.

Dietterich, T. G. (1998). Approximate statistical tests for comparing supervised classification learning algorithms. *Neural Computation*, 7(10), 1895–1924. doi:10.1162/089976698300017197

Drăghici, S. (2003). *Data analysis tools for DNA microarrays*. Boca Raton, FL: Chapman & Hall/CRC Press.

Drish, J. (2001). *Obtaining calibrated probability estimates from support vector machines.* Department of Computer Science and Engineering. San Diego, CA: University of California. Retrieved from http://cseweb.ucsd.edu/ users/ elkan/ 254spring01/ jdrishrep.pdf.

Everitt, B. (2006). *The Cambridge dictionary of statistics* (3rd ed.). Cambridge, UK: Cambridge University Press.

Fawcett, T. (2006). An introduction to ROC analysis. *Pattern Recognition Letters*, *27*(8), 861–874. doi:10.1016/j.patrec.2005.10.010

Good, P. I., & Hardin, J. W. (2006). *Common errors in statistics (and how to avoid them)* (2nd ed.). Hoboken, NJ: John Wiley & Sons. doi:10.1002/0471998524

Hand, D. J., & Till, R. J. (2001). A simple generalization of the area under the ROC curve to multiple class classification problems. *Machine Learning*, *45*(2), 171–186. doi:10.1023/A:1010920819831

Hanley, J. A., & McNeil, B. J. (1982). The meaning and use of the area under a receiver operating characteristic (ROC) curve. *Radiology*, *143*, 29–36.

Kranzler, G., Moursund, J., & Kranzler, J. H. (2006). *Statistics for the terrified* (4th ed.). Upper Saddle River, NJ: Prentice Hall/Pearson Education International.

Kuncheva, L. I. (2004). *Combining pattern classifiers: methods and algorithms.* Hoboken, NJ: John Wiley & Sons. doi:10.1002/0471660264

Niculescu-Mizil, A., & Caruana, R. (2005). Obtaining calibrated probabilities from Boosting. *Proceedings of the 21th Confrence in Uncertainty in Artificial Intelligence, Edinburgh, Scotland* (pp. 413-420). Arlington, VA: AUAI Press.

Pepe, M. S. (2004). *The statistical evaluation of medical tests for classification and prediction* (1st paperback ed.). Oxford, UK: Oxford University Press.

Sheskin, D. J. (2004). *Handbook of parametric and nonparametric statistical procedures* (3rd ed.). Boca Raton, FL: Chapman & Hall/CRC.

Swets, J. A., Dawes, R. M., & Monahan, J. (2000). Better decisions through science. *Scientific American*, *283*(4), 82–87. doi:10.1038/scientificamerican1000-82

Toothaker, L. E. (1993). *Multiple comparison procedures.* Newbury Park, CA: Sage Publications.

van Emden, H. (2008). *Statistics for the terrified biologists.* Hoboken, NJ: John Wiley & Sons.

Zadrozny, B., & Elkan, C. (2001). Obtaining calibrated probability estimates from decision trees and naive Bayesian classifiers. In C. E. Brodley, & A. Pohoreckyj Danyluk (Ed.), *Proceedings of the 18th International Conference on Machine Learning, Williamstown, MA* (pp. 609-616). San Francisco, CA: Morgan Kaufmann.

Zar, J. H. (1999). *Biostatistical analysis* (4th ed.). Upper Saddle River, NJ: Prentice Hall/Pearson Education International.

ENDNOTES

[1] Analysis of Variance may be a strange name for the procedure testing the equality of means, but, in fact, this equality (or inequality) is established based on the ratio of variances; hence, the name for the test.

[2] A simple rule for approximating the integral of a function, $f(x)$, between two limits, using the formula $\int_{a}^{a+h} f(x)\,dx = \frac{1}{2}h\left[f(a) + f(a+h)\right]$ (Everitt, 2006).

[3] This is a two-sided rank sum test of the null hypothesis that data in vectors x and y are independent samples from identical continuous distributions with equal medians, against the alternative that they do not have equal medians. If each vector is composed of probabilistic outputs of one of the two classes, then the larger a difference in medians is, the better two classes are separated. Since the AUC is equivalent to the Wilcoxon rank sum test, it can be used as a measure of class separation in binary classification tasks.

[4] By classification algorithm we will understand in this chapter both single classifiers and ensembles of classifiers. Feature selection is assumed was done before classification.

[5] Although the chi-square distribution (continuous distribution) is employed to evaluate the McNemar's test statistic, in reality this distribution provides an approximation for the exact distribution, which is the binomial distribution (discrete distribution). Hence, subtracting 1 in the numerator 'corrects' such a continuous approximation (Sheskin, 2004). Sheskin pointed out that for small sample sizes, it may be good to compute the exact binomial distribution.

[6] Significance level of a test is the level of probability at which it is agreed that the null hypothesis will be rejected (Everitt, 2006). It should be specified before the test is conducted.

[7] F-test is a statistical test for the equality of the variances of two populations having normal distributions, based on the ratio of the variances of a sample of observations (Everitt, 2006).

[8]   It should be noted that the classical one-way ANOVA is rather robust to deviations from normality, especially if the number of observations in a group is the same for all groups (Zar, 1999), (Sheskin, 2004). Outliers (a small number of data that are much more extreme than the rest of observations) can significantly distort the data distribution, thus severely violating the normality assumption. In such a case, it is better to conduct the distribution free Kruskal-Wallis test.

[9]   MATLAB® function *summary* from Statistics Toolbox™ displays such a summary.

[10]   Inter-quartile range is computed by MATLAB® function *iqr*.

[11]   These observations are regarded as being extreme enough to be potential outliers – observations that appear to markedly deviate from the other observations in a data set (Everitt, 2006).

Chapter 23
Bolstered Resubstitution Error

ALTERNATIVE TO TRADITIONAL ERROR ESTIMATORS

Unbiased (objective) classification error estimation is the important issue in evaluating performance of machine learning algorithms applied to high dimensional gene expression data. Its importance comes from the fact that it is very easy to get fooled by the inflated (over-optimistic) error rate due to overfitting caused by scarcity of training instances. Braga-Neto and Dougherty recently proposed a new error estimator less prone to error bias. It is called bolstered error, because it is based on "bolstering" the original empirical data distribution by suitable kernels placed at each training data point (Braga-Neto & Dougherty, Bolstered error estimation, 2004).

The bolstered error estimate is, in general, computed by Monte-Carlo sampling. However, a very small number of Monte-Carlo samples are needed in practice, which results in a very fast error estimator, which is in contrast to other resampling techniques, such as the bootstrap. If a classifier is linear (the decision boundary separating two classes is a hyperplane) and a Gaussian bolstering kernel is used, the bolstered error can be analytically computed.

Bolstering has the effect of reducing the variance of the error estimation[1]. Variance reduction is desirable if the trustworthy error estimation is sought. By selecting the appropriate parameters of the bolstering kernels, one can also reduce bias. Of

DOI: 10.4018/978-1-60960-557-5.ch023

course, reducing both bias and variance increases the overall accuracy of the classification error estimation.

Since our interest is in an unbiased error estimator, we need to define what is meant by 'unbiased'. Let us recall that bias is a deviation of results or inferences from the truth or the extent to which the quantity to be estimated is not estimated (Everitt, 2006). Therefore the bias is measured by the difference between a parameter estimate $\hat{\theta}$ (pay attention that θ is a parameter to be estimated while $\hat{\theta}$ is the estimate of this parameter) and its expected value $E\left(\hat{\theta}\right)$. An estimator, for which $E\left(\hat{\theta}\right) = \theta$, is called unbiased (Everitt, 2006).

The error estimator is unbiased, with respect to the unconditional error, i.e. the error not conditioned on any particular training set, if the test data come from independent instances not used to design a classifier (Braga-Neto & Dougherty, Bolstered error estimation, 2004). But where can one get these new data apart from the small training set? The answer is provided in (Braga-Neto & Dougherty, Bolstered error estimation, 2004) and it is to generate artificial samples based on the training data. These artificial or virtual instances can then serve as an independent test set. Thus, any reduction of the available data observed in the cross-validation techniques can be avoided.

Experiments carried out in (Braga-Neto & Dougherty, Bolstered error estimation, 2004) convincingly demonstrated the superiority of the bolstered error estimation over such traditional error estimation techniques as resubstitution, leave-one-out, 10-fold cross-validation, and the 0.632 bootstrap. Due to this fact, we do not discuss these traditional techniques in detail here. Interested readers can consult, e.g., (Dudoit & Fridlyand, 2003).

Main Idea

The origins of the bolstered error estimation lay in the technique known as smoothed error estimation and proposed for Linear Discriminant Analysis (LDA) (linear classifier in a 1D space) in (Glick, 1978). Although this technique is exclusively limited to LDA, it inspired to use bolstering kernels similarly to smoothing functions.

Let S_N be the training data (N instances x_i and their labels y_i) and $g\left(S_N, \cdot\right): \mathbb{R}^D \rightarrow \{0,1\}$ be a classifier that maps a D dimensional unlabeled instance into one of the two classes, based on the training data.

In this chapter, we will consider bolstered variants of resubstitution error (Braga-Neto & Dougherty, Bolstered error estimation, 2004), (Braga-Neto, Fads and fallacies in the name of small-sample microarray classification - A highlight of misunderstanding and erroneous usage in the applications of genomic signal processing, 2007). Thus, it is good to define this kind of error before proceeding further. Suppose that a classifier $g(S_N, \cdot)$ was designed by using S_N. The resubstitution error is the proportion of training instances that were misclassified by this classifier. In other words, the same set of instances is used in training and testing. It is easy to guess that the resubstitution error will be very optimistic, compared to the true error: for example, a 1-nearest neighbor classifier has zero resubstitution error as each training instance has itself as the nearest neighbor.

The resubstitution error for two-class problems can be defined as

$$\hat{\varepsilon} = \frac{1}{N} \sum_{i=1}^{N} \left(g(S_N, x_i) I_{y_i=0} + \left(1 - g(S_N, x_i)\right) I_{y_i=1} \right),$$

where x_i is the ith training instance, and I_C is an indicator function, which is equal to 1 if the statement C is true and zero otherwise.

The function g is a sharp 0-1 step function (i.e. its values are either 0 or 1). The idea behind smoothed estimators (Glick, 1978) is to replace g by a suitably chosen function taking values in the interval $[0,1]$. This replacement intends to reduce the variance of the original estimator, which can be quite large for small values of N. Due to a number of reasons stated in (Braga-Neto & Dougherty, Bolstered error estimation, 2004), the extension of smoothing to classifiers other than LDA is not straightforward. Hence, Braga-Neto and Dougherty went further by replacing a smoothing function with a pointwise bolstering kernel $f_i^{\diamond}$, which is, in fact, a D variate probability density function defined as

$$f^{\diamond}(x) = \frac{1}{N} \sum_{i=1}^{N} f_i^{\diamond}(x - x_i) I_{y=y_i}.$$

Let $A_j = \left\{ x \in \mathbb{R}^D \mid g(S_N, x) = j \right\}$ for $j = 0,1$ be two decision regions for the designed classifier. That is, A_0 contains all instances classified as class 0 and A_1 contains all instances classified as 1. Then the bolstered resubstitution error estimator is

$$\hat{\varepsilon}^{\diamond} = \frac{1}{N} \sum_{i=1}^{N} \left(\int_{A_1} f_i^{\diamond}(x - x_i) \, dx I_{y_i=0} + \int_{A_0} f_i^{\diamond}(x - x_i) \, dx I_{y_i=1} \right)$$

where the integrals are the error contributions made by the data points, according to whether $y_i = 0$ or 1. Therefore, the bolstered resubstitution error is equal to the sum of all error contributions divided by the number of training instances.

When a classifier is linear, it is usually possible to find analytical expressions for the integrals in the formula above. However, in the general case, the Monte-Carlo integration has to be applied:

$$\hat{\varepsilon}^{\Diamond} \approx \frac{1}{N} \sum_{i=1}^{N} \left(\sum_{j=1}^{M} I_{x_{ij} \in A_1} I_{y_i=0} + \sum_{j=1}^{M} I_{x_{ij} \in A_0} I_{y_i=1} \right),$$

where $\left\{ x_{ij} \right\}_{j=1,...,M}$ are samples randomly drawn from the distribution $f_i^{\Diamond}$. The value of M does not have to be large; experiments in (Braga-Neto & Dougherty, Bolstered error estimation, 2004) showed that $M = 10$ will be enough in order to approximate the integrals. That is, each training point serves as a "seed" for generating (producing) the M artificial test points.

In order to perform random sampling from a certain distribution (the exact sampling procedure will be described in the next section), the bolstering kernels have to be defined. A good choice for the bolstering kernel is a zero-mean, spherical D variate normal distribution with diagonal covariance matrix:

$$f_i^{\Diamond} = \frac{1}{(2\pi)^{D/2} \sigma_i^D} exp\left(-\frac{x^2}{2\sigma_i^2} \right),$$

where $\sigma_i, \ldots, \sigma_N$ are the standard deviations. This choice allows not making complicated inferences from a limited amount of microarray data.

The choice of the standard deviations is a critical issue. The larger σ_i's, i.e. the "wider" bolstering kernels, the more variance reduction can be achieved, though after a certain point this advantage becomes offset by increased bias (recall bias-variance dilemma!). The larger σ_i's, the more test points are spread around a training point that served as their generator.

The advantage of the bolstered error is that the σ_i's are automatically computed rather than being set up by a user. In addition, the σ_i's are not static; they vary from point to point in order to be robust to the data. In (Braga-Neto & Dougherty, Bolstered error estimation, 2004), the following formula is given:

$$\sigma_i = \frac{\hat{d}(y_i)}{F_{d_i}^{-1}(0.5)},$$

where $\hat{d}(y_i)$ is the sample-based estimate of the mean minimum distance between training points belonging to class y_i; $F_{d_i}^{-1}(0.5)$ is the inverse of the chi-square cdf (cumulative distribution function[2]) with parameter D (the number of degrees of freedom) for probability 0.5.

The motivation for such a definition of σ_i's is that the median distance of a test point to the origin is equal to the estimated $\hat{d}(y_i)$, so that half of the test points will be farther from the center than $\hat{d}(y_i)$ while the other half will be nearer. In the Gaussian case, distances between D dimensional points are distributed as a chi random variable with D degrees of freedom[3]. The quantity $F_{d_i}^{-1}(0.5)$ then implies the median distance, which needs to be multiplied by σ_i due to the bolstering kernel $f_i^{\diamond}$. As a result, one has $\sigma_i F_{d_i}^{-1}(0.5) = \hat{d}(y_i)$, from which one obtains the expression for σ_i's.

$F_{d_i}^{-1}(0.5)$ can be pre-computed in advance. On the other hand, $\hat{d}(y_i)$ is computed as follows: first, the minimum distance of x_i to all other points x_j ($i \neq j$) belonging to the same class as x_i is found; after that the average of such minimum distances is calculated.

In situations where resubstitution is heavily biased due to overfitting, it may not be a good idea to spread incorrectly classified points as this further increases a bias of the error estimate. Bias can be reduced if one assigns $\sigma_i = 0$ (which corresponds to no bolstering) to incorrectly classified training instances. This, however, increases variance. Such a variant of bolstered error estimator is called a semi-bolstered resubstitution error estimator. It may be useful to apply instead of the bolstered resubstitution estimator for such classifiers as Naïve Bayes that tend to overfit, given high dimensional data and small sample size. Such classifiers often have low variance and high bias. Hence increasing variance of error estimation may not harm, especially if bias can be simultaneously lowered.

Generation of Gaussian Distributed Random Samples

One aspect that was left untouched in the previous section is how to generate random samples from the Gaussian distribution. Although there are several sampling methods suitable for this task, I chose the Marsaglia polar method, which is the polar form of the Box-Müller transform (Box & Müller, 1958) intended to gener-

Figure 1. Gaussian distributed random samples drawn from a uniform distribution inside the unit circle

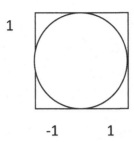

ate Gaussian pseudo-random numbers from the uniform pseudo-random numbers. The polar form does not require to compute trigonometric functions and hence, it is faster than the basic form of the Box-Müller transform.

The polar form of the Box-Müller transform starts from generating a pair of uniformly distributed random numbers $u_1, u_2 \in [0,1]$. These numbers are then transformed to a new interval $[-1,1]$ by $u_i = 2u_i - 1$, $i = 1, 2$. Let $r = u_1^2 + u_2^2$. If $r \geq 1$ or $r = 0$ (i.e. either a point with coordinates (u_1, u_2) lies outside a circle with unit radius or we got the trivial solution: $u_1 = u_2 = 0$), another pair of uniformly distributed $[0,1]$ numbers is drawn, these are again mapped into the $[-1,1]$ interval, and a new point location is checked as above. If $r < 1$, the generated point lies inside the unit circle (see Figure 1) and this is what we need. By repeating this process again and again, we draw uniformly distributed points lying inside the unit circle.

The next task is to obtain normally distributed points from these uniformly distributed points. By omitting the exact derivation[4], two standard normal (i.e. they have zero mean and unit standard deviation) numbers are found as

$$z_1 = u_1 \sqrt{\frac{-2\ln r}{r}}; \qquad z_2 = u_2 \sqrt{\frac{-2\ln r}{r}}.$$

Since we need not simply standard normal numbers but those with mean μ (μ is unnecessary zero) and standard deviation σ (σ is unnecessary equal to 1), a simple transformation lets us to reach this objective:

$$z_1' = \mu + z_1\sigma, \; z_2' = \mu - z_2\sigma.$$

In the last formulas, the estimate of σ is pre-computed for each training point as explained in the previous section, while the mean is given by coordinates of the training point in question. In Figure 1, this training point is located at the center of the circle. Different signs in front of the second summand in the equations above intend to evenly distribute generated points around the mean.

Thus, the whole process of test data creation can be seen as consisting in the following steps:

- Compute the estimate of the standard deviation for the current training point. The mean is automatically defined by the training point itself.
- Generate two normally distributed numbers z_1 and z_2.
- Obtain the desired normally distributed numbers $z_1^{'}$ and $z_2^{'}$ based on the current mean and standard deviation.

Given M points to be generated, this process is repeated $M / 2$ times, since each time two points are produced.

Algorithm Description

Based on the description of every step of bolstered resubstitution error estimation, we are now ready to summarize the whole procedure. I provide pseudo code for both bolstered and semi-bolstered resubstitution error estimation in Figure 2 and Figure 3, respectively.

In either case, training is done only if a classifier needs it (e.g., k-nearest neighbors classifier does not need this step if k is pre-set). The entire training set is used in the classifier training in order to have as much information as possible. Once the classifier has been trained, artificial test instances are created for each training instance and the contribution of their misclassification to the total error rate is accumulated. After this process finished, the number of accumulated errors needs to be normalized by the product of the number of training instances and the number of Monte-Carlo samples per each training instance. This is because the overall number of test instances is equal to the number of training instances times the number of samples generated per training instance.

Small changes to pseudo code in Figure 2 are necessary if one wants to estimate the semi-bolstered resubstitution error rate. In this case, before deciding whether to apply bolstering or not, one needs to test each training instance with the trained classifier. If there is a misclassification error, then no bolstering is used and it is assumed that one incurred M errors as if all M artificial test instances were misclassified. This is done in order to keep the total error from becoming over-optimistic.

Figure 2. Pseudo code of the bolstered resubstitution error estimation

Bolstered resubstitution error estimation

Let N and M be the number of training instances and the number of Monte-Carlo samples, respectively;
Set the total bolstered resubstitute error, *err*, to zero;

Train a classifier, if necessary, on the entire training data;

For each training instance x_i do
    Generate M test points;
    Estimate the test error rate, e_i, of the trained classifier on them;
    Accumulate e_i into the total error count: $err = err + e_i$;
End

Total bolstered resubstitution error is computed as $err / (MN)$.

Figure 3. Pseudo code of the semi-bolstered resubstitution error estimation

Semi-bolstered resubstitution error estimation

Let N and M be the number of training instances and the number of Monte-Carlo samples, respectively;
Set the total bolstered resubstitute error, *err*, to zero;

Train a classifier, if necessary, on the entire training data;

For each training instance x_i do
    Test the trained classifier on x_i;
    If x_i is not misclassified, then
        Generate M test points;
        Estimate the test error rate, e_i, of the trained classifier on them;
        Accumulate e_i into the total error count: $err = err + e_i$;
    Else /* no bolstering */
        Increase the total error count by M: $err = err + M$;
    End
End

If no misclassification of a given training instance occurs, we follow the same steps as in the case of bolstered resubstitution error estimation.

Another Algorithm for Generation of Extra Data

An interesting complementary method for generation of artificial samples for small-sample size problems is proposed in (Li, Fang, Lai, & Hu, 2009). I found that readers could benefit from knowing its principles as it has similar roots in random sample generation as the method of Braga-Neto and Dougherty. Since I chose not to encode it, only general details are outlined below.

The method of Li et al. treats each gene independently of the others. It consists of three following steps done for each gene separately from the others:

- Gene selection by making use of a simple statistic, e.g., t-statistic.
- Instance clustering into k groups in 1D space of gene expression levels. Each group contains instances, all belonging to one of the classes. The mean and standard deviation for each group are computed, too.
- Generation of virtual samples for each group based on the basic Box-Müller transform and the mean/standard deviation estimates found in the previous step. The number of virtual samples is inversely proportional to the number of real samples in a group.

Li et al. utilized extra samples for augmenting the training set. However, nothing precludes collecting virtual instances for a separate test set. So, the method of Li et al. can also be used to assess test error in parallel with the method of Braga-Neto and Dougherty. The only drawback of Li's method may be the treatment of genes independently of each other. It seems that it may be the reason why increasing the training set size did not always lead to a better classification accuracy in experiments with several gene expression data sets, carried out in (Li, Fang, Lai, & Hu, 2009). Moreover, experimental results of Li et al. show that the classification accuracy first increased and then decreased as the number of virtual samples grew. It may be that gene independence is not a problem for small data sets. However, as the number of instances rises, the independence assumption is violated, thus leading to poorer performance of a classifier. These are, however, only hypotheses that require theoretical and experimental validation, which is beyond the scope of this book.

The more information about virtual sample generation by Li et al. can be obtained from (Li & Lin, Using virtual sample generation to build up management knowledge in the early manufacturing stages, 2006), (Li, Hsu, Tsai, Lu, & Hu, 2007), (Li, Yeh, Tsai, Fang, & Hu, 2007).

MATLAB Code

I tied the implementation of bolstered and semi-bolstered resubstitution error estimators with particular classifiers such as *k*-nearest neighbors and support vector machine. Therefore the names of the functions contain the reference to this fact. For example, MATLAB® function *bolsterederror_Neareset_Neighbor* implements the bolstered resubstitution error estimation for *k*-nearest neighbors, semi-bolstered resubstitution error estimation is encoded in *semibolsterederror_Nearest_Neighbor*, while *bolsterederror_SVM* is given as an example of the bolstered resubstitution error estimation when a classifier requires training.

The Marsaglia polar sampling is implemented as a C routine *marsagliapolar* that is invoked from MATLAB® code. Although the number of Monte-Carlo samples per point is rather small, I decided to deviate from my preference and to implement the sampler in C. As a by-product, the C programming style for MATLAB® is provided as an example.

In *marsagliapolar*, *rand*() is the function for $[0, RAND\_MAX]$ uniform random number generation (hence, division of the result returned by *rand*() by $RAND\_MAX$ in order to obtain numbers within the interval $[0,1]$), *p* is the data dimensionality, *n* is equal to *M* (the number of Monte-Carlo samples), log() is the function computing the natural logarithm of its argument, *m* and *s* are the mean and the estimate of the standard deviation, respectively. The mean is given by coordinates of the current training point x_i while the estimate of the standard deviation is the σ_i, computed in the calling MATLAB® function. After each loop cycle over *i* two samples are generated and stored in matrix *X* so that after $n/2$ cycles we have $2M/2 = M$ samples.

Please notice that although *marsagliapolar* is called from MATLAB® code, the actual interface between MATLAB® and C is based on using a special function – *mexFunction*. More information about this interface can be found in MATLAB® documentation and a pocket-size book (Davis & Sigmon, 2005).

```
function [err, targets, probabilities] = bolsterederror_
NearestNeighbor(train_patterns, train_targets, params, nsam-
ples)

% Bolstered resubstitution error estimation for k-Nearest Neighbors
% Original source:
% Ulisses Braga-Neto, Edward R. Dougherty. Bolstered error estimation.
% Pattern Recognition 37(6), 1267-1281, 2004.
%
```

```
% Inputs:
%    train_patterns      - Train patterns
%    train_targets       - Train targets
%    params              - Parameters of k-Nearest Neighbors
%    nsamples            - The number of times to sample the
%                           multivariate normal distribution
%
% Output:
%
%    e                   - Bolstered error estimate
%    targets             - Predictions
%    probabilities       - Class probability estimates for predictions


targets = [];
probabilities = [];

% Get the number of training patterns
L           = length(train_targets);
% Get the number of classes
u           = unique(train_targets);
c           = length(u);

if mod(nsamples,2)
    warning('The number of times to sample the multivariate
normal distribution must be even!');
end

% Find indices of patterns belonging to each class
for i = 1:c
    ind{i} = find(train_targets == u(i));
end

% Get data dimensionality
D           = size(train_patterns, 1);

% Get the correction factor (see pp.1272-1273 of the original paper)
% for computing the estimates of the standard deviation
cp = sqrt(chi2inv(0.5,D));
```

```
% Set bolstered error to zero
err = 0;


% Compute the mean minimum distance between patterns belonging to each
% class. After that, compute the estimates of the standard deviation.
% Finally, sample points from the multivariate normal distribution and
% test a classifier on the sampled points
for i = 1:c
    % Get patterns of class i
    X = train_patterns(:,ind{i});
    % Get their number
    n = length(ind{i});
    % Compute a distance matrix
    d2 = sum(X.^2,1);
    distance = repmat(d2,n,1) + repmat(d2',1,n) - 2*X'*X;
    % Sort distances along columns of the distance matrix in the
    % ascending order of magnitude
    distance = sort(distance,1,'ascend');
    % Find the mean among all minimum distances ignoring the first row
    % consisting of all zeroes (distance of a pattern to itself)
    dhat(i) = mean(sqrt(distance(2,:)));
    % Compute the estimate of the standard deviation
    s = dhat(i)/cp;
    % Make a D-dimensional vector
    sig{i} = repmat(s,D,1);
    % Now test a classifier on the points sampled from the multivariate
    % normal distribution
    for j = 1:n
        fprintf(1,'Point %d of class %d is being classified...\
n',j,i);
        for k = 1:nsamples/2% sample nsamples/2 times (each
time two D-dimensional points)
            % Mean = X(:,j), i.e., the ith point itself
            % Sigma = sig{i}
            Y = marsagliapolar(X(:,j),sig{i},2,D); % Y is of Dx2
            % Test a classifier
            [test_targets, prob] = NearestNeighbor(train_pat-
terns, train_targets, Y, params);
            % Compute test error and increase the bolstered
error
```

```
            % counter if necessary
            err = err + sum(test_targets ~= repmat(train_
targets(ind{i}(j)),1,2));
            if nargout > 1
                % Store predicted labels for the pattern with
index in
                % ind{i}(j)
                targets(2*k-1:2*k,ind{i}(j)) = test_targets;
            end
            if nargout > 2
                % Store class probabilities for the pattern
with index
                % in ind{i}(j)
                % 'prob' is a 2x2 matrix
                probabilities(4*k-3:4*k,ind{i}(j)) = prob(:)';
            end
        end
    end
end

% Compute bolstered error
err = err/(nsamples*L);

return
function err = semibolsterederror_NearestNeighbor(train_pat-
terns, train_targets, params, nsamples)

% Semi-bolstered resubstitution error estimation for k-Nearest
% Neighbors
% Original source:
% Ulisses Braga-Neto, Edward R. Dougherty. Bolstered error
estimation.
% Pattern Recognition 37(6), 1267-1281, 2004.
%
% Inputs:
%   train_patterns      - Train patterns
%   train_targets       - Train targets
%   params              - Parameters of k-Nearest Neighbors
%   nsamples            - The number of times to sample the
%                           multivariate normal distribution
```

```
%
% Output:
%
%    e                      - Bolstered error estimate


% Get the number of training patterns
L           = length(train_targets);
% Get the number of classes
u           = unique(train_targets);
c           = length(u);

if mod(nsamples,2)
    warning('The number of times to sample the multivariate
normal distribution must be even!');
end

% Find indices of patterns belonging to each class
for i = 1:c
    ind{i} = find(train_targets == u(i));
end

% Get data dimensionality
D           = size(train_patterns, 1);

% Get the correction factor (see pp.1272-1273 of the original
paper)
% for computing the estimates of the standard deviation
cp = sqrt(chi2inv(0.5,D));

% Set bolstered error to zero
err = 0;

% Compute the mean minimum distance between patterns belonging to each
% class. After that, compute the estimates of the standard deviation.
% Finally, sample points from the multivariate normal distribution and
% test a classifier on the sampled points
for i = 1:c
    % Get patterns of class i
    X = train_patterns(:,ind{i});
```

```
% Get their number
n = length(ind{i});
% Compute a distance matrix
d2 = sum(X.^2,1);
distance = repmat(d2,n,1) + repmat(d2',1,n) - 2*X'*X;
% Sort distances along columns of the distance matrix in the
% ascending order of magnitude
distance = sort(distance,1,'ascend');
% Find the mean among all minimum distances ignoring the first row
% consisting of all zeroes (distance of a pattern to itself)
dhat(i) = mean(sqrt(distance(2,:)));
% Compute the estimate of the standard deviation
s = dhat(i)/cp;
% Make a D-dimensional vector
sig{i} = repmat(s,D,1);
% Now test a classifier on the points sampled from the multivariate
% normal distribution
for j = 1:n
    fprintf(1,'Point %d of class %d is being classified...\
n',j,i);
    % Test a classifier on the jth original point
    test_target = NearestNeighbor(train_patterns, train_
targets, X(:,j), params);
    if test_target == train_targets(ind{i}(j))
        % Sample points only if classification is correct
    for k = 1:nsamples/2% sample nsamples/2 times (each
time two D-dimensional points)
        % Mean = X(:,j), i.e., the ith point itself
        % Sigma = sig{i}
        Y = marsagliapolar(X(:,j),sig{i},2,D); % Y is of Dx2
        % Test a classifier
        test_targets = NearestNeighbor(train_patterns,
train_targets, Y, params);
        % Compute test error and increase the bolstered
error
        % counter if necessary
        err = err + sum(test_targets ~= repmat(train_
targets(ind{i}(j)),1,2));
    end
    else
```

```
            % if classification is incorrect, increase the
error
% counter accordingly in order to avoid the over-optimistic  %
estimate
            err = err + nsamples;
        end
    end
end


% Compute bolstered error
err = err/(nsamples*L);


return
function err = bolsterederror_SVM(train_patterns, train_tar-
gets, params, nsamples)


% Bolstered resubstitution error estimation for Support Vector Machine
% Original source:
% Ulisses Braga-Neto, Edward R. Dougherty. Bolstered error estimation.
% Pattern Recognition 37(6), 1267-1281, 2004.
%
% Inputs:
%    train_patterns      - Train patterns
%    train_targets       - Train targets
%    params              - Parameters of Support Vector Machine
%    nsamples            - The number of times to sample the
%                            multivariate normal distribution
%
% Output:
%
%    e                   - Bolstered error estimate



% Get the number of training patterns
L          = length(train_targets);
% Get the number of classes
u          = unique(train_targets);
c          = length(u);
```

```
if mod(nsamples,2)
    warning('The number of times to sample the multivariate
normal distribution must be even!');
end

% Find indices of patterns belonging to each class
for i = 1:c
    ind{i} = find(train_targets == u(i));
end

% Get data dimensionality
D           = size(train_patterns, 1);

% Train the Support Vector Machine
% X is of NxD, Y is of Nx1
[sv, ind_sv, alpha_star, b_star] = train_SVC(train_patterns,
train_targets, params);

% Get the correction factor (see pp.1272-1273 of the original paper)
% for computing the estimates of the standard deviation
cp = sqrt(chi2inv(0.5,D));

% Set bolstered error to zero
err = 0;

% Compute the mean minimum distance between patterns belonging to each
% class. After that, compute the estimates of the standard deviation.
% Finally, sample points from the multivariate normal distribution and
% test a classifier on the sampled points
for i = 1:c
    % Get patterns of class i
    X = train_patterns(:,ind{i});
    % Get their number
    n = length(ind{i});
    % Compute a distance matrix
    d2 = sum(X.^2,1);
    distance = repmat(d2,n,1) + repmat(d2',1,n) - 2*X'*X;
    % Sort distances along columns of the distance matrix in the
    % ascending order of magnitude
    distance = sort(distance,1,'ascend');
```

```
    % Find the mean among all minimum distances ignoring the first row
    % consisting of all zeroes (distance of a pattern to itself)
    dhat(i) = mean(sqrt(distance(2,:)));
    % Compute the estimate of the standard deviation
    s = dhat(i)/cp;
    % Make a D-dimensional vector
    sig{i} = repmat(s,D,1);
    % Now test a classifier on the points sampled from the multivariate
    % normal distribution
    for j = 1:n
        fprintf(1,'Point %d of class %d is being classified...\
n',j,i);
        for k = 1:nsamples/2% sample nsamples/2 times (each
time two D-dimensional points)
            % Mean = X(:,j), i.e., the ith point itself
            % Sigma = sig{i}
            Y = marsagliapolar(X(:,j),sig{i},2,D); % Y is of Dx2
            % Test a classifier
            test_targets = test_SVC(train_patterns, train_tar-
gets, Y, ind_sv, alpha_star, b_star, params);
            % Compute test error and increase the bolstered error
            % counter if necessary
            err = err + sum(test_targets ~= repmat(train_
targets(ind{i}(j)),1,2));
        end
    end
end

% Compute bolstered error
err = err/(nsamples*L);

return
#include "mex.h"
#include "matrix.h"
#include <math.h>
#include <stdlib.h>
#include <stdio.h>
#include <float.h>
#define INDEX(i,j,d) ((i) + (j)*(d)) /* d - # rows*/
```

```
/* Marsaglia polar method for normal random number generation
 * Inputs:
 * ****************************************************************
 * m - mean of the normal distribution (1xp vector)
 * s - standard deviation of the normal distribution (1xp vector)
 * n - # points to sample
 * p - data dimensionality
 *
 * Output:
 * ****************************************************************
 * X - n p-dimensional points assembled in a pxn matrix
 */


/* Prototypes of functions */
void marsagliapolar(double *X, double *m, double *s, int n, int
p);


void mexFunction(int nlhs, mxArray *plhs[], int nrhs, const
mxArray *prhs[])
{
    double *m, *s, *X;
    int n, p;


    /* Get inputs... */

    /* Check the number of input arguments */
    if (nrhs != 4)
    {
        mexErrMsgTxt("Wrong number of input arguments!");
    }

    /* Check the number of output arguments */
    if (nlhs != 1)
    {
        mexErrMsgTxt("Wrong number of output arguments!");
    }

    /* Check if the first argument is a real valued vector */
```

```
    if (!mxIsDouble(prhs[0]) || mxIsSparse(prhs[0]) ||
mxIsEmpty(prhs[0]) || mxIsComplex(prhs[0]))
    {
        mexErrMsgTxt("First argument must be a real valued
vector!");
    }
    /* Get a pointer to the real values of a vector */
    m = mxGetPr(prhs[0]);

    /* Check if the second argument is a real valued vector */
    if (!mxIsDouble(prhs[1]) || mxIsSparse(prhs[1]) ||
mxIsEmpty(prhs[1]) || mxIsComplex(prhs[1]))
    {
        mexErrMsgTxt("Second argument must be a real valued
vector!");
    }
    /* Get a pointer to the real values of a vector */
    s = mxGetPr(prhs[1]);

    /* First and second arguments must have the same number of
columns */
    if (mxGetN(prhs[0]) != mxGetN(prhs[1]))
    {
        mexErrMsgTxt("First and second arguments must have the
same number of rows!");
    }

    /* Check if the third argument is a scalar */
    if (!mxIsScalar(prhs[2]) || !mxIsDouble(prhs[2]) ||
mxIsEmpty(prhs[2]) || mxIsComplex(prhs[2]))
    {
        mexErrMsgTxt("Third argument must be a real valued
scalar!");
    }
    /* Get the number of points to sample */
    n = mxGetScalar(prhs[2]);
    /* Check if n is even */
    if (fmod(n,2) != 0)
    {
        mexErrMsgTxt("The number of points to sample must be
```

```
even!");
    }


    /* Check if the fourth argument is a scalar */
    if (!mxIsScalar(prhs[3]) || !mxIsDouble(prhs[3]) ||
mxIsEmpty(prhs[3]) || mxIsComplex(prhs[3]))
    {
        mexErrMsgTxt("Fourth argument must be a real valued
scalar!");
    }
    /* Get data dimensionality */
    p = mxGetScalar(prhs[3]);


    /* Create output... */


    plhs[0] = mxCreateDoubleMatrix(p, n, mxREAL);
    /* Get a pointer to the created vector */
    X = mxGetPr(plhs[0]);


    /* Sample n points */
    marsagliapolar(X, m, s, n, p);
}


void marsagliapolar(double *X, double *m, double *s, int n, int
p)
{
    int i, j;
    double u1, u2, r;

    for (i = 0; i < n/2; i++)
    {
        /* Generate normal random numbers */
        for (j = 0; j < p; j++)
        {
            do
            {
                u1 = 2.0*rand()/RAND_MAX - 1;
                u2 = 2.0*rand()/RAND_MAX - 1;
                r = u1*u1 + u2*u2;
```

```
        } while(r == 0 || r >= 1);
        r = sqrt(-2*log(r)/r);
        X[INDEX(j,i,p)] = m[j] + s[j]*r*u1;
        X[INDEX(j,i+n/2,p)] = m[j] - s[j]*r*u2;
    }
  }
}
```

REFERENCES

Box, G. E., & Müller, M. E. (1958). A note on the generation of random normal deviates. *Annals of Mathematical Statistics, 29*(2), 610–611. doi:10.1214/aoms/1177706645

Braga-Neto, U. M. (2007). Fads and fallacies in the name of small-sample microarray classification - A highlight of misunderstanding and erroneous usage in the applications of genomic signal processing. *IEEE Signal Processing Magazine, 24*(1), 91–99. doi:10.1109/MSP.2007.273062

Braga-Neto, U. M., & Dougherty, E. R. (2004). Bolstered error estimation. *Pattern Recognition, 36*(7), 1267–1281. doi:10.1016/j.patcog.2003.08.017

Braga-Neto, U. M., & Dougherty, E. R. (2004). Is cross-validation valid for small-sample microarray classification? *Bioinformatics (Oxford, England), 20*(3), 374–380. doi:10.1093/bioinformatics/btg419

Davis, T. A., & Sigmon, K. (2005). *MATLAB® Primer* (7th ed.). Boca Raton, FL: Chapman & Hall/CRC Press.

Dudoit, S., & Fridlyand, J. (2003). Classification in microarray experiments. In Speed, T. (Ed.), *Statistical analysis of gene expression microarray data* (pp. 93–158). Boca Raton, FL: Chapman & Hall/CRC Press. doi:10.1201/9780203011232.ch3

Everitt, B. (2006). *The Cambridge dictionary of statistics* (3rd ed.). Cambridge, UK: Cambridge University Press.

Glick, N. (1978). Additive estimators for probabilities of correct classification. *Pattern Recognition, 10*(3), 211–222. doi:10.1016/0031-3203(78)90029-8

Li, D.-C., Fang, Y.-H., Lai, Y.-Y., & Hu, S. C. (2009). Utilization of virtual samples to facilitate cancer identification for DNA microarray data in the early stages of an investigation. *Information Sciences, 179*(16), 2740–2753. doi:10.1016/j.ins.2009.04.003

Li, D.-C., Hsu, H.-C., Tsai, T.-I., Lu, T.-J., & Hu, S. C. (2007). A new method to help diagnose cancers for small sample size. *Expert Systems with Applications, 33*(2), 420–424. doi:10.1016/j.eswa.2006.05.028

Li, D.-C., & Lin, Y.-S. (2006). Using virtual sample generation to build up management knowledge in the early manufacturing stages. *European Journal of Operational Research, 175*(1), 413–434. doi:10.1016/j.ejor.2005.05.005

Li, D.-C., Yeh, C.-W., Tsai, T.-I., Fang, Y.-H., & Hu, S. C. (2007). Acquiring knowledge with limited experience. *Expert Systems: International Journal of Knowledge Engineering and Neural Networks, 24*(3), 162–170. doi:10.1111/j.1468-0394.2007.00427.x

ENDNOTES

[1]   For example, cross-validation error estimation is known to have high variance (Braga-Neto & Dougherty, Is cross-validation valid for small-sample microarray classification?, 2004).

[2]   A cumulative distribution function is a distribution showing how many values of a random variable are less than or more than given values (Everitt, 2006).

[3]   Distance between D dimensional vectors is the sum of D squared differences, where each difference is independent of the others. As will be explained in the next section by the nature of a sample generation process, each difference represents a standard normal variable, i.e. the one with zero mean and unit variance. A chi-square distribution is the probability distribution of a random variable defined as the sum of squares of a number (D) of independent standard normal variables (Everitt, 2006).

[4]   See, e.g., http://en.wikipedia.org/wiki/Box%E2%80%93Muller_transform.

Chapter 24
Performance Evaluation:
Final Check

BAYESIAN CONFIDENCE (CREDIBLE) INTERVAL

In this chapter, we address the important question frequently asked about cross-validation and bootstrap in relation to small sample classification. Are classification error estimates attained by using these techniques statistically reliable? The recent article (Isaksson, Wallman, Göransson, & Gustafsson, 2008) drew attention to this question with the striking answer "no". The value of this work is, however, not in pure critics; it also provides an alternative – Bayesian confidence (or credible) interval.

A reader who is not familiar with statistics (especially, Bayesian statistics) or who did not apply it for a long time may be puzzled: what did you mean? Those who understand the term might ask: how much of our valuable time are you going to spend on performance evaluation issues? My answer is affirmative 'very little'.

So let us begin with the definition of the Bayesian confidence interval used in statistics. As usually, I do not want to invent my own definition but instead I rely on (Everitt, 2006) as a trustworthy source of definitions. A Bayesian confidence interval is an interval of a posterior[1] distribution which is such that the density at any point inside the interval is greater than the density at any point outside and that the area under the curve for that interval is equal to a pre-specified probability level. It is calculated in essentially the same way as the frequentist confidence in-

DOI: 10.4018/978-1-60960-557-5.ch024

terval. However, in contrast to the frequentist confidence interval, which covers the true value of the parameter of interest in $100(1-\alpha)\%$ (typical value of α - significance level[2] - is 0.05) of trials on average[3], its Bayesian analogue specifies the range within which the parameter of interest lies with a certain probability (Gill, 2008), (Ramoni & Sebastiani, 2003). That is, a $100(1-\alpha)\%$ Bayesian confidence interval gives the region of the parameter space where the probability of covering the parameter of interest is equal to $1-\alpha$. If one applies this definition to the frequentist confidence interval, this would mean that the probability of coverage is either zero or one, because the interval either covers the true value of the parameter of interest or it does not in this case.

Formally, the Bayesian confidence (credible) interval is defined as follows (Gill, 2008), pp.45-46. Let C be a subset of the parameter space, Θ, such that a $100(1-\alpha)$ credible interval satisfies to the condition:

$$1 - \alpha = \int_C p(\theta|D)\,d\theta,$$

where $p(\theta|D)$ defines the posterior distribution of the unknown θ, given the observed data D. Notice that credible intervals are in general not unique; one can define C in several different ways, in each of which the equality above will still hold. It is often by convention that these intervals are centered at a mean or mode of a distribution. Differences from the given definition arises in asymmetric and multimodal distributions, and the convention here is to have so called equal tail intervals, meaning that regardless of the shape of the posterior distribution $p(\theta|D)$, the $100(1-\alpha)$ credible interval is built in such a way that $(1-\alpha)/2$ of the density is put in both the left and right tails outside of the credible interval.

The end points of the equal tail credible interval can be found by solving for the limits L and H in two integrals (Gill, 2008), p.47:

$$\frac{\alpha}{2} = \int_0^L p(\theta|D)\,d\theta,$$

$$\frac{\alpha}{2} = \int_H^\infty p(\theta|D)\,d\theta.$$

In practice, integration is replaced by summation. This is the definition we will utilize and implement in MATLAB®. But we need a formula for the posterior distribution, and it will be presented in the next section.

Why Cross-Validation and Bootstrap are Unreliable for Classification Error Estimation in Small Sample Size Problems?

Now we would like to provide the reasoning and findings of Isaksson et al. (Isaksson, Wallman, Göransson, & Gustafsson, 2008) concerning cross-validation and bootstrap for classification error estimation in small sample size problems, which is our case, too.

Cross-validation (CV) and bootstrapping (BT) were and still largely remain the main methods to assess classifier performance in terms of the error rate, especially if sample size is small and the independent test set is unavailable. However, when being applied to small sample problems, the uncertainty in error rate estimates obtained with these methods is unknown and therefore it could be quite large to be ignored.

The major conclusion and advice of Isaksson et al. are that *the final classification performance should be always reported in the form of a Bayesian confidence (credible) interval obtained from a simple holdout test*. Now we will proceed along with Isaksson et al. from the beginning till the end.

CV and BT are commonly considered as estimators of the true error rate e_{true}, i.e. the error rate of a trained classifier, provided that an independent test set is very large. That is, the true error rate is the probability of misclassification when applying the classifier to an unseen instance[4]. Isaksson et al. remarked that *a single point estimate, e_{est}, of a true performance, e_{true}, is not useful unless it is known to be close to e_{true}*. However, the problem is that in real problems, the value of e_{true} is unknown. Simulations done in (Isaksson, Wallman, Göransson, & Gustafsson, 2008) demonstrated that unless the data set size is at least as large as 1000, CV and BT estimates are poor predictors of the true error rate. Similar conclusions were drawn in (Braga-Neto & Dougherty, 2004), (Xu, Hua, Braga-Neto, Xiong, Suh, & Dougherty, 2006).

Since CV and BT estimates can be treated as unreliable, one needs to look for alternatives. One of the best solutions according to Isaksson et al. is the Bayesian confidence (credible) interval. The Bayesian posterior $p\left(e_{true}|k, N\right)$ defines the uncertainty about the true error rate after observing k errors in N tests. By means of Bayes' rule it is determined as

$$p\left(e_{true}|k,N\right) = \frac{p\left(k|e_{true},N\right)p\left(e_{true}|N\right)}{p\left(k|N\right)} = \frac{p\left(k|e_{true},N\right)p\left(e_{true}\right)}{p\left(k|N\right)},$$

where $p\left(e_{true}|N\right) = p\left(e_{true}\right)$, since knowing N does not bring new information about e_{true}, i.e. in other words, N and e_{true} are conditionally independent. The conditional probability $p\left(k|e_{true},N\right)$ comes from the binomial distribution, i.e. the probability of observing k errors in N independent trials when the probability of making an error in each trial is e_{true}. The factor $p\left(k|N\right)$ is the normalization constant in order to have the posterior between 0 and 1 as dictated by the probability laws. Thus, the posterior $p\left(e_{true}|k,N\right)$ describes our uncertainty about the true error rate after making k errors on a data set with N test examples. This posterior uncertainty is not conditioned on a particular data set, i.e. it is rather general. The only condition is k errors made on N test examples.

A Bayesian confidence (credible) interval is similar to the frequentist confidence interval. A 95% Bayesian confidence interval covers 95% of the area under the posterior $p\left(e_{true}|k,N\right)$. For the small sample size problems, the posterior and the associated confidence interval are broad, unless the prior $p\left(e_{true}\right)$ is narrow. That is, we rely on the prior knowledge of the probability of misclassification. When this probability is completely unknown, except that it is limited to the unit interval [0,1], the uniform distribution, $p\left(e_{true}\right) = 1$, is typically used. As a result, the posterior can be rewritten (by keeping in mind the formula for the binomial distribution) as

$$p\left(e_{true}|k,N\right) = \beta e_{true}^k \left(1 - e_{true}\right)^{N-k},$$

where β is the normalization constant. Examples of this posterior for different values of k and N are given in Figure 1 when e_{true} varied from 0 to 1 through 0.001 steps. One can see that though the ratio $k\,/\,N$ remains fixed for all three cases, the posterior shapes differ a lot. For smallest sample size, the posterior is rather wide. However, as sample size grows, the posterior becomes taller and narrower, which implies that our uncertainty about the true error rate decreases. The values for the Bayesian confidence interval bounds will be given in the next section.

MATLAB Code

The function *BayesianConfidenceInterval* plots the posterior for given k, N and the significance level α (default value for α is 0.05) and calculates the lower and upper bounds for the credible interval.

Figure 1. Three different Bayesian posterior distributions

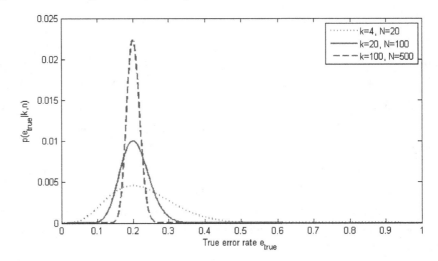

Table 1. Bounds for a Bayesian confidence interval for different values of k and N

	Lower bound	Upper bound
$k = 4, N = 20$	8.3%	42%
$k = 20, N = 100$	13.5%	29%
$k = 100, N = 500$	16.8%	23%

The true error rate runs from 0 (minimum) to 1 (maximum) through 0.001 steps. Hence, the posterior is a vector of values rather than one value. The normalization constant β is just a sum of all elements of this vector. You can verify that after normalization, each element of the posterior vector belongs to $[0,1]$. In order to find lower and upper bounds for the credible interval, the MATLAB® function *cumsum* was used, which returns the cumulative sum. This sum is the area under the posterior curve between certain limits. The 95% Bayesian confidence intervals for three pairs of k and N are presented in Table 1. As sample size N grows, the difference between the upper and lower bound (and the uncertainty about the true error rate) decreases.

```
function [el, eu]= BayesianConfidenceInterval(k, n, alpha)


% Bayesian confidence (credible) interval computation
% Original sources:
% A. Isaksson, M. Wallman, H. Göransson, M.G. Gustafsson, "Cross-
% validation and bootstrapping are unreliable in small sample
% classification", Pattern Recognition Letters 29(14), pp. 1960-1965,
% 2008
%
% Jeff Gill, "Bayesian methods: a social and behavioral sciences
% approach", in Statistics in the Social and Behavioral Sciences
% Series, Chapman & Hall/CRC, Boca Raton, FL, 2008
%
% Inputs:
%   k          - The number of errors made in n tests
%   n          - The number of tests
%   alpha      - Confidence level
%
% Outputs:
%   el         - Lower boundary of the Bayesian confidence (credible)
%                  interval
%   eu         - Upper boundary of the Bayesian confidence (credible)
%                  interval


% Default value
if nargin < 3
    alpha = 0.05;
end


% True classification errors
e = 0:0.001:1;


% Bayesian posterior p(e|k,n) according to Eq.2 on page 1965 of the
% Pattern Recognition Letters article
% Bayesian posterior defines uncertainty about the true error rate
% after observing k errors in n tests
p = e.^k.*(1-e).^(n-k);
p = p/sum(p);
% The total area under p is hence equal to 1 due to normalization
```

```
% Plot the posterior
plot(e, p, 'LineWidth', 2);
xlabel(texlabel('True error rate e_true'));
ylabel(texlabel('p(e_true|k,n)'));

% Bayesian confidence interval at (1-alpha)-level covers 100*(1-alpha)
% per cent of the area under the posterior p(e|k,n)

% We rely on the rule that no matter what the shape of the posterior
% distribution the 100(1-alpha) per cent confidence (credible) interval
% is created such that (1-alpha)/2 of the density is put in both the
% left and right tails outside of the confidence (credible) interval.
% Hence, the interval is called the equal-tail interval
% The actual calculation is done according to Eq. 2.10 on page 47 of
% Gill's book
i = find(cumsum(p) >= alpha/2);
% Lower boundary (in per cent) of the confidence (credible) interval
el = (i(1)/(length(e)-1))*100;
% Upper boundary (in per cent) of the confidence (credible) interval
i = find(cumsum(p) >= 1-alpha/2); % alpha/2+(1-alpha)
eu = (i(1)/(length(e)-1))*100;

return
```

REFERENCES

Braga-Neto, U. M., & Dougherty, E. R. (2004). Is cross-validation valid for small-sample microarray classification? *Bioinformatics (Oxford, England)*, *20*(3), 374–380. doi:10.1093/bioinformatics/btg419

Everitt, B. (2006). *The Cambridge dictionary of statistics* (3rd ed.). Cambridge, UK: Cambridge University Press.

Gill, J. (2008). *Bayesian methods: a social and behavioral sciences approach* (2nd ed.). Boca Raton, FL: Chapman & Hall/CRC.

Isaksson, A., Wallman, M., Göransson, H., & Gustafsson, M. (2008). Cross-validation and bootstrapping are unreliable in small sample classification. *Pattern Recognition Letters*, *29*(14), 1960–1965. doi:10.1016/j.patrec.2008.06.018

Ramoni, M., & Sebastiani, P. (2003). Bayesian methods. In Berthold, M., & Hand, D. J. (Eds.), *Intelligent data analysis: an introduction* (2nd ed., pp. 131–168). Berlin, Heidelberg: Springer-Verlag. doi:10.1007/978-3-540-48625-1_4

Xu, Q., Hua, J., Braga-Neto, U., Xiong, Z., Suh, E., & Dougherty, E. R. (2006). Confidence intervals for the true classification error conditioned on the estimated error. *Technology in Cancer Research & Treatment, 5*(6), 579–590.

ENDNOTES

[1]   A posterior distribution or simply a posterior is a probability distribution that summarizes information about a random variable or parameter after, or a posteriori to, having obtained new information from empirical data (Everitt, 2006).

[2]   It is the level of probability at which it is agreed that the null ('no difference') hypothesis will be rejected (Everitt, 2006).

[3]   This implies that, for instance, a 95% confidence interval covers the true value of the parameter in nineteen out of twenty trials on average.

[4]   One needs to remember that the training set size in CV is smaller than the total number of instances in a dataset, which introduces a bias in the performance estimates. However, since this bias is not large, it can be safely ignored.

Chapter 25
Application Examples

JOINING ALL PIECES TOGETHER

The purpose of this chapter is to provide several examples of MATLAB® code that would demonstrate how to combine all pieces (feature selection, classification, ensembles, and performance evaluation) together. The first three examples show feature selection followed by classification and performance evaluation, while the last example is intended to demonstrate the superiority of ensembles over individual classifiers.

All examples utilized the colon cancer data set (Alon, et al., 1999). The original data underwent preprocessing customary to this data set. This data set[1] contains expression of 2,000 genes for 62 cases (22 normal and 40 colon tumor). Preprocessing includes the logarithmic transformation to base 10, followed by normalization to zero mean and unit variance. The code implementing normalization is given in functions *preprocess* and *normalize*. If you have already preprocess colon data, these functions are unnecessary to run.

The only thing to remember is the fact that all functions expect data matrices to be $D \times N$, where D is the number of features (genes) and N is the number of data instances, while vectors of class labels need to be $1 \times N$. If your matrices/vectors do not meet these requirements, then transposition will do the job for you. The code

DOI: 10.4018/978-1-60960-557-5.ch025

of *preprocess* converts class labels to either -1 (colon tumor instance) or +1 (normal instance). A different labeling (e.g., 0/1) can be easily obtained if desired.

As you remember, there are two types of feature selection models: filters and wrappers. In all examples, I chose filter-like models that do not use a classifier to access feature importance. I recommend you to follow this choice as a safeguard against model overfitting. As for classifiers, I prefer those algorithms that do not require training (e.g., *k*-nearest neighbors). Given that we have only a few instances of high-dimensional data, classifier training will put pressure on our scarce data resources, which can easily result in overfitting. Of course, as you saw in one of the chapters, describing the bolstered error estimation, extra data may be artificially and cheaply generated but I would not abuse this option too often, too much, since artificial data are only a temporal solution to substitute real data until the latter become abundant. This way of reasoning does not make wrappers and classifiers requiring training completely useless: given enough data, they might yield better results than filters and classifiers without the need for training.

In all examples that will follow, the original (training) data are used for feature selection. After that, the classification occurs, using only selected genes. The bolstered resubstitution error is an estimate of the classification performance. It is estimated based on artificially generated data that play the role of test data. Since feature selection and classification are carried out on different data sets, there is no danger of severe overfitting, which could result in optimistically biased performance characteristics.

```
function [patterns, targets] = preprocess(filename1, filename2)

% Preprocess colon data for further analysis
% Colon data can be found at
% http://microarray.princeton.edu/oncology/affydata/index.html
% They were used by U. Alon, N. Barkai, D. A. Notterman, K. Gish,
% S. Ybarra, D.Mack, and A. J. Levine, "Broad patterns of gene
% expression revealed by clustering analysis of tumor and normal colon
% tissues probed by oligonucleotide arrays", Proceedings of National
% Academy of Sciences, Vol. 96, pp. 6745-6750, June 1999
%
% Inputs:
%   filename1        - Name of a file with data
%   filename2        - Name of a file with labels
%
% Outputs:
%   patterns         - Data matrix (DxN), where D is the number of
```

```
%                             genes and N is the number of samples
%   targets          - Label vector (1xN), where +1 and -1 stand for
%                             normal and colon samples, respectively


% Read data from a file
patterns = importdata(filename1);

% Check matrix size. It should be 62x2000, where 2000 means the
number % of genes and 62 means the number of samples
if size(patterns,2)~= 2000 || size(patterns,1)~= 62
    error('Wrong data matrix size!');
end

if size(patterns,1) < size(patterns,2) % NxD matrix?
    patterns = patterns';
end

% Check for NaNs
if isnan(isnan(patterns))
    error('There are NaNs!');
end

% Read labels from a file
targets = importdata(filename2);

% Check vector length. It should be 1x62, where 62 means the number of
% samples
if length(targets)~= 62
    error('Wrong label vector length!');
end

% Binarize targets so that normal class corresponds to +1,
while colon
% class corresponds to -1
targets(targets > 0) = +1;
targets(targets < 0) = -1;
```

```
% Check the number of normal and colon tumor samples
% Out of 62 samples, 22 should be normal and 40 should be colon tumors
% Get the number of normal samples
nn = length(find(targets > 0));
if nn ~= 22
    error('The number of normal samples must be 22!');
end
% Get the number of colon tumor samples
nc = length(find(targets < 0));
if nc ~= 40
    error('The number of colon samples must be 40!');
end

% Convert targets to a 1xN vector
if size(targets,1) > 1
    targets = targets';
end

% Check for zeroes
patterns(patterns < eps) = eps;

% Now perform gene normalization
patterns = log10(patterns);
% For each gene, subtract the mean and divide by the standard
% deviation, i.e. each gene has zero mean and unit standard de-
viation
patterns = normalize(patterns);

% Get the number of genes
N = size(patterns,1);
% Check for zero mean and unit standard deviation
for i = 1:N
    fprintf(1,'Gene no.%d...\n',i);
    g = patterns(i,:);
    fprintf(1,'Mean = %e, Std = %f\n', mean(g), std(g));
end

% For this database, the mean of each gene is slightly nonzero!

return
```

```
function X = normalize(X)

% Normalize values in X to zero mean and unit variance
%
% Input:
%   X              - DxN matrix (D - dimensionality, N - #points)
%
% Output:
%   X              - normalized X
%
% Note: each of D dimensions is normalized independently of others!


[D,N] = size(X);

s = std(X,0,2);
s(s < eps) = eps;
sX = (X - repmat(mean(X,2),[1 N]))./repmat(s,[1 N]);

return
```

Example 1

In this example, feature selection is done with *BAHSIC* algorithm (Song, Bedo, Borgwardt, Gretton, & Smola, 2007), (Song, Smola, Gretton, Borgwardt, & Bedo, 2007) while classification results are obtained with 3-nearest neighbors. The number of genes to select is pre-set to 50.

The input parameters of *example*1 include the matrix of training instances, the vector of class labels for these instances, and the number of Monte-Carlo samples per training instance, needed to estimate the bolstered resubstitution error. Therefore the size of test data is equal to the number of Monte-Carlo samples times the number of training instances. The number of Monte-Carlo samples is set to 10.

The output parameters are several performance characteristics such as the bolstered resubstitution error, Bayesian credible interval for this error, area under the ROC curve as well as the indices of selected genes. These indices can be useful in further analysis as the colon data set is accompanied by the detailed description of each gene.

The "positive" and "negative" classes are colon tumor and normal, respectively. Such an association of names for classes and classes themselves is caused by the fact that we are trying to predict the case of a cancer, i.e. diagnosing a patient who has a malignant tumor. If the test outcome is positive, such a patient is declared to have colon cancer; otherwise, he is considered to be healthy (normal).

Since the ROC curve construction and the computation of the area under it assume 0 / 1 class labeling, we first reassign labels to meet this requirement. Let me remind you that the first two parameters of the functions *ROC* and *AUC* mean the true class labels of the test instances and the class probability estimates for the test instances, respectively. As the total number of test instances is a multiple of the number of Monte-Carlo samples, we need to reshape the vector of training labels and re-adjust the numbers of "positive" and "negative" instances used in classification.

Keep in mind that the bolstered error is a number between 0 and 1 while the Bayesian credible interval contains figures in per cent. So, do not become confused if the returned error is 0.086 whereas the Bayesian credible interval is [6.7 11.1]: this does not mean our credible interval "missed" the point estimate. Multiply the error by 100 and you will get the scale of measurement in per cent.

Two figures accompany this example: ROC curve in Figure 1 and the plot of the Bayesian posterior distribution in Figure 2. The ROC curve starts at (0,0) and goes up along the Y-axis till (0,0.62), where it turns and reaches the true positive rate of 1 at the false positive rate somewhere between 0.65 and 0.7. After that, it goes horizontally till (1,1). From Figure 1 it is easy to see that a classifier performed moderately well (AUC = 0.92) as the ROC curve is much above the diagonal line corresponding to a randomly guessing classifier.

From Figure 2 one can conclude that the Bayesian credible interval ([12.5 18.1]) around the bolstered resubstitution error (15.0%) is quite narrow. You can get different figures due to random sampling present in the test data generation.

```
function [err, area, idx, el, eu] = example1(train_patterns,
train_targets, M)

% This example is based on colon data
% Inputs:
%   train_patterns    - Training patterns (DxN matrix, where D - data
%                         dimensionality, N - number of patterns)
%   train_targets     - Training targets (1xN vector)
%   M                 - The number of Monte-Carlo samples
%
% Outputs:
%   err               - Bolstered resubstitution error rate (not in
%                         percent)
```

Figure 1. ROC curve

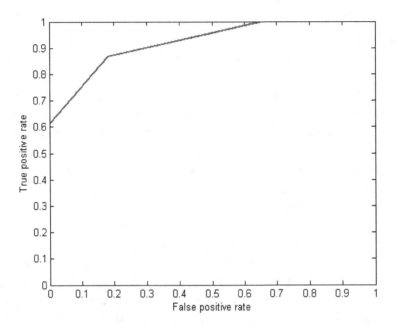

Figure 2. Bayesian posterior distribution

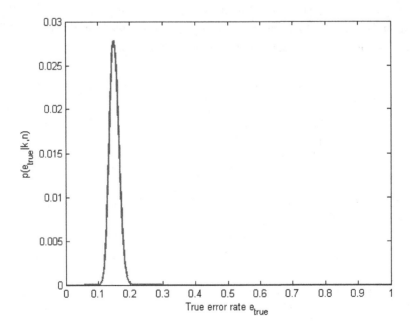

```
%    area              - Area under the ROC curve
%    idx               - Selected features (vector of feature indices)
%    el                - Lower boundary of the Bayesian confidence
%                          (credible) interval containing err
%    eu                - Upper boundary of the Bayesian confidence
%                          (credible) interval containing err


% Reassign class labels so that -1s are replaced with 0s and +1 stay
% unchanged
% These operations are specific to colon data only

% Get unique training labels
Uc = unique(train_targets);
% Find indices of "positive" (colon cancer) instances
i = find(train_targets == Uc(1));
% Get the number of "positive" instances
p = M*length(i);
if Uc(1) ~= 0
    % Replace -1s with 0s
    train_targets(i) = 0;
end
% Find indices of "negative" (normal) instances
i = find(train_targets == Uc(2));
% Get the number of "negative" instances
n = M*length(i);
if Uc(2) ~= 1
    % Replace non-1s with 1s
    train_targets(i) = 1;
end

% Gene selection with BAHSIC
idx = BAHSIC(train_patterns, train_targets, 5, 50, 'rbf', 1,
'linear', []);

% 3-nearest neighbor classification using only selected fea-
tures
[err, targets, prob] = bolstederror_NearestNeighbor(train_
patterns(idx,:), train_targets, 3, M);
```

```
% Extract the probabilities of the "positive" class
% Attention! This operation is specific to the colon data only
% For other datasets, the command could be prob =
prob(2:2:end);
prob = prob(1:2:end,:);

% Convert a matrix into a vector
prob = prob(:)';

% As the number of test patterns is a multiple of M, take into
account % this fact by modifying training labels accordingly
t = repmat(train_targets, M, 1);
t = t(:)';

% Plot the ROC curve
ROC(t, prob, p, n, '-', 'r');
% Compute the area under this curve
area = AUC(t, prob, p, n);

figure;

% Compute Bayesian confidence interval covering the given bolstered
% error
[el, eu]= BayesianConfidenceInterval(floor(err*M*length(train_
targets)), M*length(train_targets));

return
```

Example 2

For this example, the gene selection method is *EVDGeneSelection* (Li, Sun, & Grosse, 2004). The number of genes selected is automatically determined by the method itself. As a classifier, a 3-nearest neighbor was utilized. The number of Monte-Carlo samples remained 10 as in the previous example.

The ROC curve and the Bayesian posterior distribution are shown in Figure 3 and Figure 4, respectively. As you can observe, this ROC curve is located strictly above the curve in Figure 1, meaning that in example 2 the classification performance is better than that in example 1 on the entire range of false positives. The credible interval is accordingly shifted to the left, compared to that for example 1. Except

Figure 3. ROC curve

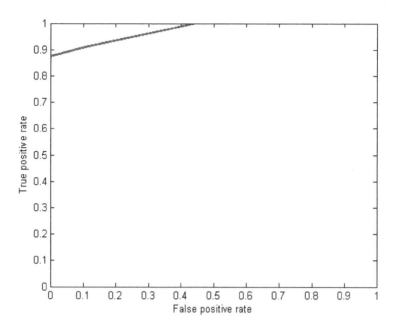

Figure 4. Bayesian posterior distribution

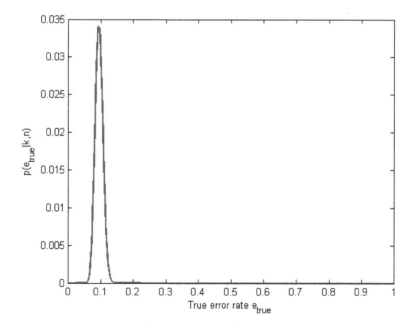

for the change of the gene selection method, the rest of code for this example 2 is the same as it is for the previous example.

The bolstered error rate was 9.5%; the Bayesian credible interval was [7.4 12.0]; the AUC was 0.97; 49 genes were selected.

```
function [err, area, idx, el, eu] = example2(train_patterns,
train_targets, M)

% This example is based on colon data
% Inputs:
%    train_patterns  - Training patterns (DxN matrix, where D - data
%                        dimensionality, N - number of patterns)
%    train_targets   - Training targets (1xN vector)
%    M               - The number of Monte-Carlo samples
%
% Outputs:
%    err             - Bolstered resubstitution error rate (not in
%                        percent)
%    area            - Area under the ROC curve
%    idx             - Selected features (vector of feature indices)
%    el              - Lower boundary of the Bayesian confidence
%                        credible) interval containing err
%    eu              - Upper boundary of the Bayesian confidence
%                        (credible) interval containing err

% Reassign class labels so that -1s are replaced with 0s and +1 stay
% unchanged
% These operations are specific to colon data only

% Get unique training labels
Uc = unique(train_targets);
% Find indices of "positive" (colon cancer) instances
i = find(train_targets == Uc(1));
% Get the number of "positive" instances
p = M*length(i);
if Uc(1) ~= 0
    % Replace -1s with 0s
    train_targets(i) = 0;
end
% Find indices of "negative" (normal) instances
```

```
i = find(train_targets == Uc(2));
% Get the number of "negative" instances
n = M*length(i);
if Uc(2) ~= 1
    % Replace non-1s with 1s
    train_targets(i) = 1;
end

% Gene selection with EVDGeneSelection
idx = EVDGeneSelection(train_patterns, train_targets);

% 3-nearest neighbor classification using only selected fea-
tures
[err, targets, prob] = bolsterederror_NearestNeighbor(train_
patterns(idx,:), train_targets, 3, M);

% Extract the probabilities of the "positive" class
% Attention! This operation is specific to the colon data only
% For other datasets, the command could be prob =
prob(2:2:end);
prob = prob(1:2:end,:);

% Convert a matrix into a vector
prob = prob(:)';

% As the number of test patterns is a multiple of M, take into
account % this fact by modifying training labels accordingly
t = repmat(train_targets, M, 1);
t = t(:)';

% Plot the ROC curve
ROC(t, prob, p, n, '-', 'r');
% Compute the area under this curve
area = AUC(t, prob, p, n);

figure;

% Compute Bayesian confidence interval covering the given bolstered
% error
[el, eu]= BayesianConfidenceInterval(floor(err*M*length(train_
```

```
targets)), M*length(train_targets));

return
```

Example 3

In this example, the gene selection method is *SVDEntropyFeatureSelection* (Varshavsky, Gottlieb, Linial, & Horn, 2006) with simple ranking as a criterion for picking genes. The number of genes selected is automatically determined by the method. The classifier is a 3-nearest neighbor. The number of Monte-Carlo samples was 10. Figure 5 and Figure 6 show the ROC curve and the Bayesian posterior distribution, respectively.

The bolstered error rate was 8.1%; the Bayesian credible interval was [6.3 10.6]; the AUC was 0.97; 316 genes were selected. That is, classification performance turned out to be similar to that in the previous example, but it was achieved with more than six times more genes.

```
function [err, area, idx, el, eu] = example3(train_patterns,
train_targets, M)

% This example is based on colon data
% Inputs:
%   train_patterns  - Training patterns (DxN matrix, where D - data
%     dimensionality, N - number of patterns)
%   train_targets   - Training targets (1xN vector)
%   M               - The number of Monte-Carlo samples
%
% Outputs:
%   err             - Bolstered resubstitution error rate (not in
%                       percent)
%   area            - Area under the ROC curve
%   idx             - Selected features (vector of feature indices)
%   el              - Lower boundary of the Bayesian confidence
%                       (credible) interval containing err
%   eu              - Upper boundary of the Bayesian confidence
%                       (credible) interval containing err


% Reassign class labels so that -1s are replaced with 0s and +1 stay
% unchanged
```

Figure 5. ROC curve

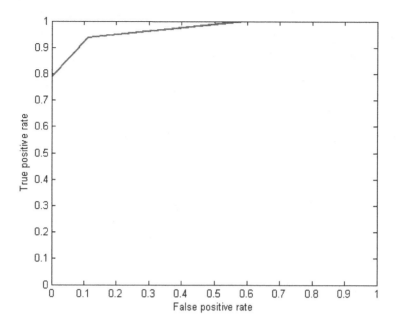

Figure 6. Bayesian posterior distribution

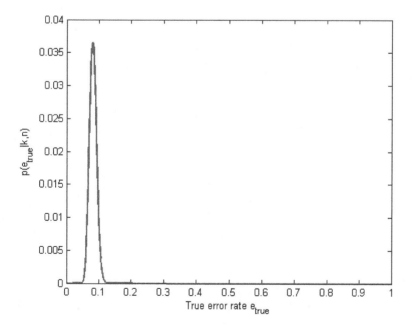

```matlab
% These operations are specific to colon data only

% Get unique training labels
Uc = unique(train_targets);
% Find indices of "positive" (colon cancer) instances
i = find(train_targets == Uc(1));
% Get the number of "positive" instances
p = M*length(i);
if Uc(1) ~= 0
    % Replace -1s with 0s
    train_targets(i) = 0;
end
% Find indices of "negative" (normal) instances
i = find(train_targets == Uc(2));
% Get the number of "negative" instances
n = M*length(i);
if Uc(2) ~= 1
    % Replace non-1s with 1s
    train_targets(i) = 1;
end

% Gene selection with SVDEntropyFeatureSelection
idx = SVDEntropyFeatureSelection(train_patterns, 'sr');

% 3-nearest neighbor classification using only selected features
[err, targets, prob] = bolstederror_NearestNeighbor(train_
patterns(idx,:), train_targets, 3, M);

% Extract the probabilities of the "positive" class
% Attention! This operation is specific to the colon data only
% For other datasets, the command could be prob =
prob(2:2:end);
prob = prob(1:2:end,:);

% Convert a matrix into a vector
prob = prob(:)';

% As the number of test patterns is a multiple of M, take into
account % this fact by modifying training labels accordingly
t = repmat(train_targets, M, 1);
```

```
t = t(:)';

% Plot the ROC curve
ROC(t, prob, p, n, '-', 'r');
% Compute the area under this curve
area = AUC(t, prob, p, n);

figure;

% Compute Bayesian confidence interval covering the given bolstered
% error
[el, eu]= BayesianConfidenceInterval(floor(err*M*length(train_
targets)), M*length(train_targets));

return
```

Example 4

The last example concerns classifier ensembles. An ensemble consists of five classifiers, each utilizing its own gene selection method. All the classifiers in the ensemble are either 3- or 5-nearest neighbors.

The first three ensemble members carry out classification with the 50, 5, and 25 genes selected by the *BAHSIC* algorithm. Pay attention that in addition to the varying number of selected genes and the value of k, kernels and kernel parameters differ from classifier to classifier so that classification results are different, too, for these ensemble members. The other two ensemble members rely on genes selected by *EVDGeneSelection* and *SVDEntropyFeatureSelection*, respectively.

Classifier combination is done with the majority voting rule, which combines predictions made by all five ensemble members into a single vote (class label). The bolstered resubstitution error and its Bayesian credible interval are also computed for the whole ensemble. The Bayesian posterior distribution is shown in Figure 7.

Errors attained by ensemble members on a single run were 13.4%, 31.1%, 16.6%, 9.2%, and 8.9%, while the ensemble error was only 6.5%, i.e. the ensemble turned out better than any ensemble member. Such a fact is not accidental: extensive studies in (Okun, Valentini, & Priisalu, 2008) demonstrated that ensembles of nearest neighbor classifiers are vastly superior to the best ensemble member on a wide range of gene expression data sets. The Bayesian credible interval for the ensemble error is [4.9 8.8]. The genes selected by each ensemble member were collected in the cell array *all_genes*. The number of such genes was equal to 50, 5, 25, 49, and 316, respectively. By looking at the last two figures obtained when the number of se-

Figure 7. Bayesian posterior distribution

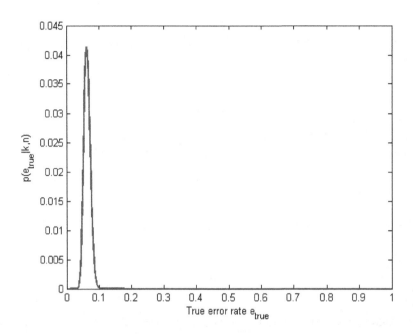

lected genes has been automatically determined rather than pre-defined by a user, it is possible to preliminarily conclude that the colon data set requires to consider rather many genes for accurate cancer prediction. For instance, only five genes chosen by *BAHSIC* yielded very poor prediction accuracy, compared to the case of 50 genes.

```
function [err, ens_err, all_genes, el, eu] = example4(train_
patterns, train_targets, M)

% This example is based on colon data
% Inputs:
%   train_patterns  - Training patterns (DxN matrix, where D - data
%     dimensionality, N - number of patterns)
%   train_targets   - Training targets (1xN vector)
%   M               - The number of Monte-Carlo samples to select
%
% Outputs:
%   err             - Bolstered resubstitution error rate (not in
%                     percent) for each ensemble member
```

```
%    ens_err     - Ensemble error
%    all_genes   - Selected features (vector of feature indices) for
%                      the entire ensemble
%    el          - Lower boundary of the Bayesian confidence
%                      (credible) interval containing ens_err
%    eu          - Upper boundary of the Bayesian confidence
%                      (credible) interval containing ens_err


% Reassign class labels so that -1s are replaced with 0s and +1 stay
% unchanged
% These operations are specific to colon data only

% Get unique training labels
Uc = unique(train_targets);
% Find indices of "positive" (colon cancer) instances
i = find(train_targets == Uc(1));
% Get the number of "positive" instances
p = M*length(i);
if Uc(1) ~= 0
    % Replace -1s with 0s
    train_targets(i) = 0;
end
% Find indices of "negative" (normal) instances
i = find(train_targets == Uc(2));
% Get the number of "negative" instances
n = M*length(i);
if Uc(2) ~= 1
    % Replace non-1s with 1s
    train_targets(i) = 1;
end
% Find unique training labels after relabelling
Uc = unique(train_targets);

% First ensemble member
% ////////////////////

% Gene selection with BAHSIC
idx = BAHSIC(train_patterns, train_targets, 5, 50, 'rbf', 1,
'linear', []);
```

```
all_genes{1} = idx;


% 3-nearest neighbor classification using only selected fea-
tures
[err(1), targets] = bolstederror_NearestNeighbor(train_
patterns(idx,:), train_targets, 3, M);
base_predictions(1,:) = targets(:)';



% Second ensemble member
% ////////////////////


% Gene selection with BAHSIC
idx = BAHSIC(train_patterns, train_targets, 5, 5, 'linear', [],
'linear', []);
all_genes{2} = idx;


% 5-nearest neighbor classification using only selected fea-
tures
[err(2), targets] = bolstederror_NearestNeighbor(train_
patterns(idx,:), train_targets, 5, M);
base_predictions(2,:) = targets(:)';



% Third ensemble member
% ////////////////////


% Gene selection with BAHSIC
idx = BAHSIC(train_patterns, train_targets, 5, 25, 'rbf', 0.1,
'linear', []);
all_genes{3} = idx;


% 5-nearest neighbor classification using only selected fea-
tures
[err(3), targets] = bolstederror_NearestNeighbor(train_
patterns(idx,:), train_targets, 5, M);
base_predictions(3,:) = targets(:)';



% Fourth ensemble member
```

```
% //////////////////

% Gene selection with EVDGeneSelection
idx = EVDGeneSelection(train_patterns, train_targets);
all_genes{4} = idx;

% 3-nearest neighbor classification using only selected fea-
tures
[err(4), targets] = bolsterederror_NearestNeighbor(train_
patterns(idx,:), train_targets, 3, M);
base_predictions(4,:) = targets(:)';


% Fifth ensemble member
% //////////////////

% Gene selection with SVDEntropyFeatureSelection
idx = SVDEntropyFeatureSelection(train_patterns, 'sr');
all_genes{5} = idx;

% 3-nearest neighbor classification using only selected fea-
tures
[err(5), targets] = bolsterederror_NearestNeighbor(train_
patterns(idx,:), train_targets, 3, M);
base_predictions(5,:) = targets(:)';


% Combine predictions of ensemble members with the majority
voting
[ensemble_predictions, ensemble_soft_predictions] = majority_
vote(base_predictions, 1:size(base_predictions,1), Uc);

% As the number of test patterns is a multiple of M, take into
account % this fact by modifying training labels accordingly
t = repmat(train_targets, M, 1);
t = t(:)';

% Compute ensemble error
ens_err = class_error(t, ensemble_predictions)/100;
```

```
% Compute Bayesian confidence interval covering the given en-
semble error
[el, eu]= BayesianConfidenceInterval(floor(ens_
err*M*length(train_targets)), M*length(train_targets));

return
function errprcnt = class_error(test_targets,targets)

% Calculate error percentage based on true and predicted test
labels
errprcnt = mean(test_targets ~= targets);
errprcnt = 100*errprcnt;

return
```

REFERENCES

Alon, U., Barkai, N., Notterman, D., Gish, K., Ybarra, S., & Mack, D. (1999). Broad patterns of gene expression revealed by clustering analysis of tumor and normal colon tissues probed by oligonucleotide arrays. *Proceedings of the National Academy of Sciences of the United States of America*, *96*(12), 6745–6750. doi:10.1073/pnas.96.12.6745

Li, W., Sun, F., & Grosse, I. (2004). Extreme value distribution based gene selection criteria for discriminant microarray data analysis using logistic regression. *Journal of Computational Biology*, *11*(2-3), 215–226. doi:10.1089/1066527041410445

Okun, O., Valentini, G., & Priisalu, H. (2008). Exploring the link between bolstered classification error and dataset complexity for gene expression-based cancer classification. In Maeda, T. (Ed.), *New signal processing research* (pp. 249–278). New York, NY: Nova Science Publishers.

Song, L., Bedo, J., Borgwardt, K. M., Gretton, A., & Smola, A. J. (2007). Gene selection via the BAHSIC family of algorithms. *Bioinformatics (Oxford, England)*, *23*(13), i490–i498. doi:10.1093/bioinformatics/btm216

Song, L., Smola, A., Gretton, A., Borgwardt, K., & Bedo, J. (2007). Supervised feature selection via dependence estimation. In Z. Ghahramani (Ed.), *Proceedings of the 24th International Conference on Machine Learning, Corvallis, OR*, (pp. 823-830).

Varshavsky, R., Gottlieb, A., Linial, M., & Horn, D. (2006). Novel unsupervised feature filtering of biological data. *Bioinformatics (Oxford, England)*, *22*(14), e507–e513. doi:10.1093/bioinformatics/btl214

ENDNOTE

[1]   It can be downloaded from http://genomics-pubs.princeton.edu/oncology/.

Chapter 26
End Remarks

A FEW WORDS IN THE END

Now when you, my dear readers, are reading these words, you got a lot of knowledge. One journey is over but another one is about to begin. I will not summarize what you have learned from this book as you can do so yourselves. I would like to talk a bit about what you may need to learn beyond this book.

First of all, you need to always remember about the potential limitations of microarray gene expression data. For example, Androulakis (Androulakis, 2005) provided a good analysis of such limitations, while Braga-Neto (Braga-Neto, 2007) described pitfalls that expect everyone dealing with small-sample microarray classification. The next logical step is to look for the integration of microarray gene expression data with other sources of data (see, e.g., (Zhang, Li, Wei, Yap, & Chen, 2007)).

It makes sense to visit the web site of Microarray Gene Expression Data Society (http://www.mged.org/) in order to be informed about the international standards that were developed or are being developed in this field. Standards are important to follow because if your products or statistical reports about drug validation do not follow the established rules, they have no market value.

Patents could also be necessary to track, independently of whether you work in academia or industry. Google Patents (http://www.google.com/patents) may be a

DOI: 10.4018/978-1-60960-557-5.ch026

starting point, but do not ignore other potential sources for patents you find. You may be aware of the patent battle in the USA on the variants of two genes BRCA1 and BRCA2 (they are responsible for breast and ovarian cancer). The fresh issue of New Scientist (10 April 2010) informs about the latest court situation (yes, 'court', do not be surprised!). Other information can be obtained from http://www.genomeweb.com if you type "gene patent" there. If you want to be your own patent engineer/lawyer, then I recommend reading the book by Ma (Ma, 2009) so that you will become familiar with the basics.

As MATLAB® is not the only programming environment useful for bioinformatics, you may be interested to read the article by Dudley and Butte (Dudley & Butte, 2009) about alternatives and programming skills required from the above-the-average researcher. Like me, they also emphasize the importance of embracing standards. Our second common point is hardware technology called Field Programmable Gate Array (FPGA). FPGA-based hardware acceleration has demonstrated its benefits for several bioinformatics applications, including sequence alignment, molecular dynamics, and proteomics (Dudley & Butte, 2009), (Storaasli, Yu, Strenski, & Maltby, 2007), (Derrien & Quinton, 2010).

I am sure you can add your own topics to study and perfect. So, let me wish you all the best in your work and life!

REFERENCES

Androulakis, I. P. (2005). Selecting maximally informative genes. *Computers & Chemical Engineering, 29*(3), 535–546. doi:10.1016/j.compchemeng.2004.08.037

Braga-Neto, U. M. (2007). Fads and fallacies in the name of small-sample microarray classification - A highlight of misunderstanding and erroneous usage in the applications of genomic signal processing. *IEEE Signal Processing Magazine, 24*(1), 91–99. doi:10.1109/MSP.2007.273062

Derrien, S., & Quinton, P. (2010). Hardware acceleration of HMMER on FPGAs. *Journal of Signal Processing Systems, 58*(1), 53–67. doi:10.1007/s11265-008-0262-y

Dudley, J. T., & Butte, A. J. (2009). A quick guide for developing effective bioinformatics programming skills. *PLoS Computational Biology, 5*(12), e1000589.. doi:10.1371/journal.pcbi.1000589

Ma, M. Y. (2009). *Fundamentals of patenting and licensing for scientists and engineers*. Singapore: World Scientific. doi:10.1142/9789812834317

Storaasli, O. O., Yu, W., Strenski, D., & Maltby, J. (2007). Performance evaluation of FPGA-based biological applications. *Proceedings of Cray Users Group, Seattle, WA.*

Zhang, X., Li, L., Wei, D., Yap, Y., & Chen, F. (2007). Moving cancer diagnostics from bench to bedside. *Trends in Biotechnology, 25*(4), 166–173. doi:10.1016/j. tibtech.2007.02.006

About the Author

Oleg Okun got an MSc degree in radiophysics and electronics in 1990 from Belarusian State University and a PhD degree in computer science in 1996 from the United Institute of Informatics Problems (Belarus). The topic of his PhD thesis was the distance transform and its application to document image analysis. From 1998 to 2008 he was with Machine Vision Group at the University of Oulu (Finland), where he taught several courses and did research on document image analysis and machine learning (with application to bioinformatics). Starting from June of 2008 he led a research team in the Swedish start-up company Precise Biometrics. His tasks there included biometric (fingerprint) algorithms, e-identity based on biometrics, biometric solutions for embedded systems, and biometric standardization. In July of 2009 he was the head of the Swedish delegation at the annual ISO/IEC JTC 1 SC 37 (Biometrics) meeting, where his contribution to one of the international standards has been approved. He has 75 publications, including two co-edited books published by Springer-Verlag in 2008 and 2009. He is the co-organizer of three international workshops on ensemble methods (SUEMA'2007, SUEMA'2008, and SUEMA'2010). His professional interests include a wide range of topics such as machine learning, data mining, statistics, bioinformatics, biometrics, extreme programming, extreme project management, strategic IT management, and cloud computing.

Index

A

adaptive boosting (AdaBoost) algo-
    rithm 314, 315, 316, 317, 318,
    319, 320, 321, 322
analysis of variance (ANOVA) 342,
    353, 381
ANOVA, nonparametric 342, 353,
    363, 369, 373, 379, 380
ANOVA, parametric 342, 353, 355,
    369, 374, 375, 380
Area Under the ROC Curve (AUC)
    338, 339

B

backward contribution selection 128,
    129, 134, 135
backward elimination using HSIC
    (BAHSIC) 147, 148, 149, 150,
    151, 156
BAHSIC algorithm 418, 421, 429,
    430, 431, 432, 434
Baron Münchausen 296, 297
Bayes error 334, 335, 336, 337, 339,
    340

Bayesian analogues 407
Bayesian confidence 406, 407, 408,
    409, 410, 411, 412
Bayes' Theorem 13, 14, 15, 17
Bayes, Thomas 13
bias-variance decomposition 255, 259
binary classification 53, 54, 56
bioinformatics 118, 119, 121
biological data 6
biological evolution 177
bolstered error estimation 383, 384,
    385, 386, 387, 392, 394, 395,
    396, 397, 398, 399, 400, 404
boosting 314, 318, 319, 320, 322, 326,
    327, 328
BoostingTrees 320, 321, 322
boostrap aggregating (BAGGING)
    296, 297, 301, 304, 308, 311,
    312
bootstrap 296, 297, 299, 302, 311, 313
bootstrapping 318, 383, 384, 406, 408
bootstrap sampling 302, 313

C

cancer 1, 3, 4, 69, 112, 114, 261, 262,

263, 264, 265, 268, 291, 293, 300, 312, 313

cancerous cells 3

CART 298, 299, 313

centroid 36, 47, 51

classification 414, 415, 418, 419, 421, 422, 425, 426, 428, 429, 432, 433, 434

classification accuracy 341, 345, 351, 352

classification algorithms 53, 341, 352

classification and regression trees (CART) 53, 54, 55, 59, 60, 67

classification error 11

classification tree 53, 55, 61, 331

classifier ensembles 260, 262, 272

classifiers 11, 12, 260-273, 277-283, 289-295, 316-323, 326-333, 342, 344-348, 352, 356, 358, 365

class labels 55, 56, 60, 64

coalition games 124

colon cancer data sets 332, 414

combiner 252, 254

complementary DNA (cDNA) 2, 4

confusion matrix 343, 356, 357, 358, 359

contribution-selection algorithm 123

covariance matrix 336, 337

cross-covariance operator 142

crossover constant (CR) 240, 241, 242, 243, 244, 245, 247

cross-validation (CV) 406, 408, 412, 413

cumulative distribution function (CDF) 258

curse of dimensionality 71, 72

D

data classification 10, 252

data mining 11, 315, 316, 327, 328

Decision Tree 11

deoxyribonucleic acid (DNA) 1, 2, 3, 4, 5, 262, 268, 291, 292

differential evolution 236, 237, 238, 241, 242, 243, 244, 250, 251

discretization 209, 221

E

e-criterion 166, 167, 168

ensemble methodology 256

ensemble pruning 265, 266, 294

ensembles 414, 429

entropy 209, 211, 214, 215, 217, 218, 219, 222, 223, 225, 226, 229, 230, 231, 232, 233, 234

error bias 383

error estimators 383, 384, 385, 387

Euclidean distance 34, 36

Euclidean space 268

EVDGeneSelection 169, 171

evolutionary algorithm for gene selection 178, 179

evolutionary optimization methods 236

extreme value distribution (EVD) 161, 162, 165, 166, 167, 168, 171, 175, 176

extreme value distribution theory 159

F

false acceptance rate (FAR) 124

False Negative (FN) rate 343, 357, 358

False Positive (FP) rate 343, 345, 346, 348, 351, 356, 357, 358

false positive rate 341, 346, 356, 358, 360

feasible region 75, 78, 94

feature extraction 121

feature selection 117, 118, 121, 123,

124, 129, 134, 139, 223, 224, 226, 228, 229, 234, 414, 415, 418, 434
feature selection methods 177
field programmable gate array (FPGA) 437, 438
forest models 331
forward contribution selection 128, 129, 130, 131
forward selection 1 (FS1) 226, 227, 228, 229, 230
forward selection 2 (FS2) 226, 227, 228, 230
forward selection using HSIC (FOH-SIC) 147, 148

G

game theory 123, 124, 129, 134, 139
Gaussian bolstering kernel 383
gene discretization 239
gene expression 2, 6, 7, 16, 18, 19, 20, 24, 29, 30, 69, 334
gene expression levels 32
gene pool (GP) 180, 184, 185, 187, 188, 189, 190, 191
generalized EVD PDFs (GEVPDF) 161
gene selection 143, 149, 159, 160, 164, 166, 167, 171, 172, 175, 300, 302, 330, 331, 333
Gini index 58
GooglePatents 119
GoogleScholar 119, 122
Gram-Schmidt orthogonalization process 224

H

heuristic sequential method 213
high dimensional gene expression data 383

Hilbert-Schmidt Independence Criterion (HSIC) 143, 144, 145, 146, 147, 148, 149, 150, 151, 153, 154
Hilbert-Schmidt norm 141
housekeeping genes 2
hyperplane distance nearest neighbor (HKNN) 35, 36, 37, 38, 39, 45

I

internal nodes 54
irrelevant genes 7

K

Karush-Kuhn-Tucker complementary condition 78
kernel function 72, 74, 93, 102, 103, 104, 106, 107, 111
kernel methods 140
kernel target alignment (KTA) 149
k-nearest-neighbor (KNN) 32, 33, 34, 35, 39, 51, 178, 200, 261, 262, 263, 264, 265, 270, 271, 293, 320, 331, 333
Kullback-Leibler distance 184

L

labels 10, 11
linear algebra 36
linear classification 70, 72
linear discriminant analysis (LDA) 384, 385
linear hyperplane 36, 255
linear programme 75
linear SVM 256
logarithmic transformation 8
logistic regression 159, 160, 163, 164, 165, 167, 169, 171, 173, 175

M

machine learning 316, 322, 326, 327, 328, 329, 333
machine learning algorithms 383
machine learning methods 6
MAGE group 7
malignant tumor 3
marginalization 16
Markov blanket 206, 207, 208, 211, 212, 213, 214, 221, 222
MATLAB® 19, 26, 27, 93, 96, 97, 150, 214, 229, 237, 242, 270, 289, 302, 304, 305, 308
MATLAB®, Classification Toolbox for 320
MATLAB® code 19, 331, 335, 414
MATLAB® programming environment 408, 410, 414, 437
maximum mean discrepancy (MMD) 149
m-class 69
McNemar's test 342, 352, 354, 381
meningitis 30
messenger RNA (mRNA) 1, 2, 5
microarray and gene expression (MAGE) 7
microarray data 33, 34, 49, 332
microarray gene expression 1, 300, 316, 319, 320, 341, 342
Microarray Gene Expression Data Society 436
misclassified instances 314, 319, 328
Monte-Carlo samples 383, 418, 419, 422, 424, 426, 430
multi-layer perceptron kernel (MLP) 74, 98, 99, 101
mutation 178, 187, 237, 238, 240, 241, 242, 243, 250, 251
mutation operation 237

N

naïve Bayes 11, 13, 255, 261, 262, 267, 271
Naïve Bayes classifiers 331
natural selection 177
nearest neighbor (NN) 11, 32, 33, 34, 35, 39, 45, 48, 49, 50, 51
negative predictive value (NPV) 344
NN algorithms 33
non-deterministic polynomial-time hard (NP-hard) 236, 251
nonempty set 34
normalize function 414, 417, 418
nucleotides 1

O

objective function 75, 76, 77, 79, 89, 91, 93, 94
oligonucleotide 8, 9
optimal parameter values 12
optimization theory 74
orthogonal eigenvectors 224
out-of-bag (OOB) 299, 300, 301, 303, 304, 313
over-expression 20
overfitting 12

P

pattern analysis 140, 141
pattern recognition 316
p-criterion 167, 168
performance evaluation 414
petal length (PL) 55, 56
petal width (PW) 55, 56
population (Pop) 237, 238, 239, 241, 242, 243, 246, 247, 248
positive predictive value (PPV) 341, 344, 351
possible class labels 342

posterior distribution formula 407, 408, 413

predictive gene selection 236

predictive power 261

predominant gene 208

preprocess function 414, 415

probability density function (PDF) 161, 162, 163, 169, 170

probability distribution 204

probably approximately correct (PAC) learning algorithms 315

prostate cancer data sets 332

Q

quadratic programme 75, 76, 77, 95

R

radial basis function (RBF) 74, 95, 98, 99, 100, 101, 103, 104, 106, 111, 149, 150, 153, 155

randomization 261, 298, 299, 302

random sampling 296, 301

RankGene software 183

receiver operating characteristic (ROC) 11, 21, 29, 60, 61, 95, 112, 124, 184, 201, 202, 263, 271, 272, 292, 322

receiver operating characteristic (ROC) curve 263, 292, 338, 339, 341, 345-350, 356, 359, 360, 361, 380, 418-429

recombination operation 237

redundancy-based feature selection 203

redundancy based filter (RBF) 211, 212, 214, 215

redundant feature 207

resubstitution error 384, 385, 386, 387, 389, 390, 391, 392, 395, 398

ribonucleic acid (RNA) 1

S

scoring function 178

selection operation 237

sepal length (SL) 55, 56

sepal width (SW) 55

serial analysis of gene expression (SAGE) 5

shapley value 123, 124, 125, 126, 130, 132, 134, 136, 139

simple ranking (SR) 226, 228, 229, 230

single-gene logistic regression model 163, 164

singular value decomposition (SVD) 223, 224, 225, 226, 229, 230, 231, 232, 233, 234

smoothed error estimation 384

specificity 344, 358

Statistics Toolbox 26, 27

support vector machine (SVM) 11, 35, 51, 68, 69, 70, 71, 72, 76, 77, 79, 80, 81, 94, 95, 96, 101, 115, 140, 141, 256, 258, 262, 291, 292, 320, 331

SVD-entropy 223, 225, 226, 229, 230, 231, 232, 233, 234

T

terminal nodes 54, 56

tissue-specific genes 2

tree models 329, 331

True Negative (TN) rate 343, 357, 358

true positive rate 341, 356, 358, 360

True Positive (TP) rate 343, 345, 346, 351, 353, 354, 356, 357, 358

twoing rule 58, 60, 62

U

unbiased error estimators 384
unbiased (objective) classification error
    estimation 383

V

Vapnik-Chervonenkis (VC) dimension
    337, 338, 340

W

weak learners 315, 318, 319, 320, 321,
    322
weighted learning 314
weighted training 314, 318, 321, 324,
    328
wrapper models 223

———